Frontiers in Ceramic Science

(Volume 2)

(Catalytic Materials for Hydrogen Production and Electro-oxidation Reactions)

Edited by

Moisés R. Cesário, Cédric Gennequin, Edmond Abi-Aad

Unit of Environmental Chemistry and Interactions with Living Organisms,
University of the Littoral Opal Coast, Dunkerque, France

&

Daniel A. de Macedo

Department of Materials Engineering, Federal University of Paraíba,
João Pessoa - PB, Brazil

Frontiers in Ceramic Science

Volume # 2

Catalytic Materials for Hydrogen Production and Electro-oxidation Reactions

Editors: Moisés R. Cesário, Cédric Gennequin, Edmond Abi-Aad & Daniel A. de Macedo

ISSN (Online): 2542-5269

ISSN (Print): 2542-5250

ISBN (Online): 978-1-68108-758-0

ISBN (Print): 978-1-68108-759-7

General:

1. Any dispute or claim arising out of or in connection with this License Agreement or the Work (including non-contractual disputes or claims) will be governed by and construed in accordance with the laws of the U.A.E. as applied in the Emirate of Dubai. Each party agrees that the courts of the Emirate of Dubai shall have exclusive jurisdiction to settle any dispute or claim arising out of or in connection with this License Agreement or the Work (including non-contractual disputes or claims).

2. Your rights under this License Agreement will automatically terminate without notice and without the need for a court order if at any point you breach any terms of this License Agreement. In no event will any delay or failure by Bentham Science Publishers in enforcing your compliance with this License Agreement constitute a waiver of any of its rights.

3. You acknowledge that you have read this License Agreement, and agree to be bound by its terms and conditions. To the extent that any other terms and conditions presented on any website of Bentham Science Publishers conflict with, or are inconsistent with, the terms and conditions set out in this License Agreement, you acknowledge that the terms and conditions set out in this License Agreement shall prevail.

Bentham Science Publishers Ltd.
Executive Suite Y - 2
PO Box 7917, Saif Zone
Sharjah, U.A.E.
Email: subscriptions@benthamscience.org

**BENTHAM
SCIENCE**

CONTENTS

FOREWORD

Hydrogen is important in many current industrial processes and is envisaged to become even more important in the future both as a raw material and also as a secondary energy carrier. At present, almost all the total hydrogen demand is met by the use of fossil fuels, (for example by steam reforming of natural gas) adding to global warming, despite the fact that pure hydrogen could be simply produced by water electrolysis. The reason for this situation is the high electrical demand for water electrolysis, which results in a hydrogen production cost that is currently uncompetitive for commercial retail when traditional technologies are used. In this respect, a distinct market opportunity is opening for alternative, energy efficient, methods for the commercial production of hydrogen. As fossil fuels become ever more scarce, hydrogen prices will continue to increase. This factor combined with increasing hydrogen demand and environmental concerns leads to an urgent need for commercialization of new hydrogen production methods. In this context, the first chapters of the book focus on alternative hydrogen production processes from a range of biomass derived precursors, by classical steam reforming, CO_2 sorption enhanced steam reforming and dry reforming, while the later chapters aim to complete the energy cycle by discussing the electrooxidation of the hydrogen and syngas ($CO + H_2$ mixture) products and their valorization for energy conversion in electrochemical devices.

The authors of the book are selected, leading world experts in their fields, from three different continents; a factor that underscores the tangible global urgency of the topic.

Dr. Duncan Paul Fagg
Nanotechnology Research Division
Centre for Mechanical Technology and Automation
Department of Mechanical Engineering
University of Aveiro, Aveiro
Portugal

PREFACE

The reduction of greenhouse gases is effectively our current focus in order to have a more sustainable life and environment. In this context, the main focus of this book is hydrogen production processes and hydrogen electro-oxidation reactions. These two joint aspects will allow to propose a catalytic/electrocatalytic solution to the environmental problems. Understanding the properties of catalytic materials and catalysis parameters governing chemical and electrochemical reactions in reforming reactions and energy conversion devices, such as solid oxide fuel cells, are the main topics of this book.

The book begins with a chapter addressing the hydrogen production by steam reforming of biomass renewable resources and biomass tar. The reactivity of different catalytic materials, mainly molybdena and vanadia-based catalystys, as well as their advantages and disadvantages in the steam reforming reaction has been addressed.

The second chapter reports the study of steam reforming coupled with CO_2 sorption. The authors focused on the development of a bifunctional material capable of sorber CO_2 and being active and stable in steam methane reforming at unusual temperatures.

The third chapter presents a review on the recent advances in the catalytic dry reforming of methane, alcohols and biomass tar. The authors propose a discussion on preparation methods and physico-chemical properties of most recent catalytic materials used in this process. A special attention has been paid to the different strategies to minimize the carbon deposition problem. Emphasis is also given to the plasma, microwaves, solar energy, and electrical current technologies associated to dry reforming reactions.

The forth chapter focuses on the applicability of bimetallic catalytic materials in the dry reforming of methane. Thermal, structural, textural, and catalytic studies have been emphasized. Moreover, the electrocatalytic performance of hydrogen and biogas electro-oxidation reactions has been studied by impedance spectroscopy.

The fifth chapter reports the history, concept, operation principle, electro-oxidation reaction mechanisms, and the anode component of SOFCs, addressing recent works in the field. Emphasis is also given to the preparation of electrocatalyst and how the synthesis parameters could affect microstructure and electrochemical properties.

The sixth chapter is also addressed to discuss anode materials, mainly perovskite-type oxides, for electro-oxidation reactions. The authors propose a comprehensive discussion on the development of electrode materials with combined properties of structural stability, electronic and ionic conductivity, thermal and chemical expansion, and electrochemical activity.

We would like to express our gratitude to all the eminent contributors for their excellent contributions and we believe that this e-book will be a reference to academic/industrial scientists from chemistry, physics, and materials science interested in the catalytic materials for hydrogen production and their valorization in electro-oxidation reactions.

Moisés R. Cesário, Cédric Gennequin, Edmond Abi-Aad
Unit of Environmental Chemistry and Interactions with Living Organisms,
University of the Littoral Opal Coast, Dunkerque,
France

&

Daniel A. de Macedo
Department of Materials Engineering,
Federal University of Paraíba, João Pessoa - PB,
Brazil

List of Contributors

Allan J. M. de Araújo	Department of Materials Engineering, Federal University of Rio Grande do Norte, Natal, Brazil
Braúlio S. Barros	Department of Mechanical Engineering, Federal University of Pernambuco, Recife, Brazil
Carlos A. Paskocimas	Department of Materials Engineering, Federal University of Rio Grande do Norte, Natal, Brazil
Cédric Gennequin	Unit of Environmental Chemistry and Interactions with Living Organisms, University of the Littoral Opal Coast, Dunkerque, France
Claire Courson	Institut de Chimie et Procédés pour l'Energie, l'Environnement et la Santé (ICPEES)-UMR CNRS 7515, University of Strasbourg, Strasbourg, France
Daniel A. de Macedo	Department of Materials Engineering, Federal University of Paraíba, João Pessoa, Brazil
Dong-Kyun Seo	School of Molecular Sciences, Arizona State University, Tempe, USA
Dulce M. de A. Melo	Department of Chemistry, Federal University of Rio Grande do Norte, Natal, Brazil
Edmond Abi-Aad	Department of Chemistry, University of the Littoral Opal Coast, Dunkerque, France
Esa Turpeinen	Environmental and Chemical Engineering, Faculty of Technology, University of Oulu, Oulu, Finland
Francisco J.A Loureiro	Department of Mechanical Engineering, University of Aveiro, Aveiro, Portugal
Gabriel M. Santos	Department of Petroleum Engineering, University of Campinas, São Paulo, Brazil
Gheorghita Mitran	Laboratory of Chemical Technology and Catalysis, Department of Organic Chemistry, Biochemistry & Catalysis, Faculty of Chemistry, University of Bucharest, Bucharest, Romania
Glageane S. Souza	Department of Materials Engineering, Federal University of Paraíba, João Pessoa, Brazil
Haingomalala L. Tidahy	Unit of Environmental Chemistry and Interactions with Living Organisms, University of the Littoral Opal Coast, Dunkerque, France
Irina E. Kuritsyna	Institute of Solid State Physics RAS, Chernogolovka, Russia
Jane Estephane	Department of Chemical Engineering, University of Balamand, El Koura, Lebanon
João Paulo de F. Grilo	Department of Materials and Ceramic Engineering, University of Aveiro, Aveiro, Portugal
Madona Labaki	Laboratory of Physical–Chemistry of Materials (LPCM/PR2N), Lebanese University, Fanar, Lebanon
Moisés R. Cesario	Unit of Environmental Chemistry and Interactions with Living Organisms, University of the Littoral Opal Coast, Dunkerque, France

Nikolay V. Lyskov Department of Chemical Engineering, Institute of Problems of Chemical Physics RAS, Russia

Octavian-Dumitru Pavel Laboratory of Chemical Technology and Catalysis, Department of Organic Chemistry, Biochemistry & Catalysis, University of Bucharest, Bucharest, Romania

Prem Seelam Environmental and Chemical Engineering, Faculty of Technology, University of Oulu, Oulu, Finland

Rubens M. Nascimento Department of Materials Engineering, Federal University of Rio Grande do Norte, Natal, Brazil

Samer Aouad Department of Chemistry, Faculty of Sciences, University of Balamand, El Koura, Lebanon

Satu Ojala Environmental and Chemical Engineering, Faculty of Technology, University of Oulu, Oulu, Finland

Vladislav A. Kolotygin Institute of Solid State Physics RAS , Chernogolovka, Russia

INTRODUCTION

The research of new sources of "clean" energy, with a significant reduction in the greenhouses gases emission like carbon dioxide (CO_2) and methane (CH_4) is strongly encouraged. In this regard, this book will consider reforming reactions for hydrogen and syngas ($CO + H_2$ mixture) production and their valorization in the electro-oxidation reaction for energy conversion in electrochemical devices.

The approach involves classical steam, CO_2 sorption enhanced steam reforming and dry reforming of hydrocarbons (Toluene, Benzene, Methane, *etc.*) or oxygenated organic compounds (Phenol, Ethanol, Glycerol, *etc.*).

The syngas derived from the dry reforming reaction can be used to develop synthetic fuels by Fischer-Tropsch process or purified to obtain hydrogen. Steam reforming converts hydrocarbon and water vapor into a gaseous mixture of H_2, CO and CO_2. These reactions are highly endothermic and thermodynamically favored by high temperature and low pressures, which require stable and active catalytic materials to induce a high conversion rate. However, deactivation catalysts caused by carbon deposition from secondary reactions, is a major obstacle for commercial purposes in the chemical industry.

Hydrogen also can be obtained by CO_2 sorption enhanced steam reforming which combines conventional reforming with in situ CO_2 sorption by sorbent materials, allowing to improve the overall hydrogen production.

A crucial step before to implement such processes is understanding the different factors controlling reforming reactions (powder synthesis method, particle size, dispersion of the active phase on the support, surface area, reducibility of multicomponent phases, and operating conditions). The development of new functional materials for electro-oxidation reactions in solid oxide Fuel Cells (SOFC) is a crucial issue regarding the improvement of their electrochemical performance.

CHAPTER 1

Impact of Molybdena and Vanadia Mixed Based Oxides on Hydrogen Production by Steam Reforming

Gheorghita Mitran[*, 1], **Dong-Kyun Seo**[2] and **Octavian-Dumitru Pavel**[1]

[1] *Laboratory of Chemical Technology and Catalysis, Department of Organic Chemistry, Biochemistry & Catalysis, Faculty of Chemistry, University of Bucharest, 4-12, Blv. Regina Elisabeta, 030018 Bucharest, Romania*

[2] *School of Molecular Sciences, Arizona State University, Tempe, AZ, 85287-1604, USA*

Abstract: Hydrogen seems to be the fuel of the future since it is clean-burning and its only by-product is water. Currently, around 95% of the hydrogen global production is accomplished by non-renewable energy sources, 4% is obtained from water and only 1% from biomass. Hydrogen production from renewable energy sources such as biomass represents an important challenge for the future. Nowadays, steam reforming is the cheapest way to produce hydrogen. This chapter summarizes data regarding hydrogen production by steam reforming of biomass renewable sources and biomass tar, emphasizing the catalysts development for this process. The development of high active catalysts with good stability and selectivity continues to be a challenge. For this purpose, the reactivity of different catalytic systems as well as their advantages and disadvantages will be discussed.

Keywords: Biomass, Biogas, Bio-oil, Hydrogen, Mixed Oxides, Molybdena, Nickel, Noble Metals, Steam Reforming, Vanadia.

INTRODUCTION

The growing global energy demand claims alternative fuel sources to replace traditional fossil fuels which are declining [1] and their price is steadily increasing.

The renewable sources which have the advantage to reduce greenhouse gas emissions represent an alternative [2]. Biomass is considered as a carbon neutral,

---

[*] **Corresponding author Gheorghita Mitran**: Laboratory of Chemical Technology and Catalysis, Department of Organic Chemistry, Biochemistry & Catalysis, Faculty of Chemistry, University of Bucharest, 4-12, Blv. Regina Elisabeta, 030018 Bucharest, Romania, Fax: 0040213159249; Tel: 0040213051464; E-mails: geta_mitran@yahoo.com, geta.mitran@chimie.unibuc.ro

Moisés R. Cesário, Cédric Gennequin, Edmond Abi-Aad & Daniel A. de Macedo (Eds.)

meeting the requirements of environmental protection and is considered as a clean source of energy.

Hydrogen represents an important alternative energy source to be used in electrochemical devices like fuel cells. However, unlike fossil fuels, hydrogen is not found free in nature and the current technology for its production is too expensive and wasteful energy consumption [3]. Hydrogen production from biomass can be achieved by thermochemical and biological methods. The biological method presents the advantage of operating under mild conditions, being environmentally friendly, but provides low yields of hydrogen [4]. The thermochemical method is fast and shows high yield of hydrogen, which makes it the most promising way to obtain hydrogen. Hydrogen can be produced from biomass by thermochemical processes such as pyrolysis, gasification, and steam reforming. Currently, steam reforming is the most used method to produce hydrogen due to its low-cost in comparison to other methods and its characteristic low CO_2 emission. However, hydrogen production cost [5], transport and storage are much more expensive compared to other liquid fuels such as ethanol, methanol, and gasoline. Nevertheless, hydrogen will be the fuel of the future for many reasons: (i) it is a clean fuel; (ii) it can be obtained from many energy sources, particularly renewable ones; (iii) it is a solution for the sustainable energy supply.

Whereas, biomass is considered the best option for derived fuels and chemical production making the knowledge of the chemical structure and organic components of biomass important [4].

In gasification and pyrolysis processes for hydrogen production the reformed gas must be converted by the water gas shift reaction. The disadvantage of these methods consists in char and tar formation as a result of the decomposition process of biomass.

In gasification reactions the quantity and quality of the products is largely influenced by the gasifying agent [6]. A multitude of gasifying agents such as oxygen, air, steam, and carbon dioxide can be used. The most used is air since the cost is almost zero, although pure oxygen leads to syngas with higher quality.

For biomass oxygen/steam gasification, hydrogen and carbon monoxide contents achieved 63-73%, since gasification with air leads to H_2 and CO contents of 52-63% [7, 8]. The steam gasification has received much attention since it produces a relatively high content of hydrogen in the gaseous fuel.

Another alternative for biofuels obtaining from biomass is the individual fraction of biomass transformation [9]. There are four main fractions: fatty acids, starch, sugars, and lignocelluloses.

There are two types of biomass feedstock that can be converted into hydrogen: (i) special bioenergy crops, and (ii) residues/organic waste from agricultural farming that are less expensive [10]. The disadvantage is that the yield of hydrogen obtained from biomass is relatively low, 16-18% from biomass weight [11].

Most researchers carried out experiments for hydrogen production from different resources and with different reactors: batch-type reactor, fluidized bed reactor, but regardless of the method was observed besides getting a small amount of hydrogen and large amounts of tar and char obtaining.

As a result, an important role for process optimization has the choice of the catalyst and the working parameters such as the gasifier temperature, steam to biomass ratio.

The hydrogen production from biomass cannot compete with natural gas steam reforming processes with well-developed technologies. However, part of the biomass can be used to produce chemicals and a residual fraction could be used for hydrogen generation, as an economically viable option, considering the key role of hydrogen for a clean and sustainable energy of the future. It is established that hydrogen will represent 11% (2025) and 34% (2050) of the total energy demand [12]. Steam reforming of natural gas and biomass gasification will become the most important processes at the end of the 21st century.

STEAM REFORMING

The most important method for hydrogen production is catalytic conversion (steam reforming) of hydrocarbons, followed by water electrolysis process. Hydrogen can be produced from natural gas, oil, coal, and biomass.

Natural and Biogas Steam Reforming

Natural gas is a nonrenewable feedstock through whose reforming generates 48% of the hydrogen production of the world. The composition of natural gas is 95 vol.% CH_4, 3.5 vol.% C_{2+}, 1 vol.% N_2, and 0.5 vol.% CO_2 [13]. The main reactions of natural gas steam reforming are:

The steam reforming of natural gas takes place at high temperatures of 800-900 °C, pressures of 1.5-3MPa and steam to carbon molar ratios of 2.5-3.

$$CH_4 + H_2O = CO + 3H_2 \qquad \Delta H^0_{298K} = 206 \text{ KJ/mol}$$
$$CO + H_2O = CO_2 + H_2 \qquad \Delta H^0_{298K} = -41 \text{ KJ/mol}$$
$$CH_4 + CO_2 = 2CO + 2H_2 \qquad \Delta H^0_{298K} = 247 \text{ KJ/mol}$$
$$C_nH_m + nH_2O = nCO + (n+m/2)H_2 \qquad \Delta H^0_{298K} > 0 \text{ KJ/mol}$$

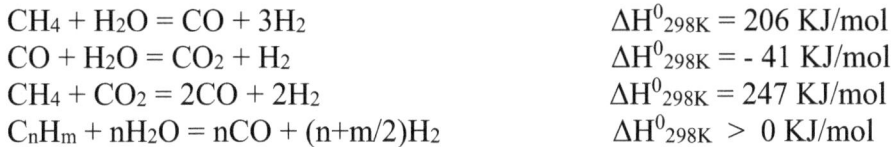

with the secondary reactions:

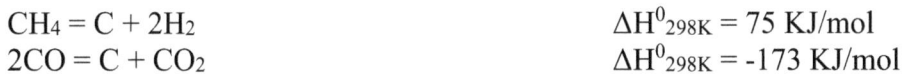

$$CH_4 = C + 2H_2 \qquad \Delta H^0_{298K} = 75 \text{ KJ/mol}$$
$$2CO = C + CO_2 \qquad \Delta H^0_{298K} = -173 \text{ KJ/mol}$$

Carbon deposits often form on the catalyst surface.

The reaction between CO_2 and methane has received much attention due to its decreasing greenhouse gas emissions. The disadvantages of this reaction are the high operating temperature, typically 900 °C. Therefore, aiming to eliminate this inconvenience, two commercial technologies that combine reforming with partial oxidation have been developed. The reaction is a very promising option for the syngas production with H_2/CO ratio close to 1, slightly below the ratio obtained in the steam reforming.

The introduction of membranes in the steam reforming process leads to the shift of the equilibrium, improving the conversion and the hydrogen production.

For the long term, renewable sources or biomass derivates are the most promising methods for hydrogen production. Biogas is a gaseous fossil fuel obtained from animals and buried plants [14] that represents an alternative to the natural gas. The composition of biogas is 60-65 vol.% CH_4, 30-35 vol.% CO_2, water vapor, hydrogen, and hydrogen sulfide. CO_2, H_2S and other impurities are removed by purification and the remaining CH_4 is used as a natural gas substitute (biomethane). The anaerobic digestion of organic wastes comprises four sequences:

(i) Hydrolysis: large organic molecules are hydrolyzed by anaerobic bacteria into smaller molecules;

(ii) Acidogenesis: smaller molecules are converted by acidogenic bacteria into carbon dioxide, ammonia, hydrogen, and organic acids;

(iii) Acetogenesis: organic acids are converted by acetogenic bacteria into hydrogen, ammonia and carbon dioxide;

(iv) Methanogenesis: acetic acid is decomposed to carbon dioxide and methane.

Another method of obtaining methane and carbon dioxide is the anaerobic decomposition (digestion) of glucose:

$$C_6H_{12}O_6 = 3CO_2 + 3CH_4$$

The main parameters which must be taken into account in the steam reforming process are temperature, pressure and molar steam to hydrocarbon ratio.

Another important factor is the choice of the catalyst which has a major influence on the reaction. The steam reforming catalysts are mainly based on nickel and cobalt. The reaction between methane and steam is strongly endothermic. The equilibrium reaction is thermodynamically favored at high temperature, low pressure and high steam to methane ratio [15].

The steam reforming of methane has been studied at high temperature (1100-1400 °C), with a complex mixture as reactants [16] and it has been noted that, no kinetic effect was found when CO was introduced in the mixture of methane and water. The addition of CO_2 into the reaction makes it easily promoted because CO2 acts as an oxidizer.

Another reforming option consists of steam and oxygen introduction, at the same time, the method known as the oxidative steam reforming (OSR) [17] which is a combination between partial oxidation and steam reforming. This method presents an advantage that the heat of endothermic steam reforming reaction is generated by exothermic oxidation of hydrocarbon. For example, in the methane OSR, the conversion is noticeable at 400 °C and reaches 99% at 700 °C. The requirements for the catalysts used for OSR are high activity and selectivity both for SR and OSR; good stability at high temperatures in order to obtain maximum of hydrogen yields; resistance toward carbon and sulfur poisoning, respectively.

Active and selective catalysts for synthesis gas by OSR reactions are those that contain group VIII metals (Pt, Rh, Ru). The main reactions of hydrocarbons oxidative steam reforming are:

The activity of transitional metals catalysts, both in steam reforming as well as in oxidative steam reforming of methane, decreased in the order Rh, Ru > Ni > Pt > Pd > Co [18].

Many researchers have focused on Ni based catalysts as low-cost, easily accessible and good catalytic activity. The steam reforming of methane over Ni based catalysts has reported a CH_4 conversion of 73% on Ni/Al_2O_3, 92% on $Ni/ZrO_2-Al_2O_3$, and 93% on $Ni/CeO_2-Al_2O_3$ and $Ni/ZrO_2-CeO_2-Al_2O_3$ [19].

$$CH_4 + 2O_2 = CO2 + 2H_2O$$

$$CH_4 + H_2O = CO + 3H_2$$

$$CO + H_2O = CO_2 + H_2$$

$$CH_4 + 2H_2O = CO_2 + 4H_2$$

$$C_mH_n + mO_2 = mCO_2 + n/2H_2$$

$$C_mH_n + mH_2O = mCO + (m+n/2)H_2$$

$$CO + H_2O = CO_2 + H_2$$

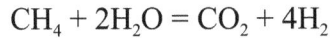

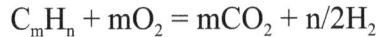

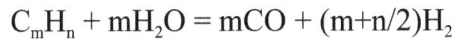

Biogas is an important raw material for hydrogen generation from reforming reactions, but it has the drawback of poisoning of the catalyst with H_2S, leading to rapid loss of catalytic activity. The advantage of the steam reforming process compared to dry reforming and dry oxidation reforming maximizes the H_2 yield.

From steam reforming of light naphtha (a mixture of light hydrocarbons from ethane to pentane) more carbon is produced and specific decoking catalysis required [20]. The catalyst based on magnesium-aluminum oxide ($MgAl_2O_4$) has been shown to be very active for the removal of coke as a result of the MgO presence. The processing of light naphtha, high steam to carbon ratios is needed and the reaction temperature is low (~ 500 °C), in contrast to the natural gas steam reforming when the reaction temperature is around 800 °C. Copper-based catalysts are also efficient to avoid coke deposition, but they cannot be used above 300 °C due to sintering.

The catalysts used for superior naphtha steam reforming are based on hydrotalcite-type materials with nickel and have been observed that their catalytic activity increases with increasing the nickel content [21].

The gaseous phase obtained from pyrolysis of agricultural waste composed of H_2, CO, CH_4, CO_2, C_2H_4, and C_2H_6 can be also used for hydrogen production. For this reaction, Ni promoted with ceria and iron have been used as catalysts due to the excellent redox properties of ceria [22] and the capacity of iron to crack coke and increase the H_2 yield. The introduction of steam to the pyrolysis gas leads to increase H_2/CO ratio and decrease CO percentage.

The hydrogen production from natural gas involves the processes illustrated in Fig. (**1**).

```
┌─────────────────────────────┐
│         Natural gas         │
└─────────────────────────────┘
               ↓
┌─────────────────────────────┐
│        Pre-reforming        │
└─────────────────────────────┘
               ↓
┌─────────────────────────────┐
│       Desulphurization      │
└─────────────────────────────┘
               ↓
┌─────────────────────────────┐
│       Steam reforming       │
└─────────────────────────────┘
               ↓
┌─────────────────────────────┐
│    Water gas shift reaction │
└─────────────────────────────┘
               ↓
┌─────────────────────────────┐
│       CO₂ elimination       │
└─────────────────────────────┘
               ↓
┌─────────────────────────────┐
│          Hydrogen           │
└─────────────────────────────┘
```

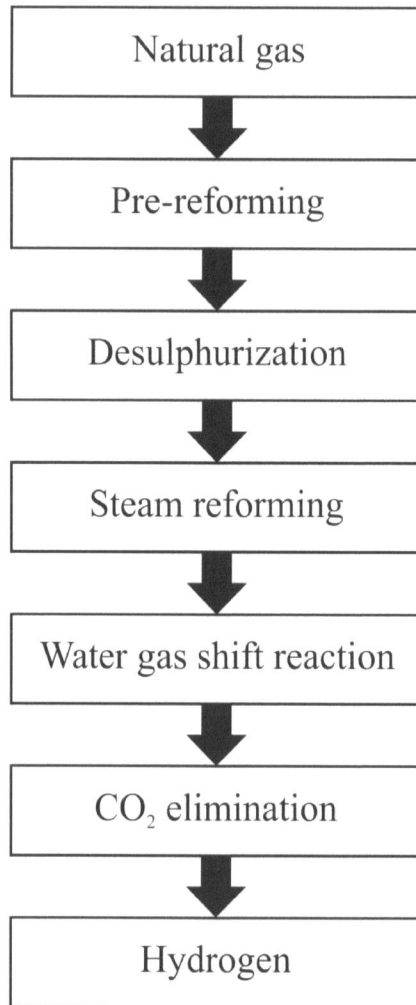

Fig. (1). Scheme of the hydrogen production by steam reforming of natural gas.

Liquefied petroleum gas (LPG) is a mixture of hydrocarbons, mainly propane (C_3H_8) and butane (C_4H_{10}), used to produce hydrogen by steam reforming. LPG can be easily transported and stored. The disadvantage of this process consists of C_2H_4 and C_2H_6 production as undesirable by-product due to decomposition of LPG. Silva [23] tested cerium- and strontium- doped $LaNiO_3$ for LPG steam and oxidative reforming and observed that in absence of oxygen they were strongly deactivated, whereas, the addition of small amount of oxygen improves the catalytic performance. Steam and autothermal reforming over ceria with high and low surface area and over Ni/Al_2O_3 were studied by Laosiripojana [24]. He observed that high surface ceria synthesized by surfactant-assisted (cetyltrimethylammonium bromide) method provided a high LPG reforming reactivity and was very resistant toward carbon deposition compared with Ni/Al_2O_3, due to the high oxygen storage capacity of ceria. The addition of oxygen in steam reforming of LPG was also explored. Oxidative steam reforming of LPG (with a propane-butane molar ratio of 1:1) has been investigated by Malaibari [25] over $Mo-Ni/Al_2O_3$ catalysts. He found that the presence of small amount of molybdenum as promoter increases both the mixture conversion as well as the hydrogen production rate, while increasing the molybdenum loading decreases the fuel conversion, probably due to active Ni species reduction to inactive Ni and Ni-Mo phases formation. Mesoporous $\gamma-Al_2O_3$ supported Ni nanoparticles and Ni-MgO catalysts [26] with large surface area, pore volumes and homogenously dispersed Ni-MgO oxides on the surface have proven to be very active catalysts for LPG pre-reforming. Pre-reforming is a steam reforming process at low temperature (300-600 °C) that converts high hydrocarbons to methane, hydrogen and carbon oxides, followed by the subsequent reforming of methane at high temperature, 800 °C, minimizing carbon deposition on the catalyst. Ni-based catalysts need to be reduced to generate metallic Ni active sites for hydrocarbons reforming. The reduction temperature has a very important influence on the catalytic activity, stability and resistance to coke deposition. Presence of NiO unreduced species on the catalyst surface not only favors hydrocarbons cracking and suppressed the water gas shift reaction, but also accentuate the coke deposition.

In the steam reforming of propane as a representative hydrocarbon of LPG, Ni-based catalysts need to be promoted with earth alkaline metals such as Mg, Ca to improve their stability and selectivity, and high S/C ratio it is also necessary to be used [27].

The oxidative steam reforming of propane was studied by Pino [28] over ceria supported platinum catalyst concluding that at low O_2/C_3H_8 molar ratio the conversion of propane is strongly influenced to the addition of water in the mixture. At a H_2O/C_3H_8 ratio equal to 2 the conversion was 80%, while by increasing the steam/fuel ratio, the conversion reaches values of 90-92%, thus the hydrogen concentration in the reaction products becomes 28% at a conversion of 92%.

Liquid Fuels Steam Reforming

Other promising sources of hydrogen are liquid fuels as gasoline and diesel. Diesel and gasoline contain higher volumetric hydrogen than other energy sources. When the hydrogen is produced from liquid fuels, several considerations have been taken into account such as carbon deposition on the catalyst surface and production of light hydrocarbons by liquid fuel reforming [29]. Another very important role has the degree of mixing between liquid fuel and gaseous reforming agents (air and steam); an incomplete mixing of reactants can destroy the reforming catalyst. Therefore, it is necessary that the reforming can be done in three simultaneous steps: atomization, evaporation and mixing with the reforming agent.

The steam reforming of a mixture with 85% pure ethanol and 15% gasoline was studied over Rh/Pt catalyst deposited on a ceramic monolith [30] and the results have confirmed that the catalyst could achieve a conversion of ethanol and gasoline around 100%. The catalyst is stable at least 110 h on stream in the absence of sulfur in mixture, while by sulfur introduction (5 ppm), the catalyst is deactivated after 22 h on steam, confirming that sulfur has been the primary cause of deactivation but not the only cause. C_2 intermediates, especially ethylene could also contribute to deactivation.

In the steam reforming of n-dodecane as a representative liquid fuel, the maximum of hydrogen was achieved at 700 °C, an excess of water makes the thermodynamic equilibrium to shift toward hydrogen, promoting the water gas shift reaction and suppressing coke formation.

Rh/CeO$_2$ catalyst was studied for steam reforming of n-dodecane as surrogate for diesel fuel, doped with tiophene as a model of organic sulfur in diesel fuel [31] in order to determine the effect of sulfur poisoning. The system is very efficient in the absence of sulfur, but it is deactivated in its presence in the liquid fuels. The catalyst stability is very good at high S/C ratio. The presence of sulfur determines the formation of Rh sulfide, which allows the hydrocarbon migration to the CeO$_2$ support, whose surface is acidic as a consequence of the appearance of oxysulfides species, stopping coke gasification reaction and the interaction between coke and oxygen is inhibited.

The resistance of noble metals supported on alumina for sulfur poisoning was studied by Xie [32] in steam reforming of liquid hydrocarbons following the order: Rh < Pt ≤ Pd ≤ Ru in catalysts deactivation.

Bio-oil Steam Reforming

The biomass pyrolysis leads to a liquid known as biomass pyrolysis oil (bio-oil or bio-crude) [33]. The components of bio-oil are naturally oxygenated compounds obtained from fragmentation and depolymerization of hemicellulose, cellulose, and lignin. They are anhydro-sugars, alcohols, phenols, carbonyl compounds, carboxylic acids, ethers, esters, and furans. The use of bio-oil in steam reforming is more suitable because the bio-oil is easier transported and with lower costs than biomass [34]. The distribution of different types of oxygenates in bio-oil depends on the feedstock and the bio-oil production conditions. The physical properties and the composition of bio-oil compared with fossil fuel oil differ through oxygen content, sulfur content and pH. Bio-oil has a high oxygen content, a high acidity and their easily polymerization determines an increase of viscosity which makes it difficult to convert. Another problem of bio-oil is its acidity making the process more expensive, but the low content of sulfur is an advantage.

The general reactions for oxygenates steam reforming are:

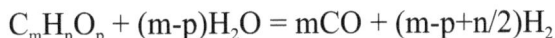

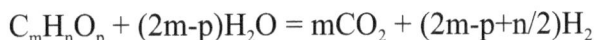

$$C_mH_nO_p + (m-p)H_2O = mCO + (m-p+n/2)H_2$$

$$C_mH_nO_p + (2m-p)H_2O = mCO_2 + (2m-p+n/2)H_2$$

Besides the main reactions, thermal decomposition of thermally unstable oxygenates may also occur:

$$C_mH_nO_p = C_xH_yO_z + gases + coke$$

The use of high temperature in steam reforming has the advantage that methane obtained from decomposition is steam reformed and water gas shift reaction is deactivated. The high ratio between steam and carbon is important because it leads to conversion increase and coke decrease.

The major components of bio-oil are phenols (38%) and acetic acid (30%) [35].

The steam reforming of *phenol* occurs according to the following reaction:

$$C_6H_5OH + 5H_2O = 6CO + 8H_2$$

and water gas shift reaction:

$$CO + H_2O = CO_2 + H_2$$

At the phenol conversion, besides active components and support, high steam to phenol ratio (to avoid coking) and temperature (complete conversion of phenol with 90% hydrogen yield can be achieved at 700 °C) have an important role. Phenol molecules are activated on metal sites whereas water molecules are activated on oxide support. There are several catalysts studied for steam reforming of phenol. The influence of support was studied by Güel [36] over Ni/K-La-ZrO$_2$ and Ni/Ce-ZrO$_2$ catalysts. They observed that Ni/Ce-ZrO$_2$ activates water gas shift reaction, being less active than Ni/K-La-ZrO$_2$.

Remon [37] studied steam reforming of different aqueous fractions obtained from biomass pyrolysis over Ni-Co/Al-Mg catalyst at 650 °C. The aqueous solutions contain different fractions of compounds: acetic acid, formic acid, propionic acid, methanol, phenol, furfural, levoglucosan, and guaiacol. As it was observed, the aqueous fractions obtained from pine have been initially converted to carbon and H_2. This is due to different reactivities of the organic compounds from solution, acetic acid and furfural being responsible for the most important differences. It has been observed that acetic acid has the lowest reactivity and low coke deposition while furfural has high reactivity and coke formation.

The steam reforming studies with model compounds such as acetic acid, furfural, phenol, levoglucosan, and guaiacol, concluded that hydrogen yield decreases in order: phenol > furfural > acetic acid > guaiacol > levoglucosan.

The hydrogen production by the sorption enhanced steam reforming (SESR) (that is an alternative for steam reforming, based on combination of reforming reaction with selective separation of CO_2 by sorption) of *acetic acid and acetone mixture*, as a model of bio-oil compounds, was studied over a Pd/Ni-Co hydrotalcite catalyst and dolomite as sorbent for CO_2 [38].

The reactions involved in steam reforming of the mixture are:

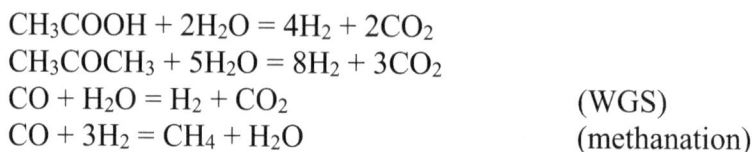

$$CH_3COOH + 2H_2O = 4H_2 + 2CO_2$$
$$CH_3COCH_3 + 5H_2O = 8H_2 + 3CO_2$$
$$CO + H_2O = H_2 + CO_2 \qquad \text{(WGS)}$$
$$CO + 3H_2 = CH_4 + H_2O \qquad \text{(methanation)}$$

The global reactions for the sorption enhanced steam reforming using CaO as a sorbent of CO_2 are:

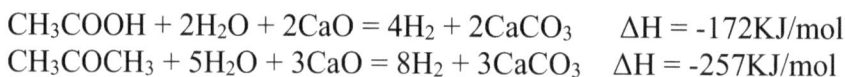

$$CH_3COOH + 2H_2O + 2CaO = 4H_2 + 2CaCO_3 \quad \Delta H = -172KJ/mol$$
$$CH_3COCH_3 + 5H_2O + 3CaO = 8H_2 + 3CaCO_3 \quad \Delta H = -257KJ/mol$$

These reactions are exothermic and supply majority of the heat required for the endothermic steam reforming reaction.

At temperatures between 525 and 675 °C, the maximum of hydrogen selectivity (> 95%) was achieved. The CH_4 concentration decreases with temperature increasing, while CO and CO_2 increase with temperature.

Rioche [39] studied acetic acid, acetone, ethanol, and phenol as bio-oil model compounds over noble metals-based catalysts and observed that the catalytic activity toward hydrogen follow the order: Rh-CeZrO$_2$ > Pt-CeZrO$_2$ ~ Rh-Al$_2$O$_3$ > Pd-CeZrO$_2$ > Pt-Al$_2$O$_3$ > Pd-Al$_2$O$_3$. As expected, ceria-zirconia supports were more efficient than alumina.

A mixture of acetic acid, ethylene glycol and acetone was chosen by Kechagiopoulos [40] for steam reforming using as a catalyst commercial nickel-based sample showing that the conversion of all compounds was complete and acetone has the higher tendency to produce coke. Ni-based catalysts supported on Ce-Zr with promoters like Mg, Ca, Y, La, or Gd were studied for steam reforming of an equimolar liquid mixture of six compounds (ethanol, propanol, butanol, lactic acid, ethylene glycol, and glycerol) [41]. Ce-Zr samples are prone to be deactivated in steam reforming because of their hydrophilic nature and this problem could be reduced by the promoter incorporation. The catalyst with Mg was the most active one (94% conversion, 80% selectivity), while that with Gd was the least active one. The increased nickel loading does not influence the activity because the surface area decreases at higher Ni loading. It was also shown that, the surfactant addition at the catalyst preparation improves the catalytic activity because it leads to enhance the surface area, specific pore volume and Ni dispersion. Wang [42] concluded that steam reforming of bio-oil or its fractions is much more difficult to be carried out than when it is working with a model compound. The reactor feeding with oil was the main problem, bio-oil cannot be totally vaporized.

Czernik [43] has been focused on catalytic steam reforming of lignocellulosic biomass-derived liquids obtained from pyrolysis. Simple oxygenated compounds like acetic acid or methanol are more reactive than hydrocarbons, while complex biomass-derived liquids need high temperatures and more steam for gasification of carbon deposition from thermal decomposition. The hydrogen yield (over commercial nickel catalyst) from hemicelluloses solution was about 70%, the lipids and lipid-derived liquids were reformed easier than lignocellulosic-based liquids (76% hydrogen from "crude glycerin" and 82% from trap grease").

Bion [44] compared crude bioethanol reforming with hydrocarbons steam reforming. The hydrogen production from hydrocarbons is thermodynamically controlled by methane formation. Compared with hydrocarbons where methane is the most stable molecule, in the case of alcohols, methanol is the most reactive being considered as a "liquid" syngas.

The *ethanol* steam reforming takes place through a series of reactions, the most important are:

$$C_2H_5OH + H_2O = 2CO + 4H_2$$
$$C_2H_5OH + 3H_2O = 2CO_2 + 6H_2$$
$$C_2H_5OH + 2H_2 = 2CH_4 + H_2O \quad \text{(hydrogenolysis to methane)}$$
$$C_2H_5OH = C_2H_4 + H_2O \quad \text{(dehydration)}$$
$$C_2H_5OH = CH_3CHO + H_2 \quad \text{(dehydrogenation)}$$
$$C_2H_5OH = CH_4 + CO + H_2 \quad \text{(cracking)}$$
$$C_2H_5OH = 1/2CO_2 + 3/2CH_4 \quad \text{(cracking)}$$

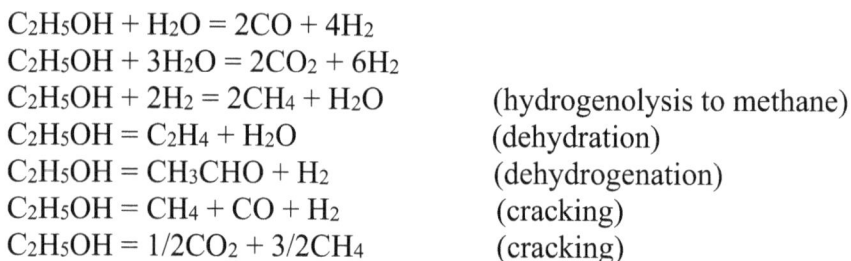

From a thermodynamic point of view, the cracking to CH_4 and CO_2 is favored at low temperatures (100-300 °C). The reforming to hydrogen and CO is the only reaction to occurs at 900 °C. The steam reforming of crude ethanol is different from that of pure ethanol because the impurities present in the crude ethanol can influence the catalyst stability and the hydrogen production. The main impurities contained by crude bioethanol are superior alcohols, esters, aldehydes, acetic acid, and amines. In order to establish which types of impurities are responsible for catalyst deactivation, the impact of different impurities on the steam reforming was studied over $Rh/MgAl_2O_4$. The following series of impurities have been studied: (i) molecules with four carbon atoms (butanal, diethyl ether, butanol, and ethyl acetate); (ii) molecules with acid and basic properties (acetic acid and diethyl amine); (iii) alcohols (methanol, 1-propanol, isopropanol, 1-butanol, and 1-pentanol). In the presence of acid-basic impurities it has been observed that diethyl amine favors the ethanol conversion compared with the presence of acetic acid as impurity that decreases ethanol conversion. This behavior can be explained by preferentially adsorption of diethyl amine on the acidic sites modifying the electronic properties of metal due to an electron transfer from the free nitrogen doublet to metal. In the presence of impurities with four carbon atoms was shown that, the presence of butanal increases the conversion of ethanol, while the presence of diethyl ether deactivates the catalyst. On the other hand, alcohols used as impurities lead to the following conclusions: methanol does not influence conversion of ethanol but improves hydrogen yield, higher alcohols decrease both the ethanol conversion and the hydrogen yield, decreasing being proportional to the number of carbon atoms of the molecule.

There are some similitudes between the steam reforming of alcohols and hydrocarbons: (i) the most active metal for both reactions is Rh; (ii) at moderate temperatures (400-500 °C) the support plays an important role while at higher temperatures (> 550 °C) reactant molecule activation is the determining step. The differences between hydrocarbons and alcohols steam reforming are: (i) in the case of alcohols the complementary reactions such as dehydration, dehydrogenation and cracking may be much faster than steam reforming; (ii) the reactivity of alcohols is higher than that of corresponding hydrocarbons for steam reforming but the hydrogen yield is lower (6 moles of H_2 per mole of ethanol and respectively 7 moles of H_2 per mole of ethane).

The ethanol steam reforming over core-shell structured Ni, Fe and Co-Pd oxides loaded on Zeolite Y catalysts was studied by Kwak [45] showing that the catalyst with Co-Pd is very active (between 70-100% conversion at 350-600 °C), and by Kim [46] over core-shell structured Si-Co-Mg oxides observing that the transfer of oxygen from the MgO to Co species plays an important role in maintaining the partially oxidized Co state resulting an increase of ethanol conversion and hydrogen yield.

Other catalysts studied for ethanol steam reforming were K-promoted Ni/ZrO_2 [47], $PtKCo/CeO_2$ [48] and Ni/TiO_2 [49]. It has been observed that the support plays an important role allowing the dispersion of the active component and improving the catalytic activity due to metal-support interactions. On the acidic supports, ethanol dehydration leads to coke formation, while by adding of alkali species the acidity is partially neutralized and coke deposition is inhibited [50]. Basic supports such as MgO, ZnO, CeO_2, and La_2O_3 are suitable for ethanol steam reforming.

Methanol steam reforming has been studied over $Cu/ZnO/Al_2O_3$ catalysts at 240-320 °C [51]. It was observed that addition of oxygen in the steam reforming reaction has been an efficient way to decrease the CO content in the products. The drawback is the fact the oxygen is firstly used in the combustion of methanol and then a reaction between unreacted methanol and water from combustion occurs. These catalysts have low stability and pyrophoric characteristics. Addition of promoters and modification of the preparation method aiming to increase the metal dispersion, surface area and decrease particle size should be done to improve the catalyst activity [52].

Other catalysts studied for steam reforming of methanol were based on noble metals such as $Pd/ZnO/Al_2O_3$ that present higher stability than copper-based catalysts [53].

Two mechanisms were proposed for the methanol steam reforming. The first one is a Langmuir-Hinshelwood mechanism [54] based on a single kind of active sites and suggests that methyl formate is an intermediate:

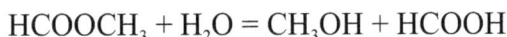

$$2CH_3OH = HCOOCH_3 + 2H_2$$

$$HCOOCH_3 + H_2O = CH_3OH + HCOOH$$

$$HCOOH = H_2 + CO_2$$

$$CO_2 + H_2 = CO + H_2O$$

The second mechanism [55] involves the existence of two types of active sites: a first one for methanol steam reforming and water gas shift reaction activation and a second one for decomposition reaction.

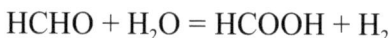

$$CH_3OH = HCHO + H_2$$

$$HCHO + H_2O = HCOOH + H_2$$

$$HCOOH = CO_2 + H_2$$

Acetic acid steam reforming has the advantage that acetic acid is safer to store and transport [56]. Thermal decomposition reaction of acetic acid takes place at high temperature even without catalyst, while the steam reforming reaction needs presence of a catalyst to break C-C and C-H bonds. Noble metals on alumina were studied as catalysts and it was observed that acetic acid conversion decreases in order (between 97% and 29%): Rh > Ru > Pd > Pt and hydrogen selectivity follows the same order Rh/Al_2O_3 (96%) > Ru/Al_2O_3 (90%) ~ Pd/Al_2O_3 (90%) > Pt/Al_2O_3 (75%).

Hu [57] studied $Cu-Zn-Co/Al_2O_3$ and found that these three metals play different roles; Cu promotes water gas shift reaction, Co is active for acetic acid steam reforming reaction and Zn enhances the catalytic performance at low temperatures. The reactions involved in the steam reforming of acetic acid are:

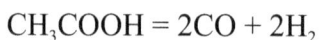

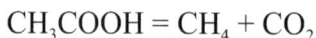

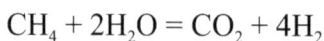

$$CH_3COOH + 2H_2O = 2CO_2 + 4H_2$$

$$CH_3COOH = 2CO + 2H_2$$

$$CH_3COOH = CH_4 + CO_2$$

$$2CH_3COOH = (CH_3)_2CO + H_2O + CO_2 \quad \text{(ketonization)}$$

$$CO + H_2O = CO_2 + H_2$$

$$CH_4 + 2H_2O = CO_2 + 4H_2$$

$$CH_4 + H_2O = CO + 3H_2$$

$$(CH_3)_2CO + 5H_2O = 3CO_2 + 8H_2$$

Glycerol is another alternative for hydrogen production by steam reforming [58]. Natural glycerol is a byproduct of biodiesel plants, between 10-30% of vegetable oils or animal fats are converted to glycerol as byproduct [59]. Synthetic glycerol is derived from propylene. The glycerol steam reforming generates 7 moles of H_2 per mole of glycerol compared with methane steam reforming for which 3 moles of H_2 can be obtained per mole of methane [60]:

$$C_3H_8O_3 + 3H_2O = 7H_2 + 3CO_2$$

The products obtained from glycerol steam reforming depend on the reaction temperature. At low temperatures, the main products are H_2, CO and CH_4 due to the fact that water gas shift reaction and CH_4 formation are favored at these temperatures. The glycerol steam reforming takes place at high temperatures. Therefore, the process is not convenient from an economic point of view. The hydrogen production is also influenced by the molar water/glycerol ratio. The hydrogen yield increases with molar ratio increasing from 4 to 15, whatever the temperature. Ratios higher than 5 inhibit carbon formation. Pressure is another factor that influences the reaction. High pressures lead to a low hydrogen yield. For this purpose, pressure decreasing could be done using a carrier gas to dilute the reactants.

The catalytic performance and operating conditions for different catalysts studied in glycerol steam reforming are summarized in Table **1**.

Table 1. Summary of the catalytic performance of different systems for glycerol steam reforming.

Catalyst	Operating Conditions	Conversion (%)	H_2 Yield/Selectivity (%)	Reference
Ni/CeO$_2$	600°C, W/G ratio = 9	100	/70	[58]
Ni/TiO$_2$	600°C, W/G ratio = 9	60	/15	[58]
Ni/Al$_2$O$_3$	500°C, 10 wt% glycerol	100	50/	[61]
Pt/SiO$_2$	350°C, 10 wt% glycerol	100	-	[62]
Pt/Al$_2$O$_3$	350°C, 30 wt% glycerol	21	-	[63]
Pt/Y$_2$O$_3$	600°C, W/G ratio = 24	100	90/	[64]
Co/CeO$_2$	425°C, G/W/He = 2/18/80 vol (%)	100	/88	[65]
Co/Al$_2$O$_3$	500°C, P_{gly}=7.4kPa, P_{steam}=57 kPa	13	77/63	[66]
Ru/Al$_2$O$_3$-CeO$_2$	650°C, W/G ratio = 12	100	86/88	[67]

(Table 1) cont.....

Catalyst	Operating Conditions	Conversion (%)	H$_2$ Yield/Selectivity (%)	Reference
Ru/Y$_2$O$_3$	600°C, S/C ratio = 3.3	100	90/	[68]
Ni/Ca/SBA-15	600°C, W/G ratio = 6	98	53/	[69]

At the same time, Zamzuri [70] investigated Ni on different supports for glycerol steam reforming at 650 °C. The glycerol conversion follows the order SiO$_2$ < ZrO$_2$ < La$_2$O$_3$ < MgO < Al$_2$O$_3$, with the H$_2$ variation following the order Al$_2$O$_3$ > La$_2$O$_3$ > ZrO$_2$ > MgO > SiO$_2$. They considered that large surface area, small crystallite size and high dispersion of Ni on the alumina support make this catalyst more active. The production of renewable energy from glycerol represents a method for increasing the profitability of biodiesel production [71].

Biomass Tar Steam Reforming

The main problem that occurs during biomass gasification consists of formation of a complex mixture of condensable compounds named tar. The steam reforming of tar is a very good technique for its removal. The equations for this reaction are:

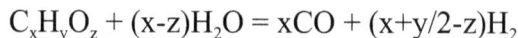

$$C_xH_y + mH_2O = xCO + (m+y/2)H_2$$

$$C_xH_yO_z + (x-z)H_2O = xCO + (x+y/2-z)H_2$$

The composition of tar is benzene (38%), toluene (14%), other 1-ring aromatic hydrocarbons (14%), naphthalene (10%), 2-rings aromatic hydrocarbons (8%), and mixed oxygenated compounds [72].

Ni-based catalysts were extensively studied for steam reforming of tar derived from cedar. Ni/Al$_2$O$_3$, Ni/ZrO$_2$ and Ni/TiO$_2$ (650 °C) [73] were very active, but their limitation is the rapid deactivation; catalyst based on BaAl$_{12}$O$_{19}$-supported Co (550 °C) [74] has been efficient for tar removing; Cu/Al$_2$O$_3$, Cu/CaO and Cu/CS [75] were also studied, and it was found that the calcined waste scallop shell (CS) support is responsible for high activity due to a strong interaction between Cu and CS support with formation of the calcium copper oxide phase that stabilizes Cu species.

Li [76] studied Ni-Cu/Mg/Al bimetallic catalysts from hydrotalcite-like compounds for the steam reforming of tar derived from the biomass pyrolysis at low temperature. The catalyst with a composition of Cu/Ni = 0.25 shows a higher catalytic performance and better coke resistance compared with monometallic

Ni/Mg/Al and Cu/Mg/Al catalysts. Catalytic performances were correlated with a high metal dispersion, high number of active sites on the surface, good oxygen affinity and surface modification by formation of small Ni-Cu alloy particles.

Chen [77] studied promoting effects of noble metals (Pt, Pd, Au, Rh, Ru and Ir) on the catalytic activity of hydrotalcite derived Ni catalysts in the steam reforming of tar derived from biomass pyrolysis. He concluded that addition of noble metals enhanced the catalytic activity. Pd was the most active promoter improving the reduction degree of Ni species, which leads to high yields of hydrogen. Ni over ceramic foam was used by Gao [78] for tar steam reforming at temperatures of 500-900 °C, with a S/C ratio from 0 to 4. It was observed that a maximum hydrogen yield was achieved at 700 °C. The structure of real tar is complex. Therefore, most researchers used the model of tar compounds such as benzene, toluene or naphthalene. Benzene steam reforming was studied by Kaisalo [79] over Ni/Al_2O_3 catalyst at 750-900 °C and by Gao [80] over Ni ceramic foam.

The steam reforming of the benzene and naphthalene mixture was studied by Furusawa [81] over Pt and Ni-based catalysts. He concluded that Pt/Al_2O_3 showed higher activity and stability. As supports, Al_2O_3 and MgO were found to be more efficient compared with other oxides such as CeO_2, ZrO_2, and TiO_2.

Steam reforming of toluene as a compound model of tar was studied over Ni/Fe/Mg zeolite-supported [82]. It has been observed that addition of Mg to Ni-Fe/zeolite catalyst enhances reaction by increasing the tolerance to carbon deposition. By adding Fe to Ni, coke deposition is suppressed improving the catalyst stability and promoting reducibility of Ni species. Ni/LaSrAlO catalysts were studied by Takise [83] for toluene steam reforming. It was observed that by increasing the support calcination temperature, both toluene conversion and hydrogen yield decrease. There is a correlation between catalytic activity and calcination temperature that influences specific surface area and lattice oxygen release rate. Toluene and naphthalene as a model tar steam reforming were studied over char supported nickel catalyst [84].

STEAM REFORMING OVER MOLYBDENA-VANADIA CATALYSTS

Vanadium and molybdenum oxides are very interesting from a chemical and catalysis point of view due to their surface mobility, Lewis acid/base sites and presence of crystal faces with different reactivity [85]. Vanadia-based compounds are used for mild oxidation, ammonoxidation and dehydrogenation, while molybdena-based catalysts are used for isomerization, polymerization and partial oxidation of hydrocarbons and alcohols. During catalytic reactions over vanadium and molybdenum oxides, the oxide surface is reduced and, as a result, oxygen vacancies are created.

Another compound of molybdenum, molybdenum carbide, is an attractive catalyst for steam reforming due to its unique surface and electronic properties. Its catalytic activity is similar to that of noble metals like Pt. The surface area and electronic properties of molybdenum carbide could increase by addition of a second metal. Molybdenum carbide was studied as catalyst for hydrogen production through the following reactions: (i) dry reforming of methane; (ii) methanol steam reforming; (iii) water gas shift reaction; and (iv) water electrolysis [86].

Methane steam reforming over supported or unsupported molybdenum carbide [87] evidenced that both catalysts are active for this reaction, the most active being the samples carburized at 700 °C. They exhibit large surface area and lowest carbon deposition on the surface.

Methanol steam reforming was studied by Szécheny [88] over Mo_2C obtained by MoO_3 carburization with C_2H_2/H_2 mixture. It has been found that methanol conversion reached 100% at 350 °C and hydrogen yield was high.

Iron group metals (Fe, Co, Ni) modified carbide catalysts were studied by Ma [89] who concluded that pure β-Mo_2C and catalyst modified with Fe and Co had poor activity and stability, while Ni modified molybdenum carbide enhanced both the activity and the stability by creation of new active sites which prevent the molybdenum carbide deactivation, promoting methanol dissociation. Mo_2C/γ-Al_2O_3 and Mo_2C/ZrO_2 were studied by Lin [90] who found that ZrO_2 improved the hydrogen yield due to its ability to retain methoxy groups on the surface compared with Al_2O_3 that restores the hydroxyl groups on the surface.

Ma [91] studied nanostructured wire-like molybdenum carbides obtained by direct carburization of aniline-intercalated Mo organic-inorganic composite and observed that α-MoC_{1-x} phase is more active than β-Mo_2C phase for methanol steam reforming. Molybdenum carbides modified with other metals were studied by Cao [92] who investigated the intermediate species generated during steam reforming of methanol. Over molybdenum carbide modified with Cu, formic acid is intermediate on the surface while on β-Mo_2C modified Ni and with Pt methyl-formate appears at intermediate. The activity and the stability are improved through the addition of Ni, Pt and Cu.

Ethanol steam reforming over nickel-molybdenum carbides doped with K was studied by Miyamoto [93]. He noted that C-O bond cleavage takes place on the oxide surface while C-C bond cleavage occurs on the carbide surface. The potassium adding improves hydrogen production being efficient for O-H bond scission and for the catalyst stability in steam reforming.

The steam reforming of the biomass tar, derived from Japanese cedar pyrolysis, over biomass char supported molybdenum carbide was investigated by Kaewpanha [94]. A good catalytic activity was observed up to 20 wt% Mo loading. Increasing Mo loading could determine the sintering of metal particles.

Molybdenum oxides were studied as catalysts for the water gas shift reaction, LPG oxidative steam reforming, bio-oil reforming, glycerol, and toluene steam reforming.

Water gas shift reaction over Cu and Au supported on molybdenum oxides [95] evidenced that bulk metallic Au is inactive while nanoparticles of Au supported on MoO_2 is a better catalyst compared with Cu nanoparticles supported on MoO_2. The catalysts characterization after reaction evidenced that Au and Cu remain in metallic state while a part of MoO_2 is oxidized to MoO_3. A mixture of 1:1 propane-butane as a component of liquefied petroleum gas (LPG) was studied for hydrogen production over Mo-Ni/Al_2O_3 catalysts [25]. It was observed that the presence of Mo improved fuel conversion, H_2 yield and resistance to coke deposition.

Bio-oil fractions steam reforming was studied over Ni-Mo supported on modified sepiolite catalysts [96] and the conclusions were that the acidified sepiolite promoted with Ni and Mo greatly improves both the catalytic activity and the hydrogen yield. The favorable conditions are temperatures of 700-800 °C and S/C ratios of 16-18 (H_2 yield of 67.5%).

MoO_3/Al_2O_3 and Mo-Ce/Al_2O_3 catalysts prepared by sol-gel method and gel combustion were studied in the glycerol steam reforming [97, 98]. The catalytic performances were studied at 400-500 °C, S/glycerol ratios of 9-20. On the Mo catalysts it has been observed that conversion increases (from 40% to 68%) with Mo loading and hydrogen yield depends on the Mo surface density (Fig. **2**).

Catalysts characterization evidenced the presence of oligomeric species octahedrally coordinated as well as polymeric molybdena species on the surface and a strong interaction of molybdena with alumina support. The combination of two metal oxides on the support improves the catalytic activity due to a good interaction between the component oxides. Ceria introduction to molybdena oxides supported on alumina provides good catalysts due to special properties of ceria named reducibility from Ce^{4+} to Ce^{3+} and creation of oxygen vacancies. A weaker interaction between molybdate species and support at ceria introduction was evidenced, while ceria appears at two different species: dispersed and bulk. The interaction between molybdate and cerium incorporated was evidenced by UV-Vis spectroscopy. On the sample without ceria, molybdenum has been presented only in tetrahedral coordinated, while the presence of ceria induces a change of molybdenum coordination in octahedral species.

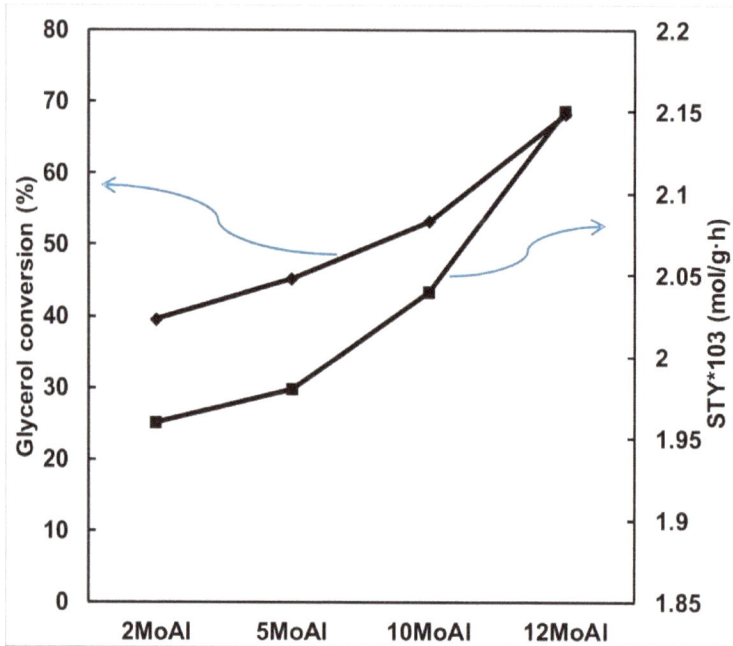

Fig. (2). Glycerol conversion and hydrogen space time yield (feed flow rate 0.06 ml/min, molar ratio steam to glycerol 15:1, P_{gly} = 4.2 kPa, P_{H2O}=62.5 kPa, P_{N2} = 33.3 kPa), temperature 500 °C).

The glycerol steam reforming over these catalysts evidenced that the glycerol conversion increases with ceria content (Fig. **3**), while the hydrogen selectivity has an optimum value at a cerium loading of 7%. At a further increasing of the Ce content above this value, its strong interaction with molybdenum and alumina was evidenced. The hydrogen selectivity was correlated with the presence of octahedral molybdenum species and with a low density on the surface.

Catalysts based on vanadia oxides have not been much studied for steam reforming reactions. They were studied in water gas shift reaction, ethanol, propane, dimethyl ether, and toluene steam reforming.

Water gas shift reaction was studied on the Pt/ZrO_2 doped with various amounts of vanadia. Polyvanadate species are present at high amounts of vanadia generating V_2O_5 and ZrV_2O_7 on the surface [99]. The presence of monovanadate species promotes the water gas shift reaction, while polyvanadate species and V_2O_5 are less active than zirconia.

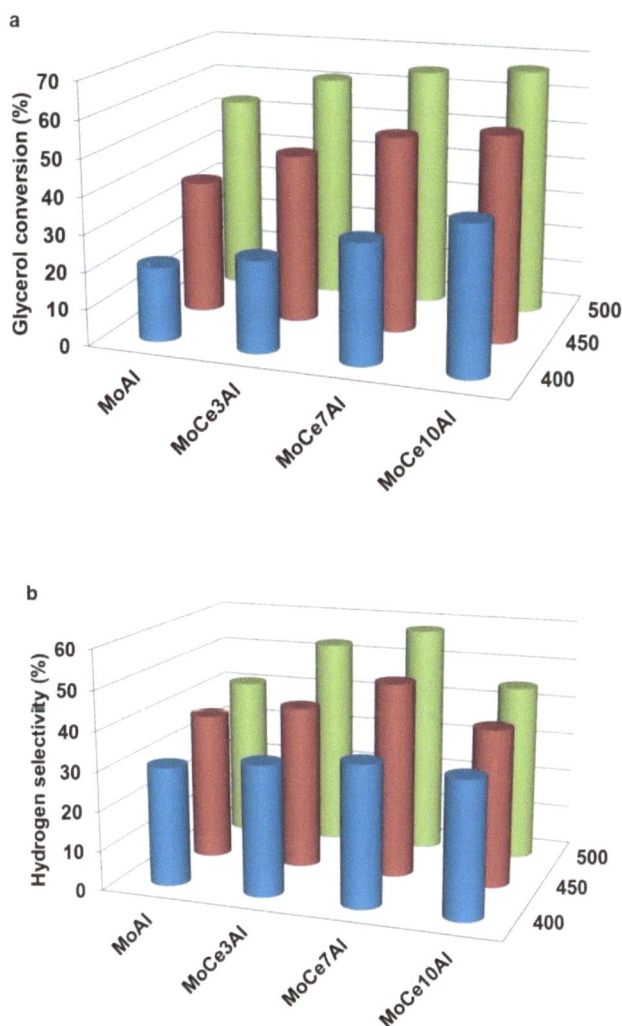

Fig. (3). Glycerol conversion (a) and H_2 selectivity (b) as a function of the temperature [99].

The dispersed vanadium species over metal oxides play an important role in water gas shift reaction [100] due to the competitive role of acid and redox sites. Therefore, the presence of vanadium over Pt/Al_2O_3 catalysts increases the activity by the formation of polymeric VO_x species that interact with alumina. On the surface was evidenced metallic Pt and V^{3+}/V^{4+} species. The presence of VO_x species favored the water gas shift reaction, but does not influence the hydrogen production.

Ethanol steam reforming reaction over bimetallic NiV-supported SiO_2 catalysts [101] was performed at 650 °C, with a water/ethanol ratio of 3:1. The presence of vanadium between the nickel particles has the role of preventing the strong agglomeration of nickel particles. However, an optimized amount of vanadium is needed to have a good catalytic activity. A further increase of the V amount leads to a drop of the acidity due to the presence of deficient coordinated V^{5+} ions that act as basic sites.

The catalyst with ratio Ni_8V_2/SiO_2 has the highest hydrogen yield (68%, at a conversion of ethanol nearly 100%) compared to other catalysts. The introduction of vanadium stabilizes the hydrogen production with a less catalytic deactivation due to poisoning with CO molecules that favor the water gas shift reaction.

The propane steam reforming was studied by Matsuka [102] with vanadium membrane reactor over nickel-based catalysts. The stability of vanadium membrane and high hydrogen selectivity has been confirmed by the reaction. The nickel catalysts supported on various oxides such as SiO_2, Al_2O_3 and CeO_2 evidenced a conversion of propane nearly 100% on alumina support, while ceria support showed high activity for water gas shift reaction, but the hydrogen production rate was higher on Ni/CeO_2 samples.

The introduction of a second metal, such as Co, Pt, Ag or Ru, was also evaluated for CeO_2 as a support material. The introduction of Pt or Ru as a second metal improved both the propane conversion and the hydrogen yield. The influence of the amount of Ag was studied on the catalysts with Ag. It was evidenced that excessive secondary metal had negative impact on the catalytic activity.

Dimethyl ether steam reforming was studied over $V-Ni/Al_2O_3$ catalysts [103] and two different mechanisms were evidenced as a function of the temperature and vanadium content. At low temperatures and low vanadium content, direct methanol decomposition to CO and H_2 was observed. At high temperatures, dimethyl ether hydrolysis and methanol reforming were observed. The vanadium presence has an influence on the nickel species on the surface enhancing the redox properties of nickel and suppressing coke formation.

Steam reforming of toluene as a model compound of the biomass tar has been studied over hydrotalcites derived vanadia catalysts [104]. The influence of different amounts of vanadium on the structure and catalytic activity has been studied. By increasing the vanadium loading it was observed that a part of Al^{3+} ions are substituted by V^{5+} ions and at high vanadia amounts their incorporation affects the crystallinity of the materials. The toluene conversion increases with vanadia content increasing while the hydrogen yield decreases in the same order. The catalytic activity is in concordance with the presence of polyvanadate species on the surface, while the hydrogen production depends by the presence of isolated species on the surface (Fig. **4**).

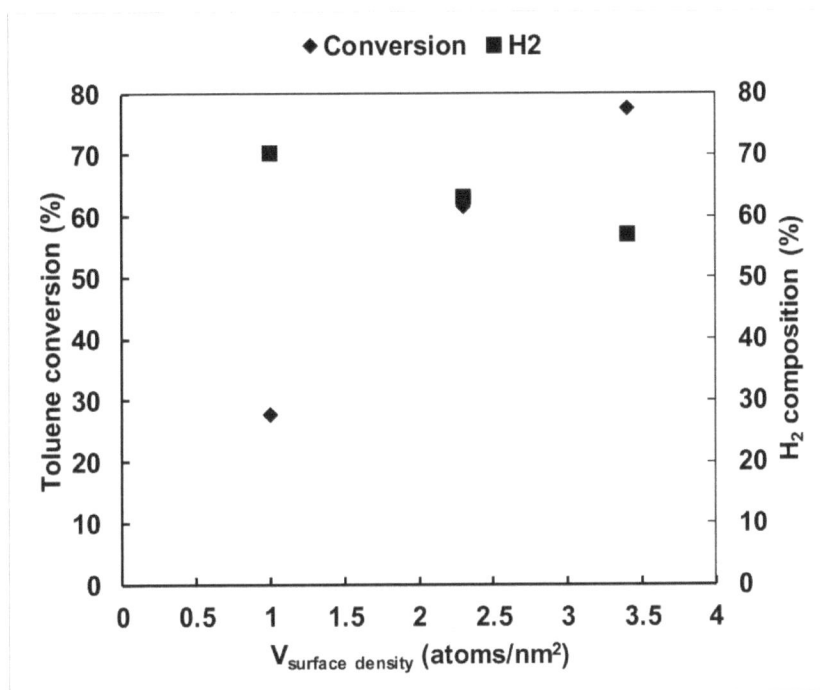

Fig. (4). Variation of toluene conversion and H_2 composition with V dispersion at 500 °C [105].

Toluene steam reforming has been intensively studied over mixed oxides derived from hydrotalcites (HT) due to their high activity for this reaction.

CONCLUSIONS

This chapter summarized some advantages and disadvantages of selected catalysts for steam reforming reactions. Nickel-based catalysts have been extensively studied due to their high activity. Their catalytic performances have been correlated with the metal dispersion, amount of active sites on the surface, oxygen affinity, and surface modification. Despite these advantages, Ni-based catalysts show the drawback of rapid deactivation due to carbon deposition. For these catalysts, it was found that the introduction of a small amounts of oxygen into the catalytic reactant mixture avoids coke deposition. Catalysts based on noble metals have high catalytic activity, good stability and resistance to coke deposition, but they are expensive. These systems are very efficient in the absence of sulfur. The presence of sulfur determines formation of sulfide on the surface, stopping coke gasification reaction and inhibiting the interaction between coke and oxygen. The catalysts based on other transition metals such as Fe, Cu, Co, V, and Mo exhibit a good performance for the steam reforming reaction, but they are also deactivated due carbon formation. The combination of two metal oxides on the support improves the catalytic activity due to a good interaction between the component oxides. Other factors that influence the steam reforming are the reaction conditions: temperature, feed composition, residence time, catalysts composition, and secondary reactions. Therefore, the selection of catalysts and the optimum parameters for reliable steam reforming reactions is a complex task.

CONSENT FOR PUBLICATION

Not applicable.

CONFLICT OF INTEREST

The authors declare no conflict of interest, financial or otherwise.

ACKNOWLEDGMENTS

We gratefully acknowledge for the use of facilities to the University of Bucharest and LeRoy Eyring Center for Solid State Science at Arizona State University.

REFERENCES

[1]   Nikolaidis, P.; Poullikkas, A. A comparative overview of hydrogen production processes. *Renew. Sustain. Energy Rev.,* **2017**, *67*, 597-611.
        [http://dx.doi.org/10.1016/j.rser.2016.09.044]

[2]   Dincer, I.; Acar, C. Innovation in hydrogen production. *Int. J. Hydrogen Energy,* **2017**, *42*, 14843-14864.
        [http://dx.doi.org/10.1016/j.ijhydene.2017.04.107]

[3]   Chaubey, R.; Sahu, S.; James, O.; Maity, S. A review on development of industrial processes and emerging techniques for production of hydrogen from renewable and sustainable sources. *Renew. Sustain. Energy Rev.,* **2013**, *23*, 443-462.
        [http://dx.doi.org/10.1016/j.rser.2013.02.019]

[4]   Balat, H.; Kirtay, E. Hydrogen from biomass - Present scenario and future prospects. *Int. J. Hydrogen Energy,* **2010**, *35*(14), 7416-7426.
        [http://dx.doi.org/10.1016/j.ijhydene.2010.04.137]

[5]   Sharma, S.; Ghoshal, S.K. Hydrogen the future transportation fuel: From production to applications. *Renew. Sustain. Energy Rev.,* **2015**, *43*, 1151-1158.
        [http://dx.doi.org/10.1016/j.rser.2014.11.093]

[6]   Guan, G.; Kaewpanha, M.; Hao, X.; Abudula, A. Catalytic steam reforming of biomass tar: Prospects and challenges. *Renew. Sustain. Energy Rev.,* **2016**, *58*, 450-461.
        [http://dx.doi.org/10.1016/j.rser.2015.12.316]

[7]   Kalinci, Y.; Hepbasli, A.; Dincer, I. Biomass-based hydrogen production: A review and analysis. *Int. J. Hydrogen Energy,* **2009**, *34*, 8799-8817.
        [http://dx.doi.org/10.1016/j.ijhydene.2009.08.078]

[8]   Lv, P.; Yuan, Z.; Ma, L.; Wu, C.; Chen, Y.; Zhu, J. Hydrogen-rich gas production from biomass air

and oxygen/steam gasification in a downdraft gasifier. *Renew. Energy,* **2007**, *32*, 2173-2185.
[http://dx.doi.org/10.1016/j.renene.2006.11.010]

[9]   Faba, L.; Díaz, E.; Ordóñez, S. Recent developments on the catalytic technologies for the transformation of biomass into biofuels: A patent survey. *Renew. Sustain. Energy Rev.,* **2015**, *51*, 273-287.
[http://dx.doi.org/10.1016/j.rser.2015.06.020]

[10]   Panwar, N.L.; Kothari, R.; Tyagi, V.V. Thermo chemical conversion of biomass – Eco friendly energy routes. *Renew. Sustain. Energy Rev.,* **2012**, *16*, 1801-1816.
[http://dx.doi.org/10.1016/j.rser.2012.01.024]

[11]   Demirbas, A. Yields of hydrogen of gaseous products *via* pyrolysis from selected biomass samples. *Fuel,* **2001**, *80*, 1885-1891.
[http://dx.doi.org/10.1016/S0016-2361(01)00070-9]

[12]   Rohland, B.; Nitsch, J.; Wendt, H. Hydrogen and fuel cells-the clean energy system. *J. Power Sources,* **1992**, *37*, 271-277.
[http://dx.doi.org/10.1016/0378-7753(92)80084-O]

[13]   Garcia, L. Hydrogen production by steam reforming of natural gas and other nonrenewable feedstocks *Compendium of Hydrogen Energy, Hydrogen Production and Purification A volume in Woodhead Publishing Series in Energy,* **2015**, 83-107.

[14]   Guo, M.; Song, W.; Buhain, J. Bioenergy and biofuels: History, status, and perspective. *Renew. Sustain. Energy Rev.,* **2015**, *42*, 712-725.
[http://dx.doi.org/10.1016/j.rser.2014.10.013]

[15]   Rostrup-Nielsen, J.R.; Bøgild Hansen, J. Steam Reforming for Fuel Cells. Fuel Cells: Technologies for Fuel Processing **2011**, 49-71.
[http://dx.doi.org/10.1016/B978-0-444-53563-4.10004-5]

[16]   Hiblot, H.; Ziegler-Devin, I.; Fournet, R.; Glaude, P.A. Steam reforming of methane in a synthesis gas from biomass gasification. *Int. J. Hydrogen Energy,* **2016**, *41*, 18329-18338.
[http://dx.doi.org/10.1016/j.ijhydene.2016.07.226]

[17]   Haynes, D.J.; Shekhawat, D. Oxidative Steam Reforming. Fuel Cells: Technologies for Fuel Processing **2011**, 129-190.
[http://dx.doi.org/10.1016/B978-0-444-53563-4.10006-9]

[18]   Jones, G.; Jakobsen, J.G.; Shim, S.S.; Kleis, J.; Andersson, M.P.; Rossmeis, J. First Principles Calculations and Experimental Insight into Methane Steam Reforming over Transition Metal Catalysts. *J. Catal.,* **2008**, *259*, 147-160.
[http://dx.doi.org/10.1016/j.jcat.2008.08.003]

[19]   Cai, X.; Cai, Y.; Lin, W. Autothermal reforming of methane over Ni catalysts support over ZrO2-CeO2-Al2O3. *J. Nat. Gas Chem.,* **2008**, *12*, 201-207.
[http://dx.doi.org/10.1016/S1003-9953(08)60052-3]

[20]   Trimm, D.L. Natural Gas: Conventional Steam-Reforming *Fuels – Hydrogen Production | Natural Gas: Conventional Steam-Reforming,* **2009**, 293-299.

[21]   Melo, F.; Morlanés, N. Naphtha steam reforming for hydrogen production. *Catal. Today,* **2005**, *107-108*, 458-466.
[http://dx.doi.org/10.1016/j.cattod.2005.07.028]

[22]   Xu, X.; Jiang, E.; Wang, M.; Xu, Y. Dry and steam reforming of biomass pyrolysis gas for rich hydrogen gas. *Biomass Bioenergy,* **2015**, *78*, 6-16.
[http://dx.doi.org/10.1016/j.biombioe.2015.03.015]

[23]   Silva, P.P.; Ferreira, R.A.R.; Noronha, F.B.; Hori, C.E. Hydrogen production from steam and oxidative steam reforming of liquefied petroleum gas over cerium and strontium doped LaNiO3 catalysts. *Catal. Today,* **2017**, *289*, 211-221.

[http://dx.doi.org/10.1016/j.cattod.2016.10.003]

[24]   Laosiripojana, N.; Assabumrungrat, S. Hydrogen production from steam and autothermal reforming of LPG over high surface area ceria. *J. Power Sources,* **2006**, *158*, 1348-1357.
[http://dx.doi.org/10.1016/j.jpowsour.2005.10.058]

[25]   Malaibari, Z.O.; Amin, A.; Croiset, E.; Epling, W. Performance characteristics of Mo–Ni/Al2O3 catalysts in LPG oxidative steam reforming for hydrogen production. *Int. J. Hydrogen Energy,* **2014**, *39*(19), 10061-10073.
[http://dx.doi.org/10.1016/j.ijhydene.2014.03.169]

[26]   Tan, M.; Wang, X.; Hu, Y.; Zou, X.; Ding, W.; Lu, X. Influence of reduction temperature on properties for pre-reforming of liquefied petroleum gas over mesoporous g-alumina supported Ni-MgO catalyst by one-pot template-free route. *Int. J. Hydrogen Energy,* **2015**, *40*, 16202-16214.
[http://dx.doi.org/10.1016/j.ijhydene.2015.08.084]

[27]   Rakib, M.; Grace, J.; Lim, C.; Elnashaie, S.; Ghiasi, B. Steam reforming of propane in a fluidized bed membrane reactor for hydrogen production. *Int. J. Hydrogen Energy,* **2010**, *35*, 6276-6290.
[http://dx.doi.org/10.1016/j.ijhydene.2010.03.136]

[28]   Pino, L.; Vita, A.; Cipitì, F.; Laganá, M.; Recupero, V. Performance of Pt/CeO2 catalyst for propane oxidative steam reforming. *Appl. Catal. A Gen.,* **2006**, *306*, 68-77.
[http://dx.doi.org/10.1016/j.apcata.2006.03.031]

[29]   Bae, J.; Lee, S.; Kim, S.; Oh, J.; Choi, S.; Bae, M.; Kang, I.; Katikaneni, S.P. Liquid fuel processing for hydrogen production: A review. *Int. J. Hydrogen Energy,* **2016**, *41*, 19990-20022.
[http://dx.doi.org/10.1016/j.ijhydene.2016.08.135]

[30]   Simson, A.; Farrauto, R.; Castaldi, M. Steam reforming of ethanol/gasoline mixtures: Deactivation, regeneration and stable performance. *Appl. Catal. B,* **2011**, *106*, 295-303.
[http://dx.doi.org/10.1016/j.apcatb.2011.05.027]

[31]   Vita, A.; Italiano, C.; Pino, L.; Laganà, M.; Recupero, V. Hydrogen-rich gas production by steam reforming of n-dodecane. Part II: Stability, regenerability and sulfur poisoning of low loading Rh-based catalyst. *Appl. Catal. B,* **2017**, *218*, 317-326.
[http://dx.doi.org/10.1016/j.apcatb.2017.06.059]

[32]   Xie, C.; Chen, Y.; Engelhard, M.H.; Song, C. Comparative Study on the Sulfur Tolerance and Carbon Resistance of Supported Noble Metal Catalysts in Steam Reforming of Liquid Hydrocarbon Fuel. *ACS Catal.,* **2012**, *2*, 1127-1137.
[http://dx.doi.org/10.1021/cs200695t]

[33]   Nabgan, W.; Tuan Abdullah, T.A.; Mat, R.; Nabgan, B.; Gambo, Y.; Ibrahim, M.; Ahmad, A.; Abdul Jalila, A.; Triwahyonod, S.; Saehe, I. Renewable hydrogen production from bio-oil derivative *via* catalytic steam reforming: An overview. *Renew. Sustain. Energy Rev.,* **2017**, *79*, 347-357.
[http://dx.doi.org/10.1016/j.rser.2017.05.069]

[34]   Trane, R.; Dahl, S.; Skjøth-Rasmussen, M.S.; Jensen, A.D. Catalytic steam reforming of bio-oil. *Int. J. Hydrogen Energy,* **2012**, *37*, 6447-6472.
[http://dx.doi.org/10.1016/j.ijhydene.2012.01.023]

[35]   Bu, Q.; Lei, H.; Ren, S.; Wang, L.; Holladay, J.; Zhang, Q.; Tang, J.; Ruan, R. Phenol and phenolics from lignocellulosic biomass by catalytic microwave pyrolysis. *Bioresour. Technol.,* **2011**, *102*, 7004-7007.
[http://dx.doi.org/10.1016/j.biortech.2011.04.025]

[36]   Matas Güell, B.; Babich, I.V.; Lefferts, L.; Seshan, K. Steam reforming of phenol over Ni-based catalysts – A comparative study. *Appl. Catal. B,* **2011**, *106*, 280-286.
[http://dx.doi.org/10.1016/j.apcatb.2011.05.012]

[37]   Remón, J.; Broust, F.; Volle, G.; García, L.; Arauzo, J. Hydrogen production from pine and poplar bio-oils by catalytic steam reforming. Influence of the bio-oil composition on the process. *Int. J.*

Hydrogen Energy, **2015**, *40*, 5593-5608.
[http://dx.doi.org/10.1016/j.ijhydene.2015.02.117]

[38] Esteban-Díez, G.; Gil, M.V.; Pevida, C.; Chen, D.; Rubiera, F. Effect of operating conditions on the sorption enhanced steam reforming of blends of acetic acid and acetone as bio-oil model compounds. *Appl. Energy,* **2016**, *177*, 579-590.
[http://dx.doi.org/10.1016/j.apenergy.2016.05.149]

[39] Rioche, C.; Kulkarni, S.; Meunier, F.C.; Breen, J.P.; Burch, R. Steam reforming of model compounds and fast pyrolysis bio-oil on supported noble metal catalysts. *Appl. Catal. B,* **2005**, *61*, 130-139.
[http://dx.doi.org/10.1016/j.apcatb.2005.04.015]

[40] Kechagiopoulos, P.N.; Voutetakis, S.S.; Lemonidou, A.A.; Vasalos, I.A. Hydrogen Production *via* Steam Reforming of the Aqueous Phase of Bio-Oil in a Fixed Bed Reactor. *Energy Fuels,* **2006**, *20*, 2155-2163.
[http://dx.doi.org/10.1021/ef060083q]

[41] Sengupta, P.; Khan, A.; Abu Zahid, Md.; Ibrahim, H.; Idem, R. Evaluation of the Catalytic Activity of Various 5Ni/Ce0.5Zr0.33M0.17O2-δ Catalysts for Hydrogen Production by the Steam Reforming of a Mixture of Oxygenated Hydrocarbons. *Energy Fuels,* **2012**, *26*, 816-828.
[http://dx.doi.org/10.1021/ef201854t]

[42] Wang, D.; Czernik, S.; Chornet, E. Production of Hydrogen from Biomass by Catalytic Steam Reforming of Fast Pyrolysis Oils. *Energy Fuels,* **1998**, *12*, 19-24.
[http://dx.doi.org/10.1021/ef970102j]

[43] Czernik, S.; French, R.; Feik, C.; Chornet, E. Hydrogen by Catalytic Steam Reforming of Liquid Byproducts from Biomass Thermoconversion Processes. *Ind. Eng. Chem. Res.,* **2002**, *41*, 4209-4215.
[http://dx.doi.org/10.1021/ie020107q]

[44] Bion, N.; Eprona, F., Dupreza, D. Bioethanol reforming for H2 production. A comparison with hydrocarbon reforming. *Catalysis,* **2010**, *22*, 1-55.

[45] Sub Kwak, B.; Kim, J.; Kang, M. Hydrogen production from ethanol steam reforming over core-shell structured NixOy-, FexOy-, and CoxOy-Pd catalysts. *Int. J. Hydrogen Energy,* **2010**, *35*, 11829-11843.
[http://dx.doi.org/10.1016/j.ijhydene.2010.08.073]

[46] Kim, K.M.; Sub Kwak, B. Im, Y.; Park, N.K.; Lee, T.J.; Lee, S.T.; Kang, M. Effective hydrogen production from ethanol steam reforming using CoMg co-doped SiO2@Co1-xMgxO catalyst. *J. Ind. Eng. Chem.,* **2017**, *51*, 140-152.
[http://dx.doi.org/10.1016/j.jiec.2017.02.025]

[47] Compagnoni, M.; Tripodi, A.; Rossetti, I. Parametric study and kinetic testing for ethanol steam reforming. *Appl. Catal. B,* **2017**, *203*, 899-909.
[http://dx.doi.org/10.1016/j.apcatb.2016.11.002]

[48] Greluk, M.; Słowik, G.; Rotko, M.; Machocki, A. Steam reforming and oxidative steam reforming of ethanol over PtKCo/CeO2 catalyst. *Fuel,* **2016**, *183*, 518-530.
[http://dx.doi.org/10.1016/j.fuel.2016.06.068]

[49] Nichele, V.; Signoretto, M.; Menegazzo, F.; Rossetti, I.; Cruciani, G. Hydrogen production by ethanol steam reforming: effect of the synthesis parameters on the activity of Ni/TiO2 catalysts. *Int. J. Hydrogen Energy,* **2014**, *39*, 4252-4258.
[http://dx.doi.org/10.1016/j.ijhydene.2013.12.178]

[50] Ni, M.; Leung, D.Y.C.; Leung, M.K.H. A review on reforming bio-ethanol for hydrogen production. *Int. J. Hydrogen Energy,* **2007**, *32*, 3238-3247.
[http://dx.doi.org/10.1016/j.ijhydene.2007.04.038]

[51] Agrell, J.; Birgersson, H.; Boutonnet, M. Steam reforming of methanol over a Cu/ZnO/Al2O3 catalyst: a kinetic analysis and strategies for suppression of CO formation. *J. Power Sources,* **2002**, *106*, 249-

257.
[http://dx.doi.org/10.1016/S0378-7753(01)01027-8]

[52]　Sá, S.; Silva, H.; Brandão, L.; Sousa, J.M.; Mendes, A. Catalysts for methanol steam reforming-A review. *Appl. Catal. B,* **2010**, *99*, 43-57.
[http://dx.doi.org/10.1016/j.apcatb.2010.06.015]

[53]　Conant, T.; Karim, A.; Lebarbier, V.; Wang, Y.; Girgsdies, F.; Schlögl, R.; Datye, A. Stability of bimetallic Pd–Zn catalysts for the steam reforming of methanol. *J. Catal.,* **2008**, *257*, 64-70.
[http://dx.doi.org/10.1016/j.jcat.2008.04.018]

[54]　Takahashi, K.; Takezawa, N.; Kobayashi, H. The mechanism of steam reforming of methanol over a copper-silica catalyst. *Appl. Catal.,* **1982**, *2*, 363-366.
[http://dx.doi.org/10.1016/0166-9834(82)80154-1]

[55]　Peppley, B.A.; Amphlett, J.C.; Kearns, L.M.; Mann, R.F. Methanol–steam reforming on Cu/ZnO/Al2O3 catalysts. Part 2. A comprehensive kinetic model. *Appl. Catal. A Gen.,* **1999**, *179*, 31-49.
[http://dx.doi.org/10.1016/S0926-860X(98)00299-3]

[56]　Chen, G.; Tao, J.; Liu, C.; Yan, B.; Li, W.; Li, X. Hydrogen production *via* acetic acid steam reforming: A critical review on catalysts. *Renew. Sustain. Energy Rev.,* **2017**, *79*, 1091-1098.
[http://dx.doi.org/10.1016/j.rser.2017.05.107]

[57]　Hu, X.; Zhang, L.; Lu, G. Steam reforming of acetic acid over Cu Zn Co catalyst for hydrogen generation: synergistic effects of the metal species. *Int. J. Hydrogen Energy,* **2016**, *41*, 13960-13969.
[http://dx.doi.org/10.1016/j.ijhydene.2016.05.066]

[58]　Lin, Y.C. Catalytic valorization of glycerol to hydrogen and syngas. *Int. J. Hydrogen Energy,* **2013**, *38*, 2678-2700.
[http://dx.doi.org/10.1016/j.ijhydene.2012.12.079]

[59]　Tamošiūnas, A.; Valatkevičius, P.; Gimžauskaitė, D.; Valinčius, V.; Jeguirim, M. Glycerol steam reforming for hydrogen and synthesis gas production. *Int. J. Hydrogen Energy,* **2017**, *42*, 12896-12904.
[http://dx.doi.org/10.1016/j.ijhydene.2016.12.071]

[60]　Silva, J.M.; Soria, M.A.; Madeira, L.M. Challenges and strategies for optimization of glycerol steam reforming process. *Renew. Sustain. Energy Rev.,* **2015**, *42*, 1187-1213.
[http://dx.doi.org/10.1016/j.rser.2014.10.084]

[61]　Iriondo, A.; Barrio, V.L.; Cambra, J.F.; Arias, P.L.; Guemez, M.B.; Sanchez-Sanchez, M.C. Glycerol steam reforming over Ni catalysts supported on ceria and ceria-promoted alumina. *Int. J. Hydrogen Energy,* **2010**, *35*, 11622-11633.
[http://dx.doi.org/10.1016/j.ijhydene.2010.05.105]

[62]　Pompeo, F.; Santori, G.F.; Nichio, N.N. Hydrogen production by glycerol steam reforming with Pt/SiO2 and Ni/SiO2 catalysts. *Catal. Today,* **2011**, *172*, 183-188.
[http://dx.doi.org/10.1016/j.cattod.2011.05.001]

[63]　Montini, T.; Singh, R.; Das, P.; Lorenzut, B.; Bertero, N.; Riello, P. Renewable H2 from glycerol steam reforming: effect of La2O3 and CeO2 addition to Pt/Al2O3 catalysts. *ChemSusChem,* **2010**, *3*, 619-628.
[http://dx.doi.org/10.1002/cssc.200900243]

[64]　Cui, Y.; Galvita, V.; Rihko-Struckmann, L.; Lorenz, H.; Sundmacher, K. Steam reforming of glycerol: the experimental activity of La1-xCexNiO3 catalyst in comparison to the thermodynamic reaction equilibrium. *Appl. Catal. B,* **2009**, *90*, 29-37.
[http://dx.doi.org/10.1016/j.apcatb.2009.02.006]

[65]　Zhang, B.; Tang, X.; Li, Y.; Xu, Y.; Shen, W. Hydrogen production from steam reforming of ethanol and glycerol over ceria-supported metal catalysts. *Int. J. Hydrogen Energy,* **2007**, *32*, 2367-2373.

[http://dx.doi.org/10.1016/j.ijhydene.2006.11.003]

[66]   Cheng, C.K.; Foo, S.Y.; Adesina, A.A. H2-rich synthesis gas production over Co/Al2O3 catalyst *via* glycerol steam reforming. *Catal. Commun.,* **2010**, *12*, 292-298.
[http://dx.doi.org/10.1016/j.catcom.2010.09.018]

[67]   Demsash, H.D.; Kondamudi, K.V.K.; Upadhyayula, S.; Mohan, R. Ruthenium doped nickel-alumin--ceria catalyst in glycerol steam reforming. *Fuel Process. Technol.,* **2018**, *169*, 150-156.
[http://dx.doi.org/10.1016/j.fuproc.2017.09.017]

[68]   Hirai, T.; Ikenaga, N.O.; Miyake, T.; Suzuki, T. Production of hydrogen by steam reforming of glycerin on ruthenium catalyst. *Energy Fuels,* **2005**, *19*, 1761-1762.
[http://dx.doi.org/10.1021/ef050121q]

[69]   Calles, J.A.; Carrero, A.; Vizcaino, A.J.; Garcia-Moreno, L. Hydrogen production by glycerol steam reforming over SBA-15- supported nickel catalysts: effect of alkaline earth promoters on activity and stability. *Catal. Today,* **2014**, *227*, 198-206.
[http://dx.doi.org/10.1016/j.cattod.2013.11.006]

[70]   Zamzuri, N.H.; Mat, R.; Amin, N.A.S.; Talebian-Kiakalaieh, A. Hydrogen production from catalytic steam reforming of glycerol over various supported nickel catalysts. *Int. J. Hydrogen Energy,* **2017**, *42*, 9087-9098.
[http://dx.doi.org/10.1016/j.ijhydene.2016.05.084]

[71]   He, Q.; McNutt, J.; Yang, J. Utilization of the residual glycerol from biodiesel production for renewable energy generation. *Renew. Sustain. Energy Rev.,* **2017**, *71*, 63-76.
[http://dx.doi.org/10.1016/j.rser.2016.12.110]

[72]   Ahmed, A.M.A.; Salmiaton, A.; Choong, T.S.Y.; Wan Azlina, W.A.K.G. Review of kinetic and equilibrium concepts for biomass tar modeling by using Aspen Plus. *Renew. Sustain. Energy Rev.,* **2015**, *52*, 1623-1644.
[http://dx.doi.org/10.1016/j.rser.2015.07.125]

[73]   Miyazawa, T.; Kimura, T.; Nishikawa, J.; Kado, S.; Kunimori, K.; Tomishige, K. Catalytic performance of supported Ni catalysts in partial oxidation and steam reforming of tar derived from the pyrolysis of wood biomass. *Catal. Today,* **2006**, *115*, 254-262.
[http://dx.doi.org/10.1016/j.cattod.2006.02.055]

[74]   Li, D.; Ishikawa, C.; Koike, M.; Wang, L.; Nakagawa, Y.; Tomishige, K. Production of renewable hydrogen by steam reforming of tar from biomass pyrolysis over supported Co catalysts. *Int. J. Hydrogen Energy,* **2013**, *38*, 3572-3581.
[http://dx.doi.org/10.1016/j.ijhydene.2013.01.057]

[75]   Kaewpanha, M.; Karnjanakom, S.; Guan, G.; Hao, X.; Yang, J.; Abudula, A. Removal of biomass tar by steam reforming over calcined scallop shell supported Cu catalysts. *J. Energy Chem.,* **2017**, *26*, 660-666.
[http://dx.doi.org/10.1016/j.jechem.2017.03.012]

[76]   Li, D.; Koike, M.; Chen, J.; Nakagawa, Y.; Tomishige, K. Preparation of Ni-Cu/Mg/Al catalysts from hydrotalcite-like compounds for hydrogen production by steam reforming of biomass tar. *Int. J. Hydrogen Energy,* **2014**, *39*, 10959-10970.
[http://dx.doi.org/10.1016/j.ijhydene.2014.05.062]

[77]   Chen, J.; Tamura, M.; Nakagawa, Y.; Okumura, K.; Tomishige, K. Promoting effect of trace Pd on hydrotalcite-derived Ni/Mg/Al catalyst in oxidative steam reforming of biomass tar. *Appl. Catal. B,* **2015**, *179*, 412-421.
[http://dx.doi.org/10.1016/j.apcatb.2015.05.042]

[78]   Gao, N.; Liu, S.; Han, Y.; Xing, C.; Li, A. Steam reforming of biomass tar for hydrogen production over NiO/ceramic foam catalyst. *Int. J. Hydrogen Energy,* **2015**, *40*, 7983-7990.
[http://dx.doi.org/10.1016/j.ijhydene.2015.04.050]

[79]   Kaisalo, N.; Simell, P.; Lehtonen, J. Benzene steam reforming kinetics in biomass gasification gas cleaning. *Fuel,* **2016**, *182*, 696-703.
[http://dx.doi.org/10.1016/j.fuel.2016.06.042]

[80]   Gao, N.; Wang, X.; Li, A.; Wu, C.; Yin, Z. Hydrogen production from catalytic steam reforming of benzene as tar model compound of biomass gasification. *Fuel Process. Technol.,* **2016**, *148*, 380-387.
[http://dx.doi.org/10.1016/j.fuproc.2016.03.019]

[81]   Furusawa, T.; Saito, K.; Kori, Y.; Miura, Y.; Sato, M.; Suzuki, N. Steam reforming of naphthalene/benzene with various types of Pt- and Ni-based catalysts for hydrogen production. *Fuel,* **2013**, *103*, 111-121.
[http://dx.doi.org/10.1016/j.fuel.2011.09.026]

[82]   Ahmed, T.; Xiu, S.; Wang, L.; Shahbazi, A. Investigation of Ni/Fe/Mg zeolite-supported catalysts in steam reforming of tar using simulated-toluene as model compound. *Fuel,* **2018**, *211*, 566-571.
[http://dx.doi.org/10.1016/j.fuel.2017.09.051]

[83]   Takise, K.; Imori, M.; Mukai, D.; Ogo, S.; Sugiura, Y.; Sekine, Y. Effect of catalyst structure on steam reforming of toluene over Ni/La0.7Sr0.3AlO3−δ catalyst. *Appl. Catal. A Gen.,* **2015**, *489*, 155-161.
[http://dx.doi.org/10.1016/j.apcata.2014.10.014]

[84]   Qian, K.; Kumar, A.; Zhang, H.; Bellmer, D.; Huhnke, R. Recent advances in utilization of biochar. *Renew. Sustain. Energy Rev.,* **2015**, *42*, 1055-1064.
[http://dx.doi.org/10.1016/j.rser.2014.10.074]

[85]   Haber, J. in: Ruiz P., Delmon B. (Eds.,), "New Developments in Selective Oxidation by Heterogeneous Catalysis. *Studies in Surface Science and Catalysis, Elsevier Science Publishers,* **1992**, *72*, 279-304.
[http://dx.doi.org/10.1016/S0167-2991(08)61679-1]

[86]   Ma, Y.; Guan, G.; Hao, X.; Cao, J.; Abudula, A. Molybdenum carbide as alternative catalyst for hydrogen production – A review. *Renew. Sustain. Energy Rev.,* **2017**, *75*, 1101-1129.
[http://dx.doi.org/10.1016/j.rser.2016.11.092]

[87]   Christofoletti, T.; Assaf, J.M.; Assaf, E.M. Methane steam reforming on supported and non-supported molybdenum carbides. *Chem. Eng. J.,* **2005**, *106*, 97-103.
[http://dx.doi.org/10.1016/j.cej.2004.11.006]

[88]   Széchenyi, A.; Solymosi, F. Production of hydrogen in the decomposition of ethanol and methanol over unsupported Mo2C catalysts. *J. Phys. Chem. C,* **2007**, *111*, 509-515.
[http://dx.doi.org/10.1021/jp072439k]

[89]   Ma, Y.; Guan, G.; Shi, C.; Zhu, A.; Hao, X.; Wang, Z.; Kusakabe, K.; Abudula, A. Low temperature steam reforming of methanol to produce hydrogen over various metal-doped molybdenum carbide catalysts. *Int. J. Hydrogen Energy,* **2014**, *39*, 258-266.
[http://dx.doi.org/10.1016/j.ijhydene.2013.09.150]

[90]   Lin, S.S.Y.; Thomson, W.J.; Hagensen, T.J.; Ha, S.Y. Steam reforming of methanol using supported Mo2C catalysts. *Appl. Catal. A Gen.,* **2007**, *318*, 121-127.
[http://dx.doi.org/10.1016/j.apcata.2006.10.054]

[91]   Ma, Y.; Guan, G.; Phanthong, P.; Li, X.; Cao, J.; Hao, X.; Wang, Z.; Abudula, A. Steam reforming of methanol for hydrogen production over nanostructured wire-like molybdenum carbide catalyst. *Int. J. Hydrogen Energy,* **2014**, *39*, 18803-18811.
[http://dx.doi.org/10.1016/j.ijhydene.2014.09.062]

[92]   Cao, J.; Ma, Y.; Guan, G.; Hao, X.; Ma, X.; Wang, Z.; Kusakabe, K.; Abudula, A. Reaction intermediate species during the steam reforming of methanol over metal modified molybdenum carbide catalysts. *Appl. Catal. B,* **2016**, *189*, 12-18.
[http://dx.doi.org/10.1016/j.apcatb.2016.02.021]

[93]   Miyamoto, Y.; Akiyama, M.; Nagai, M. Steam reforming of ethanol over nickel molybdenum carbides

for hydrogen production. *Catal. Today,* **2009**, *146*, 87-95.
[http://dx.doi.org/10.1016/j.cattod.2008.12.033]

[94]   Kaewpanha, M.; Guan, G.; Ma, Y.; Hao, X.; Zhang, Z.; Reubroychareon, P.; Kusakabe, K.; Abudula, A. Hydrogen production by steam reforming of biomass tar over biomass char supported molybdenum carbide catalyst. *Int. J. Hydrogen Energy,* **2015**, *40*, 7974-7982.
[http://dx.doi.org/10.1016/j.ijhydene.2015.04.068]

[95]   Rodríguez, J.A.; Liu, P.; Hrbek, J.; Pérez, M.; Evans, J. Water–gas shift activity of Au and Cu nanoparticles supported on molybdenum oxides. *J. Mol. Catal. Chem.,* **2008**, *281*, 59-65.
[http://dx.doi.org/10.1016/j.molcata.2007.07.032]

[96]   Liu, S.; Chen, M.; Chu, L.; Yang, Z.; Zhu, C.; Wang, J.; Chen, M. Catalytic steam reforming of bio-oil aqueous fraction for hydrogen production over Ni-Mo supported on modified sepiolite catalysts. *Int. J. Hydrogen Energy,* **2013**, *38*, 3948-3955.
[http://dx.doi.org/10.1016/j.ijhydene.2013.01.117]

[97]   Mitran, G.; Pavel, O.D.; Florea, M.; Mieritz, D.G.; Seo, D.K. Hydrogen production from glycerol steam reforming over molybdena–alumina catalysts. *Catal. Commun.,* **2016**, *77*, 83-88.
[http://dx.doi.org/10.1016/j.catcom.2016.01.029]

[98]   Mitran, G.; Pavel, O.D.; Mieritz, D.G.; Seo, D.K.; Florea, M. Effect of Mo/Ce ratio in Mo–Ce–Al catalysts on the hydrogen production by steam reforming of glycerol. *Catal. Sci. Technol.,* **2016**, *6*, 7902-7912.
[http://dx.doi.org/10.1039/C6CY00999A]

[99]   Nguyen-Thanh, D.; Duarte de Farias, A.; Fraga, M. Characterization and activity of vanadia-promoted Pt/ZrO2 catalysts for the water–gas shift reaction. *Catal. Today,* **2008**, *138*, 235-238.
[http://dx.doi.org/10.1016/j.cattod.2008.05.036]

[100]   Kokumai, T.; Cantane, D.; Melo, G.; Paulucci, L.; Zanchet, D. VOx-Pt/Al2O3 catalysts for hydrogen production. *Catal. Today,* **2017**, *289*, 249-257.
[http://dx.doi.org/10.1016/j.cattod.2016.09.021]

[101]   Sub Kwak, B.; Kim, K.M.; Jo, S.W.; Do, J.Y.; Kang, S.; Park, M.; Park, N.K.; Lee, T.J.; Lee, S.T.; Kang, M. Characterizations of bimetallic NiV-supported SiO2 catalysts prepared for effectively hydrogen evolutions from ethanol steam reforming. *J. Ind. Eng. Chem.,* **2016**, *37*, 57-66.
[http://dx.doi.org/10.1016/j.jiec.2016.03.002]

[102]   Matsuka, M.; Shigedomi, K.; Ishihara, T.T. Comparative study of propane steam reforming in vanadium based catalytic membrane reactor with nickel-based catalysts. *Int. J. Hydrogen Energy,* **2014**, *139*(27), 14792-14799.

[103]   González-Gil, R.; Herrera, C.; Larrubia, M.; Kowalik, P.; Pieta, I.; Alemany, L. Hydrogen production by steam reforming of DME over Ni-based catalysts modified with vanadium. *Int. J. Hydrogen Energy,* **2016**, *41*, 19781-19788.
[http://dx.doi.org/10.1016/j.ijhydene.2016.05.074]

[104]   Mitran, G.; Mieritz, D.; Seo, D.K. Hydrotalcites with vanadium, effective catalysts for steam reforming of toluene. *Int. J. Hydrogen Energy,* **2017**, *42*, 21732-21740.
[http://dx.doi.org/10.1016/j.ijhydene.2017.07.097]

CHAPTER 2

Ni/CaO.Ca$_{12}$Al$_{14}$O$_{33}$ Based-Materials for Hydrogen Production by CO$_2$ Sorption Enhanced Steam Methane Reforming

Moisés R. Cesario[1,*], **Braúlio S. Barros**[2], **Claire Courson**[3,*] and **Dulce M.A. Melo**[4]

[1] *Unité de Chimie Environnementale et Interactions sur le Vivant (UCEIV, E.A. 4492), MREI, Université du Littoral Côte d'Opale (ULCO), 59140, Dunkerque, France*

[2] *Department of Mechanical Engineering, Federal University of Pernambuco, 50670-901, Recife, Brazil*

[3] *Institut de Chimie et Procédés pour l'Energie, l'Environnement et la Santé (ICPEES)-UMR CNRS 7515, University of Strasbourg, 67087, Strasbourg, France*

[4] *Department of Chemistry, Federal University of Rio Grande do Norte, 59072-970, Natal, Brazil*

Abstract: Sorbents with different CaO/Ca$_{12}$Al$_{14}$O$_{33}$ ratios were prepared by microwave-assisted self-combustion. Catalysts were prepared by Ni wet impregnation of sorbents and a catalyst with an optimized composition was also prepared by microwave-assisted self-combustion method. Then, the CO$_2$ sorption capacity and catalytic activity in sorption enhanced steam methane reforming (SE-SMR) for production of hydrogen were evaluated. X-ray diffraction (XRD) analysis of reduced catalysts confirmed Ni, CaO and Ca$_{12}$Al$_{14}$O$_{33}$ phases. Catalysts prepared by wet impregnation favor high surface area, and consequently the CO$_2$ sorption. The CO$_2$ sorption by CaO modifies the equilibrium of the water gas shift reaction (WGS) and consequently the hydrogen production is improved at 650 °C. The time of breakthrough for H$_2$, CO, CO$_2$, and CH$_4$ depends on both the excess of CaO and operating conditions (H$_2$O/CH$_4$ ratio). Ni-CA75 catalyst obtained by impregnation method with a CaO excess around 75% leads to the optimum activity. It exhibits high sorption capacity and hydrogen selectivity of 100% during 2 h and 16 h for H$_2$O/CH$_4$ ratios of 3 and 1, respectively. Therefore, Ni-CA75 is the most active and stable catalyst during 50 h in CO$_2$ sorption enhanced steam methane reforming, even at an unusually low temperature (650 °C).

* **Corresponding authors Moisés Cesario:** Unité de Chimie Environnementale et Interactions sur le Vivant (UCEIV, E.A. 4492), MREI, Université du Littoral Côte d'Opale (ULCO), Dunkerque, France; Tel: +33 (0)7 81 55 95 94; E-mails: moises.cesario@univ-littoral.fr, moisesrcesario@gmail.com
Claire Courson: Institut de Chimie et Procédés pour l'Energie, l'Environnement et la Santé (ICPEES)-UMR CNRS 7515, University of Strasbourg, 67087 Strasbourg, France; Tel: +33 3 68 85 27 70; E-mail: claire.courson@unistra.fr

Keywords: CO_2 Sorbent, Hydrogen, Microwave-Assisted Self-Combustion, Ni Catalyst, Sorption Capacity, Steam Methane Reforming, Wet Impregnation, Water Gas Shift.

INTRODUCTION

Research topics related to energy and environment are still important due to the need for developing clean energy technologies to produce hydrogen [1 - 3]. Hydrogen can be obtained by catalytic Steam Methane Reforming (SMR) according to the reaction (1) [4, 5]:

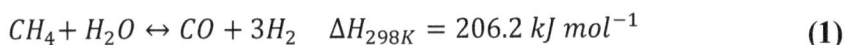

$$CH_4 + H_2O \leftrightarrow CO + 3H_2 \quad \Delta H_{298K} = 206.2 \ kJ \ mol^{-1} \tag{1}$$

This catalytic reaction occurs at high temperature (T $\geq$ 800 ° C) and low pressure (1-5 bar). The water excess promotes, in the form of CO or CO_2, the removal of carbonaceous deposits formed by the dissociation of methane. Steam reforming is favored over supported metal catalysts (Ni-alumina) in which alumina acidity can be compensated by alkaline earth oxides (MgO, CaO) dopage [6 - 9]. The steam reforming reaction is always accompanied by Water-Gas Shift Reaction (WGS) (2):

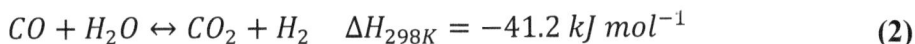

$$CO + H_2O \leftrightarrow CO_2 + H_2 \quad \Delta H_{298K} = -41.2 \ kJ \ mol^{-1} \tag{2}$$

This reaction allows increasing the global production of hydrogen and leads to the formation of undesirable CO_2 for the environment. One of the possible solutions to limit the formation of CO_2 is to trap CO_2 on a sorbent and then regenerate it by heating (Temperature Swing Absorption) before a new absorption cycle. The CO_2 produced during the desorption can be concentrated and purified for subsequent chemical upgrading to energy product (CH_4) or chemical raw material (CH_3OH). Calcium oxide can be used as a sorbent which carbonates according to the reversible reaction (3) [10 - 13]:

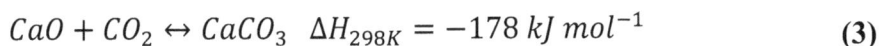

$$CaO + CO_2 \leftrightarrow CaCO_3 \quad \Delta H_{298K} = -178 \ kJ \ mol^{-1} \tag{3}$$

CO_2 sorption by CaO makes it possible to largely compensate for the very high endothermicity of the steam reforming reaction but also to shift the reaction (1) towards greater formation of hydrogen. The positive effect of CO_2 sorption is maximum at low temperatures. The thermodynamics of the sorption reaction shows that a temperature of 700 °C must not be exceeded for this effect to remain significant [14, 15].

Then, the first challenge is to have sufficient steam reforming activity at unusual temperatures of 650-700 °C, below the usual range of 800-900 °C. In this case, the choice of the catalyst and its preparation are essential. The second challenge is to maintain the CaO uptake capacity at high efficiency after several absorption-desorption cycles. The literature reports a marked decrease in the CaO sorption capacity of CO_2 depending on the time and number of cycles [12, 16 - 18]. This is related to the sintering of the CaO particles, and consequently, a decrease to specific surface area of the sorbent [18 - 21]. One of the possible solutions to improve the aging of the system is to disperse CaO on a support, taking care that the chosen support does not interact strongly with the metal and that possible interaction with CaO can be controlled.

In this context, this chapter focuses on the study of active and stable bifunctional materials for the catalytic steam reforming of methane at low temperatures as well as for the absorption of CO_2 produced during the SRM reaction. These bifunctional materials may be easily regenerated and have long-term stability.

Materials Preparation

Several synthesis methods [18, 22, 23] have been proposed to prepare Ni-based catalytic materials. The common objective is obtaining materials with high chemical homogeneity, high surface, microporosity and high reducibility. These characteristics are essential for application as catalysts in SRM. This current chapter highlights the preparation of CaO-Ca$_{12}$Al$_{14}$O$_{33}$ (noted CaO.CaAl) sorbents and simultaneous preparation of Ni/CaO-CaAl nanocatalysts/sorbents powders with different compositions x%CaO.y%CaAl by a microwave assisted self-combustion method.

Microwave Assisted Self-Combustion Synthesis

Microwave assisted self-combustion has been chosen as a principal method for the preparation of the bifunctional materials under investigation. It consists of the conventional combustion reaction of reducing and oxidizing precursors, where the heat required for ignition is provided by polar molecules exposed to the microwave radiation. The heat generated by the combustion allows the decomposition of the precursors and the formation of new crystal structures. Advantages of this synthesis method compared to the traditional combustion method using hot plate and muffle furnace are the following: reduced time required to attain the ignition temperature and uniformity of temperature distribution, which in this case are achieved within the precursor suspension. In addition, microwave heating allows better control of the synthesis conditions, since the intensity of the microwave emission can be quickly interrupted, reduced or increased, allowing to obtain materials with very specific characteristics.

Therefore, this method not only leads to a better homogeneity and purity of the final product but also saves energy and time during preparation. From an economic point of view, this process is very attractive for obtaining nanocomposites for industrial applications.

Sorbents Preparation

x%CaO.y%CaAl sorbents were prepared by microwave assisted self-combustion. This synthesis method comprises of four steps:

The first step refers to the mixture of calcium [$Ca(NO_3)_2.4H_2O$ – Merck] and aluminum [$Al(NO_3)_3.9H_2O$ – Merck] nitrates as oxidizing agents and urea fuel [$CO(NH_2)_2$ – Merck] as a reducing agent in the presence of ammonium nitrate [NH_4NO_3 – Merck]. The stoichiometric oxidizing agent/fuel molar ratio was 1/1;

The second step refers to the mixture of precursors and its constant stirring under heating at temperatures between 60 and 80 °C, obtaining a suspension with good homogenization.

The third step refers to the self-ignition of the precursor suspension under incidence of microwaves in a commercial oven (Brastemp/maximum power of 900 W) using power and frequency equal to 810 W and 2.45 GHz, respectively. The self-ignition occurs between 1 and 5 min, giving rise to the x%CaO.y%CaAl precursor powder;

The fourth step refers to the thermal treatment of the precursor powder at 900-950 °C in order to promote the crystallization of the desired material.

Fig. (1) shows the methodology for obtaining x%CaO.y%CaAl sorbents powders.

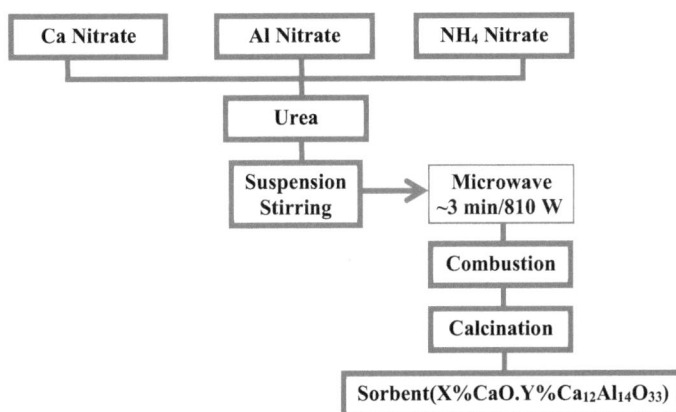

Fig. (1). Methodology for obtaining x%CaO.y%CaAl sorbents powders.

Catalyts Preparation

Ni/x%CaO.y%CaAl catalysts were prepared by microwave assisted self-combustion, adding the nickel nitrate [Ni(NO$_3$)$_2$.6H$_2$O – Merck] in first step of the procedure of sorbent synthesis. Fig. (**2**) shows the methodology for obtaining Ni/x%CaO.y%CaAl catalyst powders.

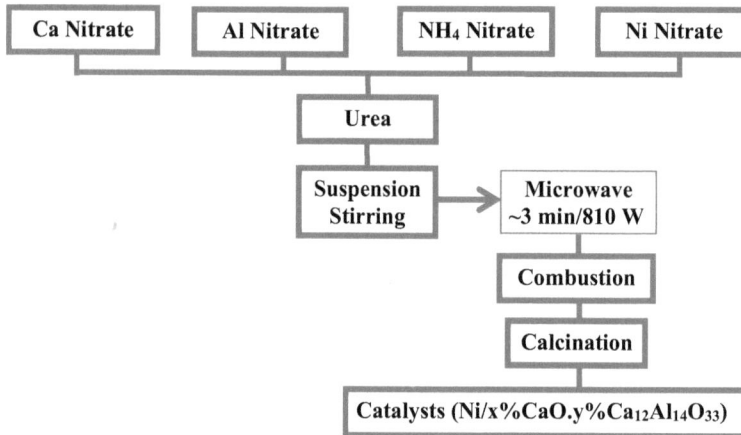

```
 ┌─────────────┐  ┌─────────────┐  ┌──────────────┐  ┌─────────────┐
 │  Ca Nitrate │  │  Al Nitrate │  │  NH₄ Nitrate │  │  Ni Nitrate │
 └─────────────┘  └─────────────┘  └──────────────┘  └─────────────┘

                      ┌──────────┐
                      │   Urea   │
                      └──────────┘

               ┌─────────────┐      ┌──────────────┐
               │ Suspension  │ ───► │  Microwave   │
               │  Stirring   │      │ ~3 min/810 W │
               └─────────────┘      └──────────────┘

                                    ┌──────────────┐
                                    │  Combustion  │
                                    └──────────────┘

                                    ┌──────────────┐
                                    │  Calcination │
                                    └──────────────┘

              ┌──────────────────────────────────────────────┐
              │ Catalysts (Ni/x%CaO.y%Ca₁₂Al₁₄O₃₃) │
              └──────────────────────────────────────────────┘
```

Fig. (2). Methodology for obtaining Ni/x%CaO.y%CaAl catalyst powders.

Another method to obtain catalysts was adopted. It consists of wet impregnation using an aqueous solution of nickel nitrate as a precursor (5% wt of Ni of the final weight of the sorbent sample). The suspension of sorbents in the nickel nitrate solution was kept under stirring at 110 °C during 30 min to induce the evaporation of the solvent. For the comparison of the sorption capacity, a sample of CaO was obtained by calcination of calcite at 900 °C for 2 h and then impregnated with 5 wt% Ni.

Table **1** shows the nomenclature of the prepared sorbents and catalysts as well as the conditions for their synthesis before calcination.

Table 1. Nomenclature of prepared samples.

Samples	Synthesis Method	Nomenclature
4%CaO.96%Ca$_{12}$Al$_{14}$O$_{33}$	Microwave assisted self-Combustion	CA4
8%CaO.92% Ca$_{12}$Al$_{14}$O$_{33}$	Microwave assisted self-Combustion	CA8
27%CaO.73%Ca$_{12}$Al$_{14}$O$_{33}$	Microwave assisted self-Combustion	CA27
48%CaO.52%Ca$_{12}$Al$_{14}$O$_{33}$	Microwave assisted self-Combustion	CA48
65%CaO.35%Ca$_{12}$Al$_{14}$O$_{33}$	Microwave assisted self-Combustion	CA65

Samples	Synthesis Method	Nomenclature
75%CaO.25%Ca$_{12}$Al$_{14}$O$_{33}$	Microwave assisted self-Combustion	CA75
5%Ni-CA4	Ni impregnation of CA4 then drying	Ni-CA4
5%Ni-CA8	Ni impregnation of CA8 then drying	Ni-CA8
5%Ni-CA27	Ni impregnation of CA27 then drying	Ni-CA27
5%Ni-CA48	Ni impregnation of CA48 then drying	Ni-CA48
5%Ni-CA65	Ni impregnation of CA65 then drying	Ni-CA65
5%Ni-CA75	Ni impregnation of CA75 then drying	Ni-CA75
5%Ni-CaO	Ni impregnation of CaO then drying	Ni-CaO
5%Ni-75%CaO.25%Ca$_{12}$Al$_{14}$O$_{33}$	Microwave assisted self-Combustion	Ni-CA75M

All the samples were further calcined at 750 °C for 4 h with a heating rate of 3 °C min^{-1}.

Materials Characterization

The sorbents and catalysts were characterized by X-ray Diffraction (XRD) using a Bruker D8-Advance diffractometer, with Cu Kα radiation (λ=1.5406 Å) at room temperature. The diffraction powder patterns were obtained in the angular range of 10–90° using step-scanning mode (0.02°/step) with a counting time of 2 s/step.

The crystallite sizes of metallic nickel and calcium oxide for catalysts which underwent a reduction treatment or the reactivity test (spent catalysts) were calculated from the broadening of the main diffraction rays using the Scherrer equation (Equation 4) [24].

$$L = \frac{0.9\lambda}{\beta \cos\theta} \qquad (4)$$

Where L is the crystallite size, λ is the X-ray wavelength (λ=1.5406 Å), β is the line broadening and θ is the Bragg angle.

The content of the active nickel phase was determined by Inductively Coupled Plasma (ICP - *Service Central d'Analyse*, CNRS, France).

The specific surface area of the sorbents and catalysts by means of BET method (Brunauer, Emmett and Teller) was determined by N$_2$ adsorption-desorption at -196 °C in a Micrometics Tri Star 3000 analyzer. The materials were degassed overnight at 250 °C before being analyzed.

FTIR spectra were measured using a Nicolet iS10 spectrometer in the range of 400-4000 cm^{-1} with a total of 32 scans and 4 cm^{-1} resolutions. Prior to analysis, samples were diluted with KBr powder and then pelletized.

The surface morphology and microstructure of the catalysts were examined using a cold field-emission gun scanning electron microscope FEG-SEM (JEOL 6700F). The microstructure and particle size of the active phase for the best catalyst was also evaluated by Transmission Electron Microscopy (TEM, TOPCOM EM-002B).

Temperature-Programmed Reduction (TPR) was carried out on a Micromeritics AutoChem II apparatus to study the reducibility of the catalytic materials and the temperatures at which reduction takes place. 50 mg was placed in a quartz U-tube (6.6 mm internal diameter) and submitted at a total gas flow of 50 ml min^{-1}, consisting of a mixture of 90% argon and 10% hydrogen. The heating rate, from room temperature to 900 °C, was 15 °C min^{-1}. A Thermal Conductivity Detector (TCD) permitted the quantitative determination of hydrogen consumption.

The Temperature-Programmed Desorption (TPD) and Oxidation (TPO) tests were performed on a Micromeritics AutoChem II coupled with a mass spectrometer (Pfeiffer Vacuum, Omnistar) in order to evaluated the amount of CO_2 absorbed (carbonates) and the amount of carbon deposited, respectively. 50 mg of spent sample was placed in a quartz U-tube (6.6 mm internal diameter) and subjected to a pure helium flow (50 ml min^{-1}) and then heated from room temperature to 900 °C with heating rate of 15 °C min^{-1}. Similar conditions were used for TPO analysis, but under helium and oxygen (1%) mixture.

A TGA Q500 Thermogravimetric (TG) analysis apparatus was used for the carbonation (sorption) and calcination (desorption) experiments. 5–10 mg of sorbents and catalysts were placed in a platinum/rhodium sample cup and heated at 800 °C, under helium (10 ml min^{-1}) for 10 minutes in order to remove water and CO_2 accumulated on the sample surface during their storage. Then, the temperature was decreased to 650 °C and the valve was switched to a 5 ml min^{-1} CO_2 flow (10% in He). The sorption duration was 30 min, following by a desorption step at 800 °C for 10 min under a pure He flow (10 ml min^{-1}). Multiple cycles, consisting of sorption and desorption steps, were repeated to test the ability of sorbents to preserve their CO_2 sorption capacity.

The experiments of CO_2 sorption enhanced steam methane reforming (SE-SMR) were carried out at 650 °C with duration controlled by the sorption capacity of the samples. The five different operating conditions for cyclic stepwise sorption-enhanced steam methane reforming over Ni/x%CaO.y%CaAl were as follows (feed flow rates under normal conditions):

Water was injected using a syringe pump. The outlet gas was analyzed by means of two micro gas chromatographs: the first one, for the separation of CH$_4$, H$_2$ and CO, use a molecular sieve column and the second a HayeSep column for the separation of CH_4 and CO_2. Before reaction test, materials were reduced in a 30%H_2/Ar flow at 800 °C for 1 h at a constant heating rate of 10 °C min^{-1}. The flow rate of H_2 was then cut and the temperature decreased to 650 °C for the addition of water and CH_4.

	Ar (ml min^{-1} g^{-1}$_{cat}$)	CH$_4$ (ml min^{-1} g^{-1}$_{cat}$)	H$_2$O (ml min^{-1} g^{-1}$_{cat}$)	H$_2$O/CH$_4$	m$_{catalyst}$ (g)
1	26	2.5	7.5	3	2.5
2	26	1.0	1.0	1	2.5
3	26	1.0	3.0	3	2.5
4	26	1.0	3.0	3	1.0
5	26	1.0	1.0	1	1.0

For the study of the Water Gas Shift reaction, the CH_4 flow was replaced by a CO flow (H_2O/CO ratio = 1) after 20 h of SE-SMR tests.

Catalytic performances were evaluated by CH_4 conversion and H_2, CO, CO_2, CH_4 molar fraction calculated as follows:

$$\text{Conversion } (CH_4)(\%) = \frac{(CH_4)_{in} - (CH_4)_{out}}{(CH_4)_{in}} \cdot 100 \tag{5}$$

$$\text{Molar Fraction } X_n = \frac{(AX_n * f_n)}{(AX_1 * f_1) + (AX_2 * f_2) + (AX_3 * f_3) + (AX_4 * f_4)} \tag{6}$$

X: product; A: peak area; f: response factor and n: variation 1-4 corresponding to H_2, CO, CO_2, CH_4, respectively.

Preliminary Study with Low CaO/CaAl Ratio

The structural study of low CaO/CaAl ratio was performed as well as the influence of this ratio on the catalytic performance.

X-ray diffraction patterns (XRD) of as-prepared x%CaO.y%CaAl powders (before calcination) are shown in Fig. (3). $CaAl_2O_4$ (JCPDS n° 23-1036, monoclinic), $Ca_{12}Al_{14}O_{33}$ (JCPDS n° 48-1743, cubic), CaO (JCPDS n° 78-0649, cubic), and $Ca_3Al_2O_6$ (JCPDS n° 34-1429, cubic) phases were identified.

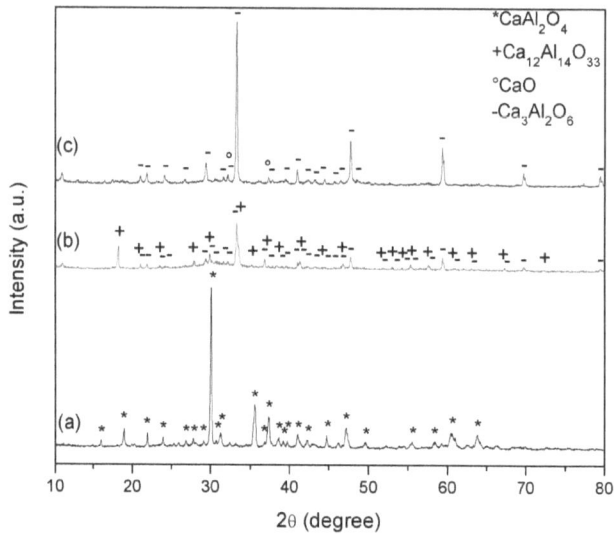

Fig. (3). XRD patterns of as-prepared samples: (a) CA4, (b) CA8 and (c) CA27.

According to these results, the mixture of desired phases (CaO + Ca$_{12}$Al$_{14}$O$_{33}$) was not obtained. Only the CA27 sample showed the presence of low intensity CaO rays. This sample was then calcined and the XRD pattern is shown in Fig. (**4a**). The calcined CA27 sample contains Ca$_3$Al$_2$O$_6$, Ca$_{12}$Al$_{14}$O$_{33}$ and CaO phases. The calcination led to the partial segregation of CaO from the Ca$_3$Al$_2$O$_6$ phase, leading to the formation of Ca$_{12}$Al$_{14}$O$_{33}$. It is likely that the increase of calcination time should lead to greater CaO segregation and total conversion of Ca$_3$Al$_2$O$_6$ into Ca$_{12}$Al$_{14}$O$_{33}$.

Fig. (**4b**) shows the XRD pattern of CA27 impregnated with nickel nitrate solution (5 wt% Ni). CaO, Ca$_{12}$Al$_{14}$O$_{33}$ and NiO (JCPDS n° 73-1519, cubic) crystalline phases were identified. No diffraction ray corresponding to any other Ca-Al phases, such as Ca$_3$Al$_2$O$_6$, CaAl$_2$O$_4$, CaAl$_4$O$_6$ or CaAl$_2$O$_{19}$ was observed. After impregnation with nickel nitrate followed by calcination, no spinel-like structure (NiAl$_2$O$_4$) or hydroxides (Ca(OH)$_2$) were formed. The absence of these crystal phases is important to assure the largest availability of CaO for CO$_2$ sorption and reducible Ni species for activity in SMR [25]. However, the amount of CaO (4, 8 and 27 wt%) in the samples is not sufficient for the sorption of CO$_2$ produced during the process. Ni-CA4, Ni-CA8 and Ni-CA27 catalysts exhibited H$_2$ yield of 80%, 77% and 74%, respectively.

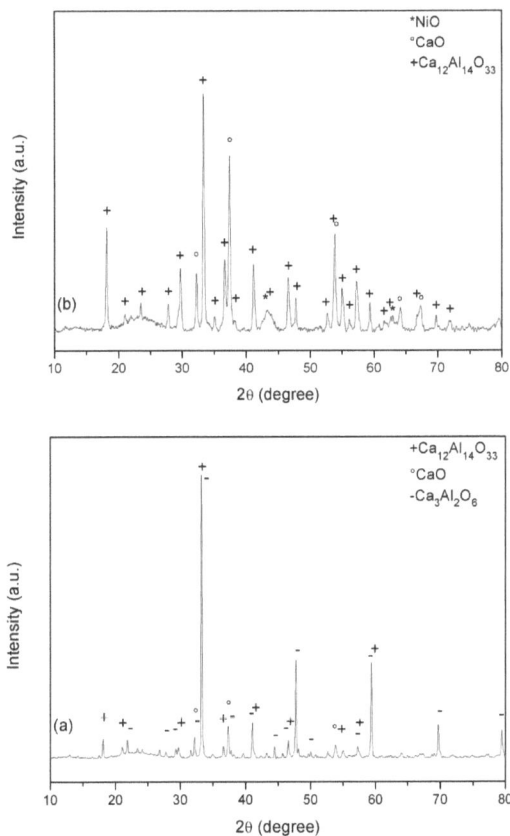

Fig. (4). XRD patterns of samples: (a) calcined CA27 and (b) calcined Ni-CA27.

Structural, Microstructural and Textural Study with High CaO/CaAl Ratio

X-ray diffraction patterns of sorbents with high CaO/CaAl ratio and corresponding Ni catalysts are shown in Figs. (**5** and **6**), respectively. Cubic structure phases of CaO, $Ca_{12}Al_{14}O_{33}$, and NiO were identified.

The preparation methods and calcination conditions ensure the selective formation of $Ca_{12}Al_{14}O_{33}$ by reacting CaO and Al_2O_3. No diffraction ray corresponding to any other Ca-Al phases was detected. A Ni/CaO standard catalyst has been prepared. After Ni nitrate impregnation or Microwave assisted self-Combustion of the catalyst followed by calcination, neither spinel-type ($NiAl_2O_4$) nor hydrated structures ($Ca(OH)_2$) was observed.

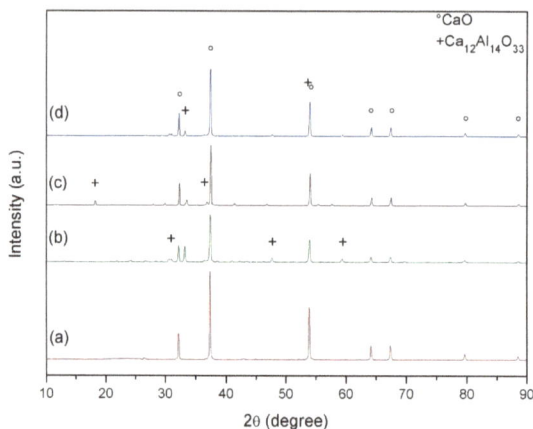

Fig. (5). XRD patterns of samples: (a) CaO, (b) CA48, (c) CA65, and (d) CA75.

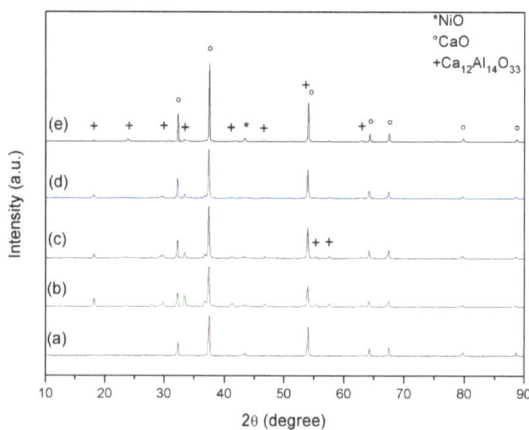

Fig. (6). XRD patterns of samples: (a) Ni-CaO, (b) Ni-CA48, (c) Ni-CA65, (d) Ni-CA75, and (e) Ni-CA75M.

Table **2** shows the surface area values of samples. Surface areas of CA48, CA65 and CA75 sorbents (supports) are lower than those of Ni-CA48, Ni-CA65 and Ni-CA75 catalysts, due to the hydration stage in the Ni impregnation. In fact, Ni-CA48, Ni-CA65 and Ni-CA75 catalysts prepared *via* wet impregnation of the support have significantly higher surface area than the Ni-CA75M catalyst prepared by the microwave assisted self-combustion method. These results are in agreement with those previously reported in literature [26]. The addition of water in the catalyst preparation step may be responsible for producing regular hexagonal crystalloid Ca(OH)$_2$ which becomes porous CaO during calcination at 900 °C leading to an increase in surface area. Such phenomenon occurs in our case during the impregnation of the support with the Ni salt. The surface areas of the Ni-CA48, Ni-CA65 or Ni–CA75 catalysts are similar within experimental error.

Table 2. Specific surface areas of sorbents and catalysts and Ni content of catalysts.

Catalysts	BET Surface Area ($m^2 g^{-1}$)	Ni Content (Weight %)
CA48	2.3	---
CA65	0.7	---
CA75	1.2	---
Ni-CA48	14.8	3.8
Ni-CA65	13.0	3.5
Ni-CA75	11.5	3.6
Ni-CA75M	0.7	4.3

Fig. (7) shows the TPR profile of Ni catalysts. Ni-CA48, Ni-CA65 and Ni-CA75 catalysts have higher reducibility than Ni-CaO and Ni-CA75M. A slight shoulder appears between 450 and 510 °C which can be considered as the result of the reduction of isolated or weakly bounded NiO. The maximum reduction peak is observed between 510 and 620 °C. For Ni-CaO and Ni-CA75M, the reduction of NiO continues until 650 °C, while for Ni-CA48, Ni-CA65 and Ni-CA75, the reduction of NiO continues until 750 °C which is an indication of different types of metal-support interactions. NiO-$Ca_{12}Al_{14}O_{33}$ interactions are stronger than those in NiO-CaO, which can be evidenced by Ni-CaO curve where the maximum reduction of NiO is located at 550 °C. In addition, for an overall preparation of the catalyst by microwave assisted self-combustion (Ni-CA75M), the amount of reducible Ni species is much lower than for the impregnation method (Ni-CA48, Ni-CA65 and Ni-CA75).

It can be seen also at 750-850 °C a small reduction peak, probably assigned to the spinel-like structure $NiAl_2O_4$. This phase may have been formed during preparation (not detected by XRD) or during TPR. The diffraction patterns of catalysts after TPR (Fig. 8) showed well defined diffraction rays of metallic nickel (JCPDS File No. 87-0712, cubic), CaO and $Ca_{12}Al_{14}O_{33}$. The average crystallite size of metallic nickel (Table 3) is about 19-28 nm.

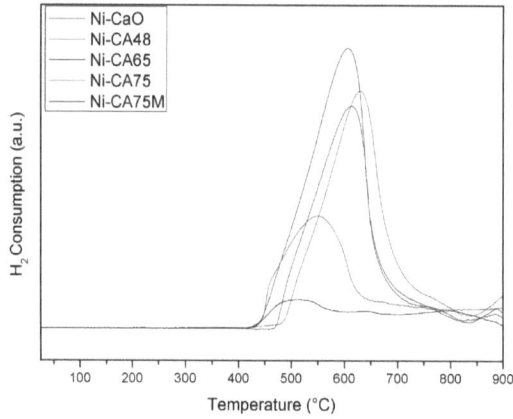

Fig. (7). Temperature-programmed reduction (TPR) profiles of the catalysts.

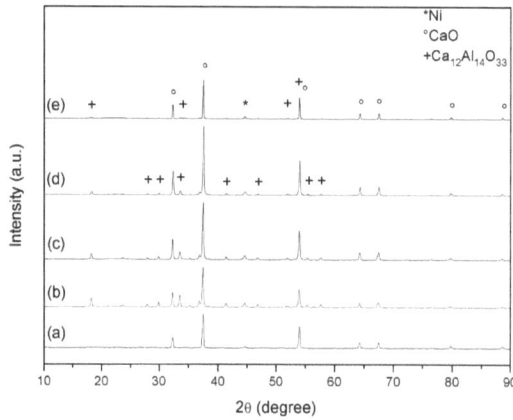

Fig. (8). X-ray diffraction patterns of the samples upon reduction: (a) Ni-CaO, (b) Ni-CA48, (c) Ni-CA65, (d) Ni-CA75, and (e) Ni-CA75M.

Table 3. Ni crystallite sizes of the catalyst upon reduction and SE-SMR reaction from XRD.

Catalysts	Crystallite Size after Reduction (nm)	SE-SMR Duration (h)	Crystallite Size after SE-SMR (nm)
Ni-CaO	19.5	25h	29.1
Ni-CA75	22.3	50h	27.9
Ni-CA75M	27.6	50h	30.3

Fig. **(9)** shows SEM images of catalysts before SE-SMR tests. Samples exhibited similar microstructures with large porous agglomerates composed of fine particles.

Fig. (9). SEM images of (a) Ni-CA48, (b) Ni-CA65, (c) Ni-CA75, and (d) Ni-CA75M before tests of CO_2 sorption-enhanced steam methane reforming.

A TEM micrograph of the Ni-CA75 sample is shown in Fig. (**10**). This sample consists of NiO crystallites dispersed in the $CaO-Ca_{12}Al_{14}O_{33}$ support. The NiO crystallite size is about 20 nm, which is in good agreement with published values of similar materials [27 - 29].

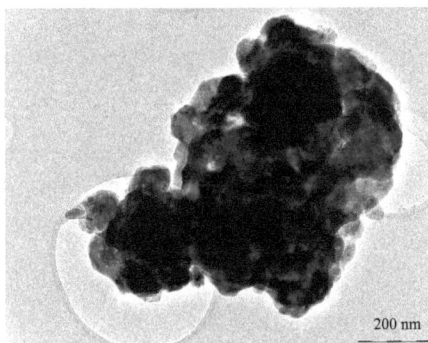

Fig. (10). TEM micrograph of the Ni-CA75 sample.

CO₂ Sorption

Fig. (**11**) shows the CO_2 sorption capacity of the catalysts after five cycles of carbonation/ calcination cycles. The sorption capacity was computed as the fraction of the total carbonation for free CaO. After five sorption/calcination cycles the samples Ni-CA75M, Ni-CaO and Ni-CA75 have sorption capacity of 8.4, 13.3, and 44.3%, respectively. Clearly, the preparation method has an effect on this property. The catalyst obtained by wet impregnation of the sorbent (Ni-CA75) allowed higher specific surface area (Table **2**) and better Ni species reducibility (Fig. **7**) probably due to a better accessibility. The $Ca_{12}Al_{14}O_{33}$ had a null CO_2 sorption capacity (not shown) but its presence enhances the sorption capacity of catalysts (Ni-CaO compared to Ni-CA75).

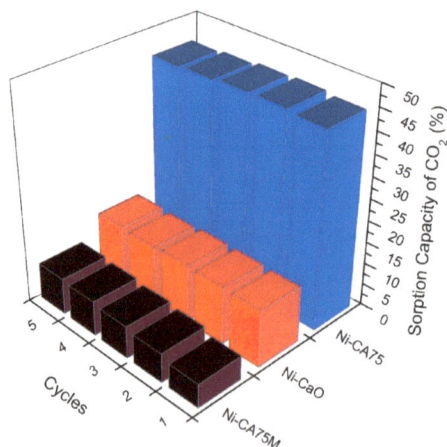

Fig. (11). Comparison of cyclic CO_2 sorption capacity on the Ni-CaO, Ni-CA75 and Ni-CA75M catalysts (CO_2 sorption: 650°C, 30 min, 10% CO_2/He; desorption: 800°C, 10 min, 100% He).

CO₂ Sorption Enhanced Steam Methane Reforming

Fig. (**12**) shows the performance of the catalysts in CO_2 sorption enhanced steam methane reforming for a H_2O/CH_4 ratio equal to 3. As it can be clearly noted the process here reported is divided into two steps, in which the total time of the first one depends on the characteristics and properties of the sorbent (CaO/CaAl ratio, sorption capacity) as well as the operating conditions (in our case H_2O/CH_4 ratio).

In the first step, only hydrogen is observed with a total conversion of CH_4. All CO produced during steam reforming is oxidized into CO_2 which is totally absorbed throughout. CO_2 sorption enhances CH_4 and CO conversions and H_2 concentration is higher than the value calculated at the thermodynamic equilibrium of SMR

(78%) because of the shift of the equilibrium of the steam reforming and water gas reactions. Then a drop in the formation of hydrogen, higher for the lowest CaO excess (Ni-CA48), and associated with the increase in CO_2 and CH_4 concentrations is observed (second step). Then the efficiency of the sorption enhancement is longer with the higher CaO excess (Ni-CA75).

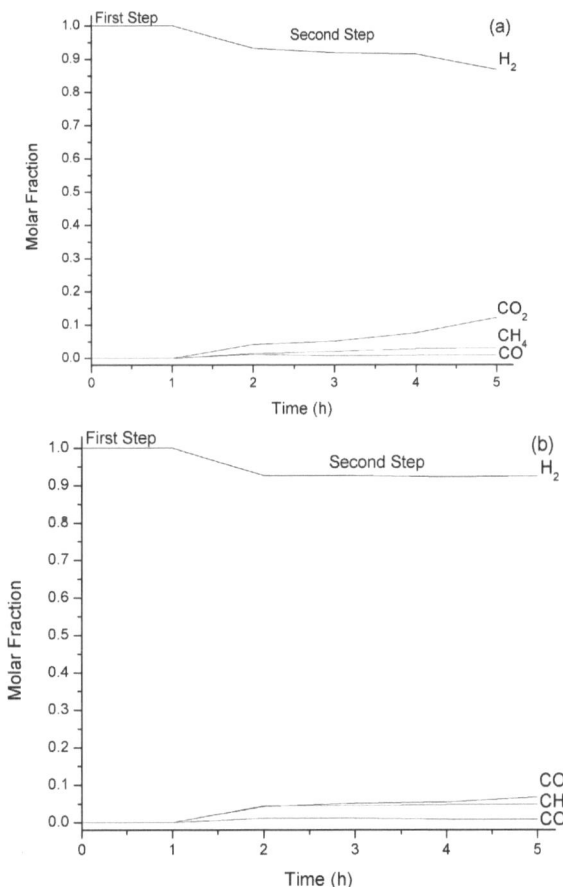

Fig. (12). CO_2 sorption enhanced steam methane reforming: molar fraction of methane and products *versus* time (a) Ni-CA48 and (b) Ni-CA75. Reaction conditions: total flow rate = 30 mL min^{-1} (Ar = 26 mL min^{-1}, CH_4 = 2.5 mL min^{-1}, H_2O = 7.5 mL min^{-1}), mass of catalyst = 2.5 g, H_2O/CH_4 molar ratio = 3, temperature 650 °C.

The H_2/CO ratio observed for the catalysts Ni-CA48, Ni-CA65, and Ni-CA75 were much higher (between 9 and 10) than that observed (between 4 and 6) for the catalysts with lower CaO contents (Ni-CA4, Ni-CA8, and Ni-CA27). This highlights the positive effect of the $CaO/Ca_{12}Al_{14}O_{33}$ ratio not only on the sorption capacity (breakthrough time) but also on the catalytic properties (H_2/CO ratio) at a temperature of 650 °C.

The slight activity decay observed by the increase in CH_4 concentration is due to the end of sorption enhancement because clearly associated to the increase in CO_2 concentration. No catalyst deactivation is observed and is corroborated with the absence of carbon residue (main cause of deactivation in steam reforming) highlighted by SEM and TPO.

Fig. (**13**) shows the performance of the NiO-CaO catalyst in SE-SMR for various H_2O/CH_4 ratios (1 or 3). These results, compared to the Fig. (**12**), clearly shows the importance of the $Ca_{12}Al_{14}O_{33}$ phase. Regardless of the operating conditions, the catalytic performance is low and decrease with time due to CaO agglomeration and loss of CO_2 sorption capacity. In fact, this catalyst presents CaO particles size growing during the catalytic test (31.1 *vs.* 39.7 nm) not observed on the other catalysts.

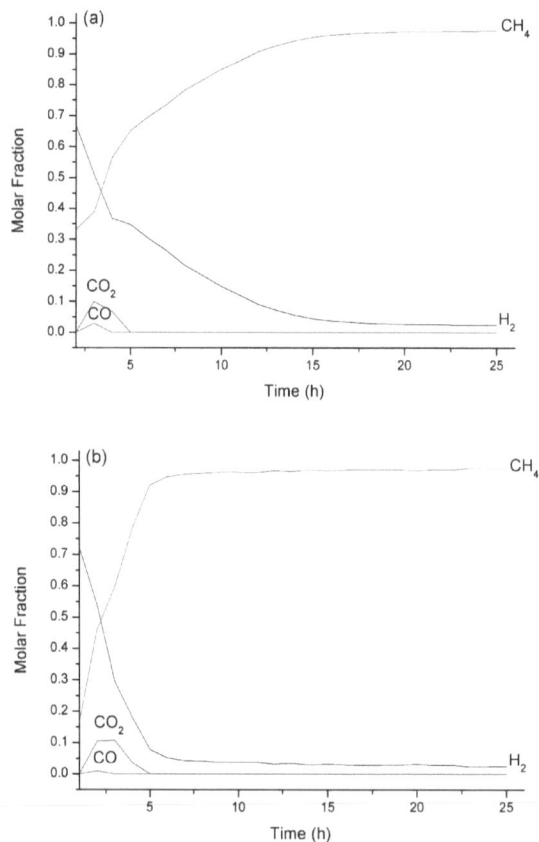

Fig. (13). CO_2 Sorption enhanced steam methane reforming: molar fraction of methane and products *versus* time (a) Ni-CaO, H_2O/CH_4 molar ratio= 1 and (b) Ni-CaO, H_2O/CH_4 molar ratio= 3. Reaction conditions:

total flow rate= 30 mL min^{-1} (Ar = 26 mL min^{-1}, CH$_4$ = 1.0 mL min^{-1}, H$_2$O = 1.0 or 3.0 mLmin^{-1}), mass of catalyst = 1.0 g, temperature 650 °C.

According to the above results (Fig. **12**) the Ni-CA75 catalyst exhibited the most promising results. This catalyst has been compared with one obtained by microwave assisted self-combustion (Ni-CA75M). Fig. (**14**) shows the performance of Ni-CA75 and Ni-CA75M catalysts in SE-SMR for H$_2$O/CH$_4$ ratio equal to 3. Ni-CA75 and Ni-CA75M showed a breakthrough time for CO$_2$ equal to 2 and 1 h, respectively. This can be justified by a greater sorption capacity of Ni-CA75 (44.3 *vs* 8.4 after five cycles of sorption-desorption).

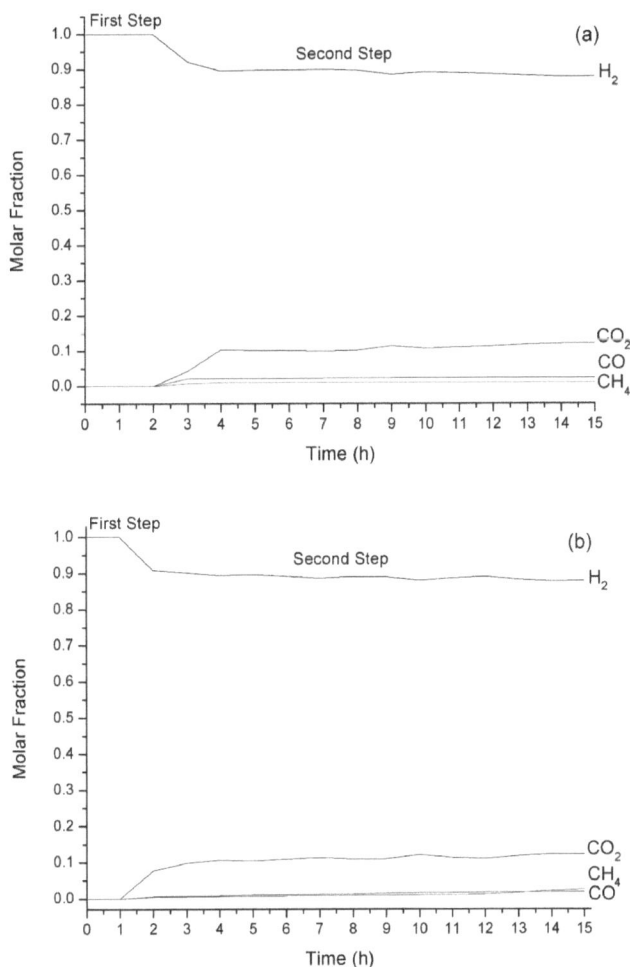

Fig. (14). CO$_2$ sorption enhanced steam methane reforming: molar fraction of methane and products *versus* time (a) Ni-CA75 and (b) Ni-CA75M. Reaction conditions: total flow rate= 30 mL min^{-1} (Ar = 26 mL min^{-1},

CH_4 = 1.0 mL min⁻¹, H_2O = 3.0 mL min⁻¹), mass of catalyst = 1.0 g, H_2O/CH_4 molar ratio= 3, temperature 650 °C.

The influence of the H_2O/CH_4 ratio was also studied. With H_2O/CH_4 ratio = 1, Ni-CA75 showed a breakthrough time for CO_2 equal to 16h (not shown). With more water (H_2O/CH_4 ratio = 3) the breakthrough time decreases significantly to 2 h (Fig. **14a**). Due to the greater amount of water, competition between water and CO_2 occurs at the sorbent surface leading to the formation of hydroxides and carbonates. Thus, smaller amount of CaO are available to absorb CO_2 and consequently, breakthrough time is shorter.

The behavior of Ni-CA75 catalyst was then studied in the WGS reaction (Fig. **15**). As soon as CH_4 has been cut (after 20h of SE-SMR), H_2 decreases and CO_2 increases abruptly, then stabilize for a long period. The entire CO is consumed and the H_2/CO_2 ratio near 1 leads to the conclusion that the WGS reaction is catalyzed by this catalyst.

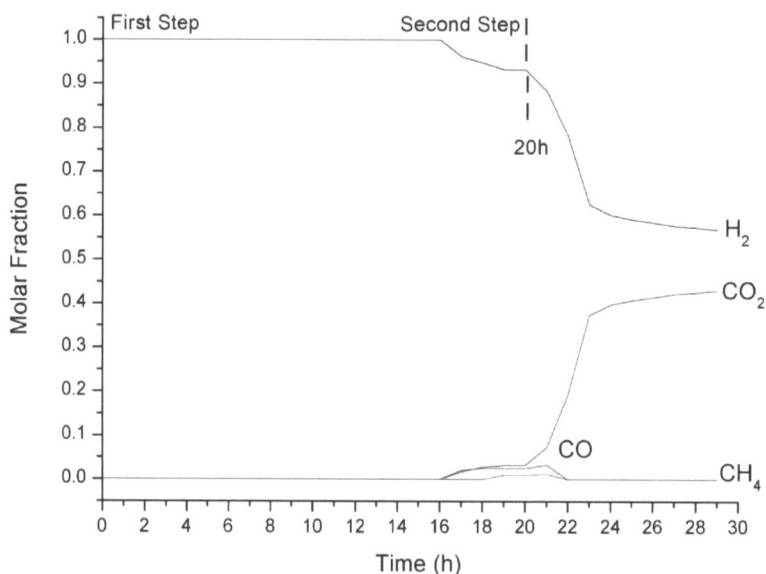

Fig. (15). CO_2 sorption enhanced steam methane reforming (SE-SMR) and Water Gas Shift reaction (WGS) with Ni-CA75. SE-SMR reaction conditions: total flow rate = 30 mL min⁻¹ (Ar = 26 mL min⁻¹, CH_4 = 1.0 mL min⁻¹, H_2O = 1.0 mL min⁻¹), mass of catalyst = 1.0 g, H_2O/CH_4 molar ratio = 1, temperature 650 °C. WGS reaction conditions: H_2O/CO molar ratio= 1.

Long-term Reactivity

Fig. (**16**) shows the performance of Ni-CaO, Ni-CA75 and Ni-CA75M over 30 cycles of carbonation/calcination in order to verify the stability of their sorption capacity over a long period. After 30 sorption/calcination cycles the samples Ni-CA75 and Ni-CA75M reached values of 33.9 and 11.9%, respectively.

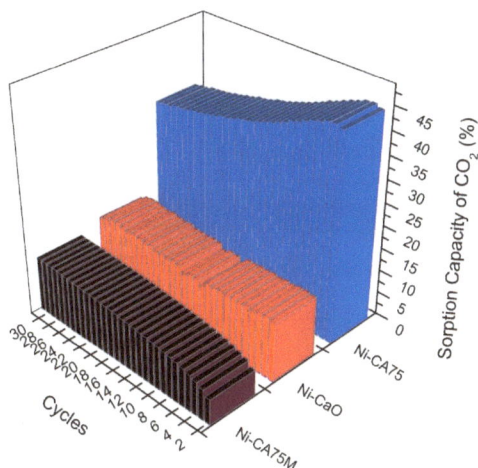

Fig. (16). CO_2 sorption capacity of Ni-CaO, Ni-CA75 and Ni-CA75M during 30 cycles of sorption/desorption. (CO_2 sorption: 650 °C, 30 min, 10% CO_2/He; desorption: 800 °C, 10 min, 100% He).

The $Ca_{12}Al_{14}O_{33}$ stabilizes the sorption capacity of catalysts. Previous studies have already showed the stability of CaO in repeated sorption–desorption cycles in presence of $Ca_{12}Al_{14}O_{33}$ [30, 31]. Here, a decay of the sorption capacity was observed after 24 cycles, sorption stabilizes and reaches 35% for Ni-CA75. The literature reported a similar CO_2 sorption capacity of 31% [27] for a sample prepared by a process involving hydration followed by calcination with the same CaO to $Ca_{12}Al_{14}O_{33}$ ratio. Regarding the Ni-CA75M catalyst, the sorption capacity increases with the increase in the number of cycles, but it has values significantly lower than the catalyst prepared by impregnation of the support due to the unrealized hydration process.

Fig. (**17**) shows the behavior of the Ni-CA75 and Ni-CA75M catalyst during 50 h of reaction. Ni-CA75 had always the best performance. Ni-CA75 and Ni-CA75M showed a breakthrough time for CO_2 equal to 7 and 1 h, respectively. After the first step, hydrogen production was stable for a period of 50 h (second step) when the reaction is carried out using the Ni-CA75 catalyst. CO and CO_2 formation and methane conversion also remain constant during this time. Regarding the Ni-CA75M catalyst, the hydrogen concentration in the second step decreases after 26h followed by an increase in CO_2 concentration probably due to the formation of a $CaCO_3$ layer that limits the diffusion of CO_2 and its sorption on the available CaO. This activity decay is not due an increase in Ni particles size as confirmed by XRD (Table **3**).

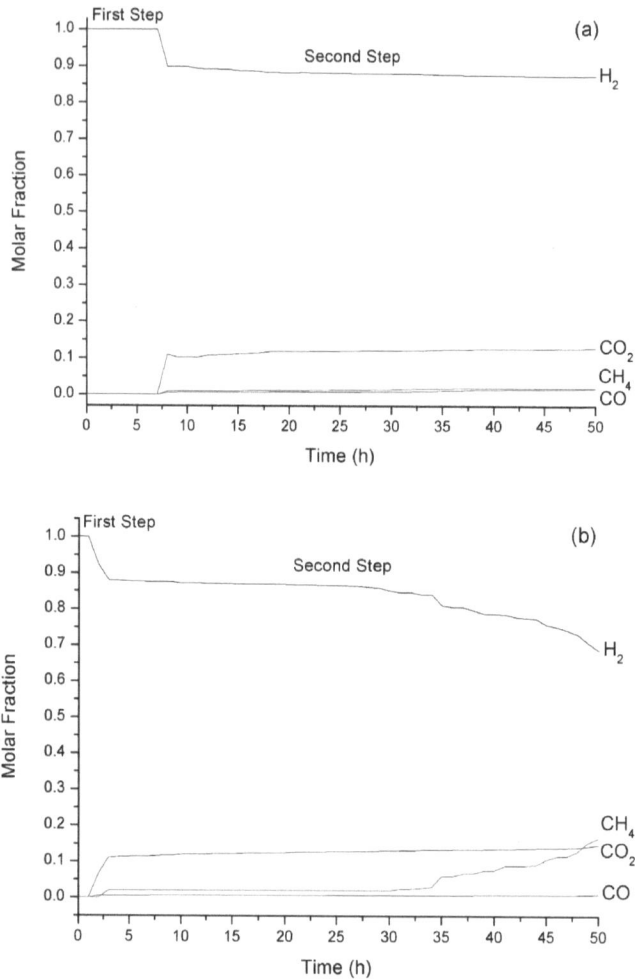

Fig. (17). CO$_2$ sorption enhanced steam methane reforming: molar fraction of methane and products *versus* time (a) Ni-CA75 and (b) Ni-CA75M. Reaction conditions: total flow rate= 30 mL min^{-1} (Ar = 26 mL min^{-1}, CH$_4$ = 1.0 mL min^{-1}, H$_2$O = 3.0 mL min^{-1}), mass of catalyst = 2.5 g, H$_2$O/CH$_4$ molar ratio= 3, temperature 650 °C.

Characterizations after Tests of CO$_2$ Sorption Enhanced Steam Methane Reforming

XRD profiles of catalysts after SE-SMR tests are shown in Fig. (**18**). These patterns confirm the presence of the calcium carbonate phase (JCPDS file N° 85-0849, rhombohedral), but with a more intense ray for the Ni-CA75 sample. Ni, CaO and Ca$_{12}$Al$_{14}$O$_{33}$ have also been identified.

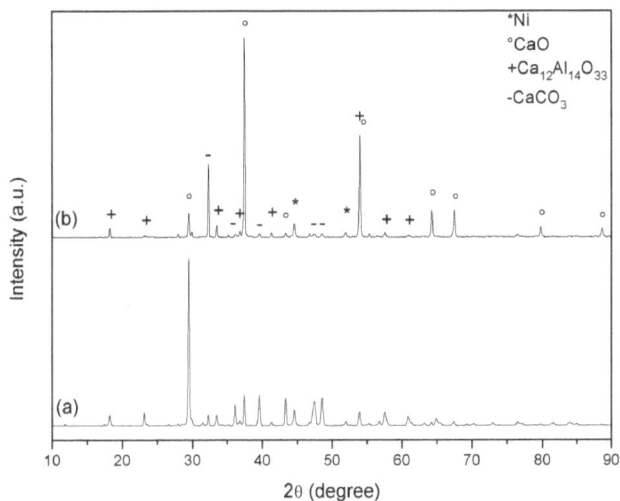

Fig. (18). XRD patterns of: (a) Ni-CA75 and (b) Ni-CA75M after SE-SMR.

Fig. (**19**) shows SEM images of catalysts after tests of SE-SMR. Both samples exhibited a typical microstructure of calcium carbonate which is formed during the carbonation reaction, but no traces of residual carbon.

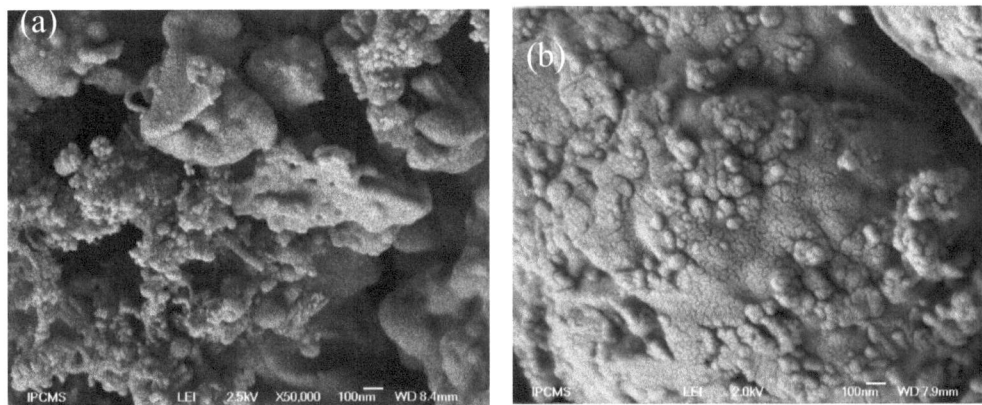

Fig. (19). SEM images of: (a) Ni-CA75 and (b) Ni-CA75M after SE-SMR.

FTIR spectra after SE-SMR tests are shown in Fig. (**20**). The band located at 840 cm^{-1} is attributed to the Al-O (metal-oxygen) stretch [32]. While the vibrations at 1407 cm^{-1} are characteristic of carbonate groups [33].

The profiles of temperature-programmed desorption of CO_2 are given in Fig. (**21**). Maximum of CO_2 desorption occurs at 702–756 °C corresponding to the decomposition of calcium carbonates [14]. Ni-CA75 catalyst showed a more intense peak of CO_2 (~ 0.1 mol released g^{-1} cat) compared to Ni-CA75M (~ 0.01 mol) after 50 h of catalytic activity. This result confirms that CO_2 sorption for Ni-CA75 was more effective than for Ni-CA75M as firstly seen with the results of CO_2 sorption enhanced steam methane reforming and XRD.

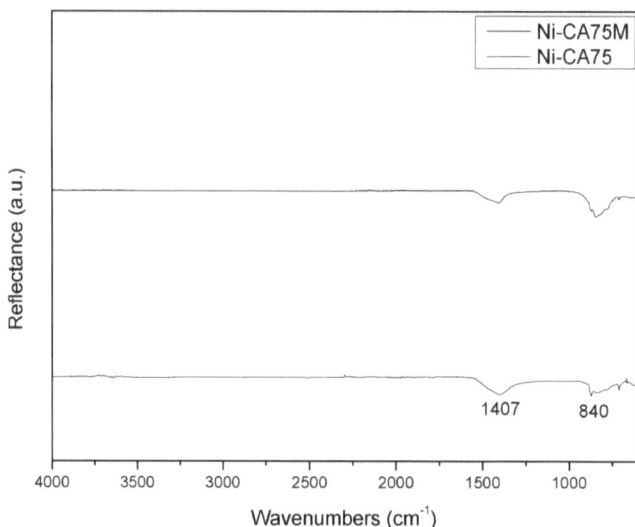

Fig. (20). FTIR spectra of: (a) Ni-CA75 and (b) Ni-CA75M after SE-SMR.

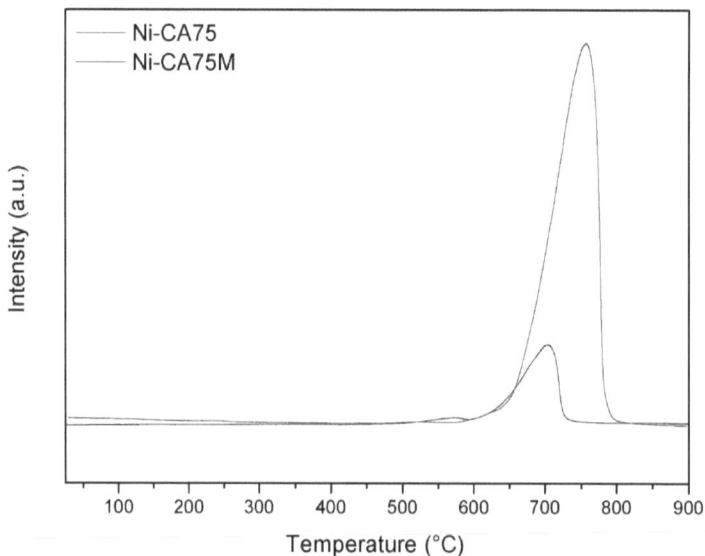

Fig. (21). TPD profiles of CO_2 for Ni-CA75 and NiCA75M after SE-SMR (ratio $H_2O/CH_4 = 3$).

CONCLUDING REMARKS

$CaO.Ca_{12}Al_{14}O_{33}$ (CaO.CaAl) sorbents and Ni catalysts were successfully prepared by microwave-assisted self-combustion. Ni catalysts have also been prepared by wet impregnation method. This method improves the surface area and consequently the CO_2 sorption. The catalyst obtained by wet impregnation (Ni-CA75) showed the highest sorption capacity during the 30 cycles of absorption-desorption. The presence of $Ca_{12}Al_{14}O_{33}$ enhances and stabilizes the sorption capacity.

$Ni-CaO.Ca_{12}Al_{14}O_{33}$ catalysts have been tested in enhanced-sorption steam methane reforming. The CaO/CaAl ratio was optimized and a large excess of CaO is required (75% CaO, 25% CaAl) for the CO_2 sorption. The CO_2 sorption by CaO shifts the equilibrium of the water gas shift reaction and improves both CH_4 conversion and selectivity to hydrogen. The breakthrough time for H_2, CO, CO_2 and CH_4 depends both on the CaO excess and on the operating conditions (H_2O/CH_4 ratio). Steam methane reforming at 650 °C is total and the hydrogen selectivity is of 100% during 2 h or 16 h according to operational conditions (H_2O/CH_4 1 or 3), validating the concept of enhanced hydrogen production by CO_2 sorption. This hydrogen production was stable for a period of 50 h when the reaction is carried out in the presence of Ni-CA75 catalyst obtained by wet impregnation. The Ni-CA75 catalyst was the most active and stable in CO_2 sorption enhanced steam methane reforming. Its stability was possible because carbon fouling was avoided by the presence of $Ca_{12}Al_{14}O_{33}$ with high oxygen mobility.

CONSENT FOR PUBLICATION

Not applicable.

CONFLICT OF INTEREST

Declare None.

ACKNOWLEDGEMENT

The authors acknowledge CAPES (DS and PDEE program BEX 0324/10-8) and Eiffel program (721398L) for their financial support.

REFERENCES

[1]   Baykara, S.Z. Hydrogen: A brief overview on its sources, production and environmental impact. *Int. J. Hydrogen Energy,* **2018**, *43*, 10605-10614.
[http://dx.doi.org/10.1016/j.ijhydene.2018.02.022]

[2]   Hosseini, S.E.; Wahid, M.A.; Jamil, M.M.; Azli, A.A.M.; Misbah, M.F. A review on biomass-based

hydrogen production for renewable energy supply. *Int. J. Energy Res.,* **2015**, *39*, 1597-1615.
[http://dx.doi.org/10.1002/er.3381]

[3]   Dutta, S. A review on production, storage of hydrogen and its utilization as an energy resource. *J. Ind. Eng. Chem.,* **2014**, *20*, 1148-1156.
[http://dx.doi.org/10.1016/j.jiec.2013.07.037]

[4]   Farsi, A.; Mansouri, S.S. Influence of nanocatalyst on oxidative coupling, steam and dry reforming of methane: A short review. *Arab. J. Chem.,* **2016**, *9*, S28-S34.
[http://dx.doi.org/10.1016/j.arabjc.2011.08.001]

[5]   Boyano, A.; Blanco-Marigorta, A.M.; Morosuk, T.; Tsatsaronis, G. Exergoenvironmental analysis of a steam methane reforming process for hydrogen production. *Energy,* **2011**, *36*, 2202-2214.
[http://dx.doi.org/10.1016/j.energy.2010.05.020]

[6]   Artetxe, M.; Alvarez, J.; Nahil, M.A.; Olazar, M.; Williams, P.T. Steam reforming of different biomass tar model compounds over Ni/Al$_2$O$_3$ catalysts. *Energy Convers. Manage.,* **2017**, *136*, 119-126.
[http://dx.doi.org/10.1016/j.enconman.2016.12.092]

[7]   Twigg, M.V.; Richardson, J.T. Effects of alumina incorporation in coprecipitated NiO-Al$_2$O$_3$ catalysts. *Appl. Catal. A Gen.,* **2000**, *190*, 61-72.
[http://dx.doi.org/10.1016/S0926-860X(99)00269-0]

[8]   Charisiou, N.D.; Papageridis, K.N.; Tzounis, L.; Sebastian, V.; Hinder, S.J.; Baker, M.A.; Alketbi, M.; Polychronopoulou, K.; Goula, M.A. Ni supported on CaO-MgO-Al$_2$O$_3$ as a highly selective and stable catalyst for H$_2$ production *via* the glycerol steam reforming reaction. *Int. J. Hydrogen Energy,* in press

[9]   Elias, K.F.M.; Lucrédio, A.F.; Assaf, E.M. Effect of CaO addition on acid properties of Ni-Ca/Al$_2$O$_3$ catalysts applied to ethanol steam reforming. *Int. J. Hydrogen Energy,* **2013**, *38*, 4407-4417.
[http://dx.doi.org/10.1016/j.ijhydene.2013.01.162]

[10]   Di Giuliano, A.; Girr, J.; Massacesi, R.; Gallucci, K.; Courson, C. Sorption enhanced steam methane reforming by Ni-CaO materials supported on mayenite. *Int. J. Hydrogen Energy,* **2017**, *42*, 13661-13680.
[http://dx.doi.org/10.1016/j.ijhydene.2016.11.198]

[11]   Zamboni, I.; Courson, C.; Niznansky, D.; Kiennemann, A. Simultaneous catalytic H$_2$ production and CO$_2$ capture in steam reforming of toluene as tar model compound from biomass gasification. *Appl. Catal. B,* **2014**, *145*, 63-72.
[http://dx.doi.org/10.1016/j.apcatb.2013.02.046]

[12]   Zamboni, I.; Courson, C.; Kiennemann, A. Fe-Ca interactions in Fe-based/CaO catalyst/sorbent for CO$_2$ sorption and hydrogen production from toluene steam reforming. *Appl. Catal. B,* **2017**, *203*, 154-165.
[http://dx.doi.org/10.1016/j.apcatb.2016.10.024]

[13]   Di Felice, L.; Courson, C.; Jand, N.; Gallucci, K.; Foscolo, P.U. Kiennemann. A. Catalytic biomass gasification: Simultaneous hydrocarbons steam reforming and CO$_2$ capture in a fluidised bed reactor. *Chem. Eng. J.,* **2009**, *154*, 375-383.
[http://dx.doi.org/10.1016/j.cej.2009.04.054]

[14]   García-Labiano, F.; Abad, A.; de Diego, L.F.; Gayán, P.; Adánez, J. Calcination of calcium-based sorbents at pressure in a broad range of CO$_2$ concentrations. *Chem. Eng. Sci.,* **2002**, *57*, 2381-2393.
[http://dx.doi.org/10.1016/S0009-2509(02)00137-9]

[15]   Zamboni, I.; Courson, C.; Kiennemann, A. Synthesis of Fe/CaO active sorbent for CO$_2$ absorption and tars removal in biomass gasification. *Catal. Today,* **2011**, *176*, 197-201.
[http://dx.doi.org/10.1016/j.cattod.2011.01.014]

[16]   Grasa, G.; González, B.; Alonso, M. Abanades. J.C. Comparison of CaO-based synthetic CO$_2$ sorbents under realistic calcination conditions. *Energy Fuels,* **2007**, *21*, 3560-3562.

[http://dx.doi.org/10.1021/ef0701687]

[17] Abanades, J.C.; Alvarez, D. Conversion limits in the reaction of CO_2 with lime. *Energy Fuels,* **2003**, *17*, 308-315.
[http://dx.doi.org/10.1021/ef020152a]

[18] Dou, B.; Wang, C.; Song, Y.; Chen, H.; Jiang, B.; Yang, M.; Xu, Y. Solid sorbents for in-situ CO_2 removal during sorption-enhanced steam reforming process: a review. *Renew. Sustain. Energy Rev.,* **2016**, *2016*(53), 536-546.
[http://dx.doi.org/10.1016/j.rser.2015.08.068]

[19] Alvarez, D.; Abanades, J.C. Determination of the critical product layer thickness in the reaction of CaO with CO_2. *Ind. Eng. Chem. Res.,* **2005**, *44*, 5608-5615.
[http://dx.doi.org/10.1021/ie050305s]

[20] Filitz, R.; Kierzkowska, A.M.; Broda, M.; Müller, C.R. Highly efficient CO_2 sorbents: development of synthetic, calcium-rich dolomites. *Environ. Sci. Technol.,* **2012**, *46*(1), 559-565.
[http://dx.doi.org/10.1021/es2034697] [PMID: 22129091]

[21] Alvarez, D.; Peña, M.; Borrego, A.G. Behavior of different calcium-based sorbents in a calcination/carbonation cycle for CO_2 capture. *Energy Fuels,* **2007**, *21*, 1534-1542.
[http://dx.doi.org/10.1021/ef060573i]

[22] Martavaltzi, C.S.; Pampaka, E.P.; Korkakaki, E.S.; Lemonidou, A.A. Hydrogen production *via* steam reforming of methane with simultaneous CO_2 Capture over $CaO\text{-}Ca_{12}Al_{14}O_{33}$. *Energy Fuels,* **2010**, *24*, 2589-2595.
[http://dx.doi.org/10.1021/ef9014058]

[23] Li, C.; Hirabayashi, D.; Suzuki, K. A crucial role of O_2^- and O_2^{2-} on mayenite structure for biomass tar steam reforming over $Ni/Ca_{12}Al_{14}O_{33}$. *Appl. Catal. B,* **2009**, *88*, 351-360.
[http://dx.doi.org/10.1016/j.apcatb.2008.11.004]

[24] Azaroff, L.V.; Buerger, M.J. *The powder method in X-ray crystallography,* 1st ed; McGraw-Hill Book Company: USA, **1958**.

[25] Lemonidou, A.A.; Vasalos, I.A. Carbon dioxide reforming of methane over 5 wt.% $Ni/CaO\text{-}Al_2O_3$ catalyst. *Appl. Catal. A Gen.,* **2002**, *228*, 227-235.
[http://dx.doi.org/10.1016/S0926-860X(01)00974-7]

[26] Martavaltzi, C.S.; Lemonidou, A.A. Parametric Study of the $CaO\text{-}Ca_{12}Al_{14}O_{33}$ Synthesis with Respect to High CO_2 Sorption Capacity and Stability on Multicycle Operation. *Ind. Eng. Chem. Res.,* **2008**, *47*, 9537-9543.
[http://dx.doi.org/10.1021/ie800882d]

[27] Martavaltzi, C.S.; Lemonidou, A.A. Hydrogen production *via* sorption enhanced reforming of methane: Development of a novel hybrid material-reforming catalyst and CO_2 sorbent. *Chem. Eng. Sci.,* **2010**, *65*, 4134-4140.
[http://dx.doi.org/10.1016/j.ces.2010.04.038]

[28] Christensen, K.O.; Chen, D.; Lødeng, R.; Holmen, A. Effect of supports and Ni crystal size on carbon formation and sintering during steam methane reforming. *Appl. Catal. A Gen.,* **2006**, *314*, 9-22.
[http://dx.doi.org/10.1016/j.apcata.2006.07.028]

[29] Vagia, E.Ch.; Lemonidou, A.A. Hydrogen production *via* steam reforming of bio-oil components over calcium aluminate supported nickel and noble metal catalysts. *Appl. Catal. A Gen.,* **2008**, *351*, 111-121.
[http://dx.doi.org/10.1016/j.apcata.2008.09.007]

[30] Martavaltzi, C.S.; Lemonidou, A.A. Development of new CaO based sorbent materials for CO_2 removal at high temperature. *Microporous Mesoporous Mater.,* **2008**, *110*, 119-127.
[http://dx.doi.org/10.1016/j.micromeso.2007.10.006]

[31] Li, Z-s.; Cai, N-s.; Huang, Y-y.; Han, H-j. Synthesis, experimental studies, and analysis of a new

calcium-based carbon dioxide absorbent. *Energy Fuels,* **2005**, *19*, 1447-1452.
[http://dx.doi.org/10.1021/ef0496799]

[32]   Gaki, A.; Chrysafi, R.; Perraki, Th.; Kakali, G. Synthesis of calcium aluminates through the polymeric precursor route. *Chem. Ind. Chem. Eng. Q.,* **2006**, *12*, 137-140.
[http://dx.doi.org/10.2298/CICEQ0602137G]

[33]   Ramasamy, V.; Ponnusamy, V.; Sabari, S.; Anishia, S.R.; Gomathi, S.S. Effect of grinding on the crystal structure of recently excavated dolomite. *Indian J. Pure Appl. Phy.,* **2009**, *47*, 586-591.

CHAPTER 3

A Review on the Dry Reforming Processes for Hydrogen Production: Catalytic Materials and Technologies

Samer Aouad[1,*], Madona Labaki[2], Satu Ojala[3], Prem Seelam[3], Esa Turpeinen[3], Cédric Gennequin[4], Jane Estephane[5] and Edmond Abi Aad[4]

[1] *Department of Chemistry, Faculty of Sciences, University of Balamand, El Koura, Lebanon*

[2] *Laboratory of Physical–Chemistry of Materials (LPCM/PR2N), Lebanese University, Fanar, Lebanon*

[3] *Environmental and Chemical Engineering, Faculty of Technology, University of Oulu, Oulu, Finland*

[4] *Unité de Chimie Environnementale et Interactions sur le Vivant (UCEIV, E.A. 4492), MREI, Université du Littoral Côte d'Opale (ULCO), Dunkerque, France*

[5] *Department of Chemical Engineering, Faculty of Engineering, University of Balamand, El Koura, Lebanon*

Abstract: Dry reforming (DR) processes consist of a reaction between an adequate feedstock and carbon dioxide to produce syngas. In the case of a renewable feedstock (biogas, bioalcohols, wood tar,…), the DR processes become very interesting since they consume greenhouse gases (CO_2, CH_4,…) and produce hydrogen and syngas mixtures that can be considered as renewable alternatives to fossil fuels. The DR processes involve endothermic reactions accompanied by side reactions that decrease the overall process efficiency. The use of a catalytic material is expected to reduce the energy required for the process and to favor the selectivity towards syngas production. Thus, in the last decades, many studies considered the synthesis of catalytic materials that are active, selective and stable in DR reactions. This chapter considers the recent advances in the catalytic DR of methane, alcohols and biomass tar. The most recent catalytic materials are discussed in terms of their preparation, physico-chemical characteristics, and intrinsic properties that serve the purpose of the DR reactions. A special attention is paid to the carbon deposition problem and the different strategies that are adopted to minimize it. A final part of the chapter discusses the most recent developments in plasma, microwaves, solar energy and electrical current technologies for dry reforming reactions. Some examples of the developed reactor technologies are also presented including chemical looping reforming, membrane reactors and ceramic counter flow reactor.

* **Corresponding author Samer Aouad:** Department of Chemistry, Faculty of Sciences, University of Balamand, El Koura, Lebanon; Tel: +55(84) 32153826; E-mail: samer.aouad@balamand.edu.lb

Moisés R. Cesário, Cédric Gennequin, Edmond Abi-Aad & Daniel A. de Macedo (Eds.)

Keywords: Alcohols, Biomass Tar, Carbon Deposition, Carbon Dioxide, Catalyst, Dry Reforming, Hydrogen, Methane, Reactor Technologies, Syngas.

INTRODUCTION

The main process technologies related to the reforming reactions are those developed for dry reforming (DR), autothermal reforming (ATR) and steam reforming (SR). The technologies for autothermal reforming and steam reforming are already adopted at the industrial scale while dry reforming technologies are still mainly at research stage even though few commercial examples exist related to methane dry reforming developed in Germany and in Denmark [1, 2]. The advantage of dry reforming resides in the possibility to use carbon dioxide as an oxidant and thereby reduce the environmental impacts of carbon dioxide emissions [3]. Concerning the feedstocks for the reforming processes, any hydrocarbon containing sufficient amounts of carbon and hydrogen could be used. In the case of dry reforming, the most studied one is methane [3], followed by ethanol [4, 5] and glycerol [6, 7]. The use of higher hydrocarbons in dry reforming feedstock has been earlier reviewed by Shah [2]. During recent years, reports have been found also related to dry reforming of more exotic feedstocks, such as diesel [8], lignin [9] and gasified waste plastic [10].

All the three mentioned reforming techniques involve the presence of an oxidant that converts the organic carbon to CO and at the same time produces H_2. The CO/H_2 ratio of the product gas depends on the oxidant (*e.g.*, CO_2, O_2, H_2O) and certain other parameters including type and presence of a catalyst, pressure, air/fuel ratio, temperature and possible on–line hydrogen separation. The selected advantages and disadvantages of the ATR, SR and DR technologies are summarized in Table **1**.

The thermodynamics of the DR reaction are not as advantageous as in the case of ATR and SR reactions which are largely considered in the literature [11 - 13]. In fact, due to the endothermic nature of the reforming (in all cases), high temperatures are needed to produce syngas or hydrogen. Since commercial applications of autothermal reforming and steam reforming already exist, this section will concentrate on the processes developed for dry reforming. There are certain reasons why the dry reforming process has not yet reached the full commercial application. These are, for example, availability of clean and steady source of CO_2 and high energy demand due to the high temperatures needed in the reaction (especially in the case of methane, which is the most stable hydrocarbon) [2, 3]. High temperatures are needed also to minimize the coking of the catalysts. These reasons increase the costs of the dry reforming process which hinders its commercialization. In practice, the energy demand of the dry reforming needs to be reduced in some way. This chapter will present selected possibilities to overcome the energy problem. Due to the endothermic nature of the DR process, the energy supply is one of the most crucial aspects of this technology. The energy for the reaction can be supplied by conventional heating (burners), but there exist other options as well.

Table 1. Comparison of steam reforming, partial oxidation and dry reforming.

Process	Main Reactions (CH$_4$ as an Example)	ΔH_{298} (kJ mol^{-1})*	Advantages	Disadvantages
Steam reforming (SR)	$CH_4 + H_2O \leftrightarrow CO + 3H_2$ $CH_4 + 2H_2O \leftrightarrow CO_2 + 4H_2$ $CO + H_2O \leftrightarrow CO_2 + H_2$	+206 +165 −41	Good availability of steam Operation temperature lower than in ATR Commercially available	Energy balance less advantageous than in ATR. High CO$_2$ emissions
Partial oxidation (used in ATR)	$CH_4 + 2O_2 \leftrightarrow CO_2 + 2H_2O$ $CH_4 + O_2 \leftrightarrow CO_2 + 2H_2$ $CH_4 + 1/2O_2 \leftrightarrow CO + 2H_2$	−803 −319 −36	No external heat required Commercially available	Need and cost of pure O$_2$ Low H$_2$/CO ratio for certain applications
Dry reforming (DR)	$CH_4 + CO_2 \leftrightarrow 2CO + 2H_2$	+247	Possibility to use CO$_2$	Constant supply of pure CO$_2$ Coking of catalyst High energy demand

*Calculated with HSC Chemistry 7.0

The most evident option is the use of a catalyst that provides a new reaction pathway that requires less energy. In the case of dry reforming reactions, the main problem is carbon deposition that can deactivate the operating catalyst [3]. From the process operation point of view, the carbon deposition can be monitored from the produced H$_2$/CO ratio. During methane reforming, in an ideal case, the H$_2$/CO ratio should be equal to 1. Any deviation from this value indicates that carbon formation is likely to occur. Furthermore, if the formation of H$_2$O occurs when H$_2$/CO ratio is 1, carbon formation may occur *via* direct reduction of CO [3]. In the case of feedstocks other than methane, the optimal H$_2$/CO ratio is different. To minimize the coke formation on the catalyst, the DR process is often run above 1000°C, which sets high demands for the thermal stability of the materials used in the process. Other approaches include adding oxygen to the feed and modifying the catalysts by rare–earth metals [1 - 4]. The development of more efficient and durable catalytic materials for dry reforming reactions has been already considered for several years, and is still a popular research topic. Typically, highly dispersed Ni–containing catalysts are considered to be the most active in DR reactions. The catalysts durability can be improved for example with potassium or potassium oxide [3]. An alternative strategy adopted to minimize carbon formation on the catalyst is the use of different hydrogen–containing additives. Kumar *et al.* [4] reported that adding 6 moles of hydrazine per 1 mole of ethanol into the feed during the dry reforming of ethanol increases the hydrogen yield and eliminates completely the carbon formation. However, using 2 moles of hydrazine led to the highest H_2/CO ratio. Glycine, urea and ammonia were also considered as possible reducing agents in the same study [4].

Fermentation of organic wastes produces biogas. Methane (CH_4) and carbon dioxide (CO_2) are the main components of biogas: 45-80% CH_4, 20–55% CO_2. CO_2 or dry reforming of methane (DRM) produces carbon monoxide (CO) and hydrogen (H_2). This reaction is interesting because it contributes to the decrease of the amount of two greenhouse gases (CH_4 and CO_2), it valorizes organic wastes (biogas from waste) and it produces a gas mixture, CO and H_2, called syngas, which could be used to synthesize chemical intermediates with higher values than methane, via the Fischer–Tropsch (FT) process. Indeed, when the H_2/CO ratio in the syngas is close to 1, the syngas could be used to produce acetic acid, aldehydes, ethanol, a wide variety of alcohols, olefins, and gasoline [2, 14]. Hence, this syngas could be considered as alternative for fossil fuels. It is also worth mentioning that natural gas deposits could contain various amounts of CO_2 and its removal can be done by the use of DRM, lowering therefore the overall operation cost by 20% compared to other reforming reactions such as steam reforming of methane (SRM) [15]. In the DRM process, steam does not figure among the reactants. This fact minimizes the energy consumption and protects from sintering and/or re–oxidation of metallic nanoparticles used as catalysts during the process. For all these reasons, many research groups are directed towards the DRM feasibility and optimization. However, the DRM reaction suffers from carbon deposition due to side reactions, and has a high endothermic character. It is known that catalysis allows to decrease the temperature required for a given process and to direct the reaction towards a desired pathway. For example, Abd Allah and Whitehead [16] demonstrated that the energy efficiency in DRM is increased by over 20% by the addition of a catalyst. Due to the industrial and environmental importance of DRM, a lot of studies were devoted to the synthesis and physico–chemical characterization of solid catalysts for this process. On January 2018, the number of scientific articles and reviews on catalytic DRM was about 953 in Sciencedirect and 1680 in Scopus. Statistical analysis on the number of publications per year is displayed in Fig. (**1**). Data for 2018 concerns only the first 15 days of January. The earliest work on DRM was published in 1973 and the number of publications was only 13 before 1999 in Sciencedirect and 24 in Scopus. An increasing interest in the catalytic DRM reaction can be clearly seen. The chemical compositions of the support, the active phase, and the promoter, the preparation method of the catalyst (impregnation, precipitation, sol–gel, ball–milling, *etc…*) and its physico–chemical properties (oxidation state, particle size...) are among the numerous parameters that influence the catalytic performance in DRM process. These parameters are discussed in this chapter based on different recent published studies with a special focus on the main findings along with the encountered limitations.

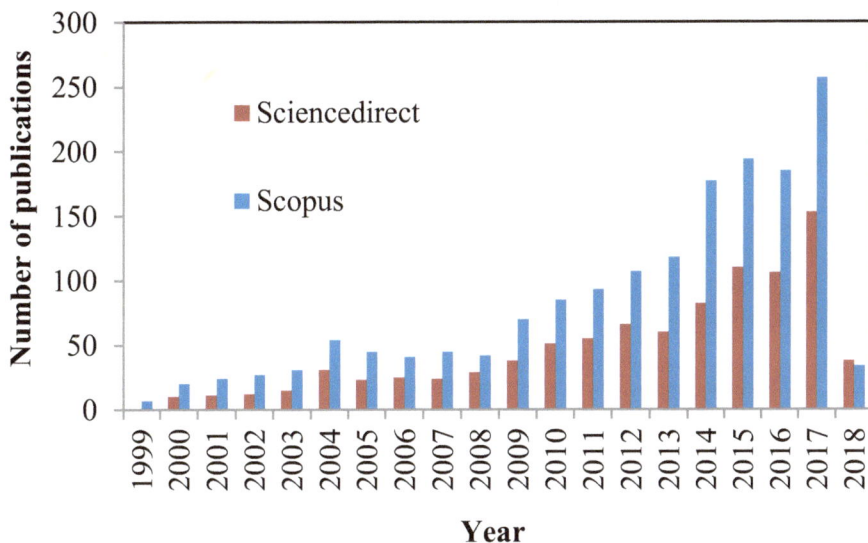

Fig. (1). Yearly number of publications studying the dry reforming of methane (search keywords: "catalysis" and "methane dry reforming"). For 2018, the first 15 days of January are only considered.

On the other hand, CO_2 or dry reforming (DR) of oxygenates such as alcohols is a promising way to produce H_2 rich stream and/or syngas (Fig. **2**). The DR technology has the strong advantage of utilizing the greenhouse gas CO_2 as an oxidant and effective reactant. The dry reforming of alcohols is a complex reaction network that involves a series of steps [17, 18]. Moreover, DR of alcohols is highly endothermic and requires high reaction temperature to achieve higher conversions compared to the steam reforming (SR) of alcohols [19, 20]. This chapter provides insights on the reactions and catalytic development in DR of short chain alcohols (methanol, ethanol, glycerol and butanol) as model compounds. Biomass derived materials and/or industrial side streams or other bio–wastes contain alcohols and are hence useful sources to generate energy in the form of H_2 rich stream. The latter can be directly fed to high temperature fuel cells (HTFC), such as solid oxide fuel cell (SOFC) and molten carbonate fuel cell (MCFC) devices [21].

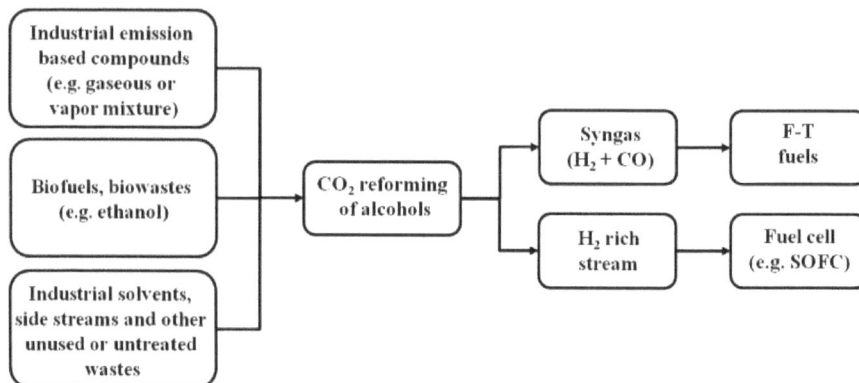

Fig. (2). Utilization of various alcohols feedstocks for H_2 and/or syngas production via CO_2 reforming.

Table **2** compares natural gas and the short chain alcohols feedstocks in terms of advantages and disadvantages.

Table 2. List of feedstocks for CO_2 reforming and their advantages and disadvantages [18, 19, 22, 23].

Feedstock	Advantages	Disadvantages
Natural gas **(CH_4)**	Availability, inexpensive and CO_2 mixed feed, existing gas infrastructure	Fossil fuel–derived feedstock, reserves depletion, high CO_2 emissions, flammable, highly volatile, energy intensive
Methanol **(CH_3OH)**	Industrial waste solvent, renewable bio–methanol, easy transportable, highest relative H content among alcohols, DR operated below 600°C	Limited, processing costs, methanol produced via syngas, volatile, toxic
Ethanol **(C_2H_5OH)**	Renewable, biofuel, biodegradable, ethanol vapors from side streams, easy transportable, non–toxic, non–volatile, non–flammable, DR operated below 800°C	Ethanol production is energy intensive, H_2*vs* syngas energy economics
Glycerol **($C_3H_3O_8$)**	Renewable, biodiesel byproduct, non–toxic, non–volatile, non–flammable, DR operated below 800°C	More difficult to process, need primary feedstock
Butanol **($C_4H_{10}O$)**	Renewable, CO_2 neutral, more safe than ethanol, high H_2O tolerance, DR operated below 900°C	Limited availability, bio–butanol production not well developed

The type of feedstock depends on the source and the geological conditions while the end application mostly depends on the H_2 to CO ratio. DR reaction is suitable for feeding fuel cell systems [24], but also to produce different H_2/CO ratios that can be used for methanol synthesis, methane production and Fischer–Tropsch fuels synthesis [25]. It is worth noting that higher alcohols present a higher C/H ratio compared to methane which increases the possibility of carbon deposition and catalyst deactivation during DR reaction.

Carbon formation/deposition is the biggest challenge in DR of hydrocarbons. It depends on the type of feedstock and the reaction conditions which influence the type of carbon formed (*e.g.*, filamentous, whiskers, carbon nanotubes, *etc*...) [2, 26, 27]. The reaction thermodynamics and the catalyst development for the dry reforming of various alcohols are reported in this chapter.

According to the World Energy Council, "bioenergy is energy from organic matter (biomass), *i.e.* all materials of biological origin that is not embedded in geological formations (fossilized). Bioenergy includes traditional biomass (example forestry and agricultural residues), modern biomass and biofuels. It represents the transformation of organic matter into a source of energy, whether it is collected from natural surroundings or specifically grown for the purpose." Bioenergy accounted for around 14% of the world's energy consumption in 2014 [28]. Biomass can be considered as a carbon neutral and renewable energy source. However, it is a low quality fuel that can be upgraded through "biomass gasification technologies" which are very important to expand its utilization [29]. Gasification occurs when biomass is heated in the presence of air, oxygen, carbon dioxide or steam in order to produce gaseous fuels especially hydrogen and syngas [30 - 33]. These gaseous products are the raw materials for the catalytic synthesis of liquid fuels (methanol, dimethyl ether, Fischer–Tropsch oils,…) and are also used to produce heat and electricity [34]. During the gasification process, the production of tar byproduct is an inevitable technical barrier [35]. Tar composition depends on the nature of the gasified biomass but a typical biomass tar contains toluene (24 wt%), other one ring aromatic hydrocarbons (22 wt%), naphthalene (15 wt%), other polycyclic aromatic hydrocarbons (20 wt%), phenolic compounds (7 wt%), heterocyclic compounds (10 wt%) and others (2 wt%) [36, 37]. For different yet important reasons, tar formation is a serious challenge to the gaseous products utilization [38]. It can account up to 40% of the total volatiles therefore it contains a large fraction of the biomass energy [39]. Moreover, tar formation is a serious problem at the gasification process level as it can clog the piping system, filters, heat exchangers, and cause important catalysts coking [40]. In addition to the technical problems, tar is a dangerous mixture because of its carcinogenic character and toxic compounds content [29]. For all the above reasons, tar content should be decreased in order to use the gaseous products from biomass gasification. More than half of the produced tar can be removed by raising the gasification process temperature to more than 800°C at the expense of energy consumption [41]. Physical removal of tar is also feasible but it decreases the overall process efficiency and may cause a secondary pollution [42, 43]. Catalytic reforming of tar is a chemical treatment technique that is more promising than thermal or physical treatments for several reasons. In fact, catalytic reforming not only reduces the risk of environmental pollution but also enhances the overall energy recovery through the production of combustible lighter gas species (CH_4) and a better quality syngas (more CO and H_2 produced) [44 - 46]. This latter solution will be considered in details in this chapter.

Finally, the production of energy (or inducing the chemical reactions) by plasma, microwaves, electrical current and solar energy are presented in the final part of this chapter. Some examples on the developed reactor technologies are also presented including chemical looping reforming, membrane reactors and ceramic counterflow reactor.

Below is a list of numbered equations of important reactions that will be referred to in several parts of the chapter. Enthalpy values are calculated at 25°C with HSC Chemistry 7.0 (all compounds except solid carbon are considered in the gaseous state)

Dry reforming of methane (DRM)

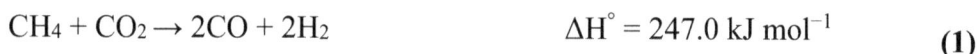

$$CH_4 + CO_2 \rightarrow 2CO + 2H_2 \qquad \Delta H^\circ = 247.0 \text{ kJ mol}^{-1} \qquad \text{(1)}$$

Steam reforming of methane reactions (SRM)

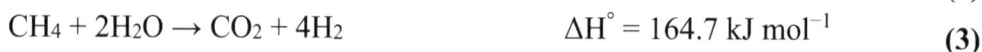

$$CH_4 + H_2O \rightarrow CO + 3H_2 \qquad \Delta H^\circ = 205.9 \text{ kJ mol}^{-1} \qquad \text{(2)}$$

$$CH_4 + 2H_2O \rightarrow CO_2 + 4H_2 \qquad \Delta H^\circ = 164.7 \text{ kJ mol}^{-1} \qquad \text{(3)}$$

Dry reforming of methanol reactions (DRMe)

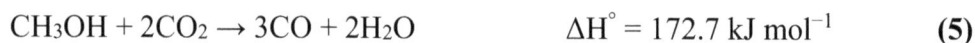

$$CH_3OH + CO_2 \rightarrow CH_2O + CO + H_2O \qquad \Delta H^\circ = 133.5 \text{ kJ mol}^{-1} \qquad \text{(4)}$$

$$CH_3OH + 2CO_2 \rightarrow 3CO + 2H_2O \qquad \Delta H^\circ = 172.7 \text{ kJ mol}^{-1} \qquad \text{(5)}$$

Methanol dehydrogenation

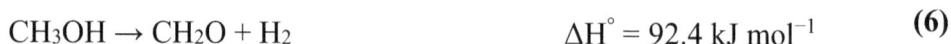

$$CH_3OH \rightarrow CH_2O + H_2 \qquad \Delta H^\circ = 92.4 \text{ kJ mol}^{-1} \qquad \text{(6)}$$

Methanol decomposition

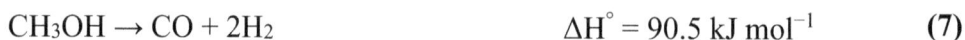

$$CH_3OH \rightarrow CO + 2H_2 \qquad \Delta H^\circ = 90.5 \text{ kJ mol}^{-1} \qquad \text{(7)}$$

Dry reforming of ethanol (DRE)

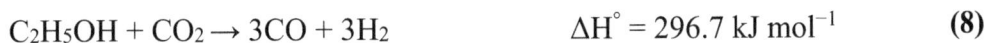

$$C_2H_5OH + CO_2 \rightarrow 3CO + 3H_2 \qquad \Delta H^\circ = 296.7 \text{ kJ mol}^{-1} \qquad \textbf{(8)}$$

Steam reforming of ethanol (SRE)

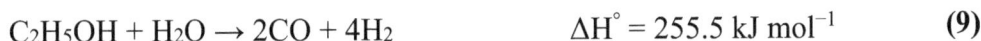

$$C_2H_5OH + H_2O \rightarrow 2CO + 4H_2 \qquad \Delta H^\circ = 255.5 \text{ kJ mol}^{-1} \qquad \textbf{(9)}$$

Ethanol decompositions

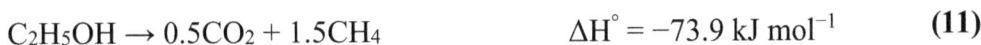

$$C_2H_5OH \rightarrow CO + CH_4 + H_2 \qquad \Delta H^\circ = 49.7 \text{ kJ mol}^{-1} \qquad \textbf{(10)}$$
$$C_2H_5OH \rightarrow 0.5CO_2 + 1.5CH_4 \qquad \Delta H^\circ = -73.9 \text{ kJ mol}^{-1} \qquad \textbf{(11)}$$

Ethanol dehydrogenation

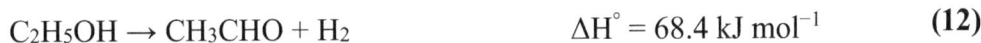

$$C_2H_5OH \rightarrow CH_3CHO + H_2 \qquad \Delta H^\circ = 68.4 \text{ kJ mol}^{-1} \qquad \textbf{(12)}$$

Acetaldehyde decomposition

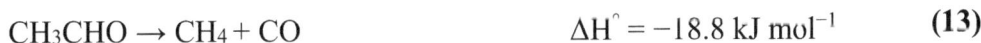

$$CH_3CHO \rightarrow CH_4 + CO \qquad \Delta H^\circ = -18.8 \text{ kJ mol}^{-1} \qquad \textbf{(13)}$$

Acetaldehyde CO_2 reforming

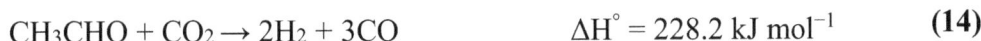

$$CH_3CHO + CO_2 \rightarrow 2H_2 + 3CO \qquad \Delta H^\circ = 228.2 \text{ kJ mol}^{-1} \qquad \textbf{(14)}$$

Dry reforming of glycerol (DRG)

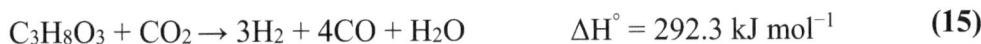

$$C_3H_8O_3 + CO_2 \rightarrow 3H_2 + 4CO + H_2O \qquad \Delta H^\circ = 292.3 \text{ kJ mol}^{-1} \qquad \textbf{(15)}$$

Glycerol dehydrogenation

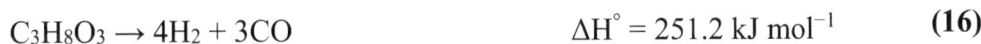

$$C_3H_8O_3 \rightarrow 4H_2 + 3CO \qquad \Delta H^\circ = 251.2 \text{ kJ mol}^{-1} \qquad \textbf{(16)}$$

Dry reforming of butanol (DRB)

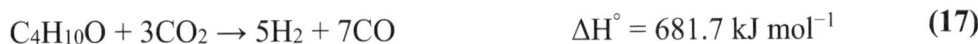

$$C_4H_{10}O + 3CO_2 \rightarrow 5H_2 + 7CO \qquad \Delta H^\circ = 681.7 \text{ kJ mol}^{-1} \qquad \textbf{(17)}$$

Butanol decomposition

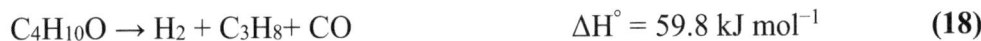

$$C_4H_{10}O \rightarrow H_2 + C_3H_8 + CO \qquad \Delta H^\circ = 59.8 \text{ kJ mol}^{-1} \qquad \textbf{(18)}$$

Butanol dehydrogenation

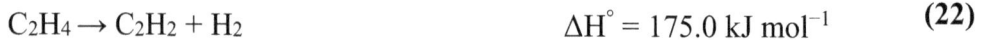

$$C_4H_{10}O \rightarrow H_2 + C_4H_8O \qquad\qquad \Delta H^\circ = 68.0 \text{ kJ mol}^{-1} \qquad \textbf{(19)}$$

Butanol dehydration

$$C_4H_{10}O \rightarrow H_2O + C_4H_8 \qquad\qquad \Delta H^\circ = 32.7 \text{ kJ mol}^{-1} \qquad \textbf{(20)}$$

$$C_3H_8 \rightarrow CH_4 + C_2H_4 \qquad\qquad \Delta H^\circ = 82.5 \text{ kJ mol}^{-1} \qquad \textbf{(21)}$$

$$C_2H_4 \rightarrow C_2H_2 + H_2 \qquad\qquad \Delta H^\circ = 175.0 \text{ kJ mol}^{-1} \qquad \textbf{(22)}$$

Carbon monoxide methanation

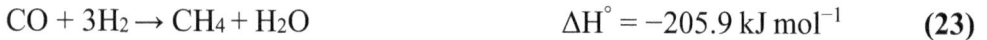

$$CO + 3H_2 \rightarrow CH_4 + H_2O \qquad\qquad \Delta H^\circ = -205.9 \text{ kJ mol}^{-1} \qquad \textbf{(23)}$$

Carbon dioxide methanation (CDM)

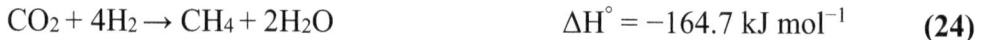

$$CO_2 + 4H_2 \rightarrow CH_4 + 2H_2O \qquad\qquad \Delta H^\circ = -164.7 \text{ kJ mol}^{-1} \qquad \textbf{(24)}$$

Water–gas shift reaction (WGSR) – Reversed Eq. 25 (RWGS)

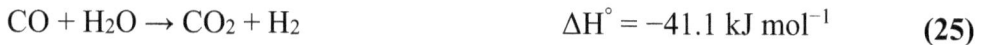

$$CO + H_2O \rightarrow CO_2 + H_2 \qquad\qquad \Delta H^\circ = -41.1 \text{ kJ mol}^{-1} \qquad \textbf{(25)}$$

Carbon gasification

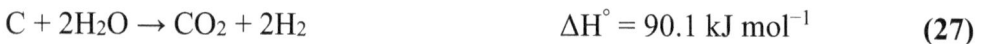

$$C + H_2O \rightarrow CO + H_2 \qquad\qquad \Delta H^\circ = 131.3 \text{ kJ mol}^{-1} \qquad \textbf{(26)}$$

$$C + 2H_2O \rightarrow CO_2 + 2H_2 \qquad\qquad \Delta H^\circ = 90.1 \text{ kJ mol}^{-1} \qquad \textbf{(27)}$$

Undesirable Carbon Formation Reactions

Ethanol dehydration/Ethylene polymerization

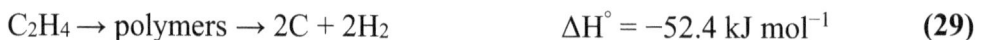

$$C_2H_5OH \rightarrow C_2H_4 + H_2O \qquad\qquad \Delta H^\circ = 45.4 \text{ kJ mol}^{-1} \qquad \textbf{(28)}$$

$$C_2H_4 \rightarrow \text{polymers} \rightarrow 2C + 2H_2 \qquad\qquad \Delta H^\circ = -52.4 \text{ kJ mol}^{-1} \qquad \textbf{(29)}$$

Methane decomposition (MD)

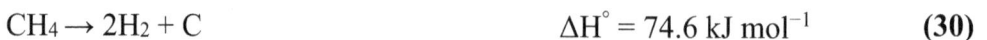

$$CH_4 \rightarrow 2H_2 + C \qquad\qquad \Delta H^\circ = 74.6 \text{ kJ mol}^{-1} \qquad \textbf{(30)}$$

$$2CO \rightarrow CO_2 + C \qquad\qquad \Delta H^\circ = -172.4 \text{ kJ mol}^{-1} \qquad \textbf{(31)}$$

Carbon monoxide hydrogenation (CMH)

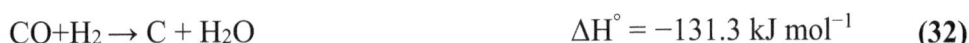

$$CO + H_2 \rightarrow C + H_2O \qquad\qquad \Delta H^\circ = -131.3 \text{ kJ mol}^{-1} \qquad \textbf{(32)}$$

Carbon dioxide hydrogenation

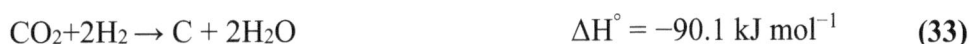

$$CO_2 + 2H_2 \rightarrow C + 2H_2O \qquad\qquad \Delta H^\circ = -90.1 \text{ kJ mol}^{-1} \qquad \textbf{(33)}$$

RECENT ADVANCES IN CO₂ REFORMING OF METHANE

Thermodynamics of the DRM Reaction

The DRM reaction is accompanied by many side reactions which are represented with equations 1–2, 24–25 and 30–33 (Eqs. 1–2, 24–25 and 30–33). The evolution of the equilibrium constant *versus* temperature is given in Fig. (**3**) for most of these reactions.

The thermodynamic calculations show that the DRM reaction is only possible, at atmospheric pressure and at temperatures higher than 633°C ($\Delta G > 0$). In fact, the C–H bond in CH_4 molecule is very stable (435 kJ mol⁻¹) and therefore a high temperature is required for a good methane conversion. However, at temperatures higher than 633°C, other reactions take place such as the SRM (Eq. 2), the methane decomposition (MD, Eq. 30) and to a less extent the reversed water–gas shift reaction (RWGS reversed Eq. 25) (Fig. **3**) which affects significantly the DRM reaction. Due to its strong endothermic character, the equilibrium constant of the DRM reaction increases sharply with temperature while for reactions with moderate endothermic character such as MD (Eq. 30) and RWGS (reversed Eq. 25), the equilibrium constants increase moderately with temperature. Conversely, exothermic reactions, are less thermodynamically favored at high temperatures (T ≥ 750°C) [47]. In real applications, the residence time in the reactor is shorter than the period needed to reach theoretical equilibrium. Therefore, the use of a catalyst is necessary to decrease the activation energy while favoring the DRM reaction pathway rather than other side reactions. Consequently, the presence of a catalyst makes the DRM process more economically sound.

Problems Encountered During the Catalytic DRM Reaction

A main problem encountered during the catalytic DRM is the deactivation of the catalyst due to several reasons and different mechanisms. Catalyst deactivation in reforming reactions is mainly due to, carbon deposition or hydrocarbons deposition –known as coking–, sintering, poisoning, and oxidation. Carbon formation could also lead to the reactor blockage and pressure drop problems.

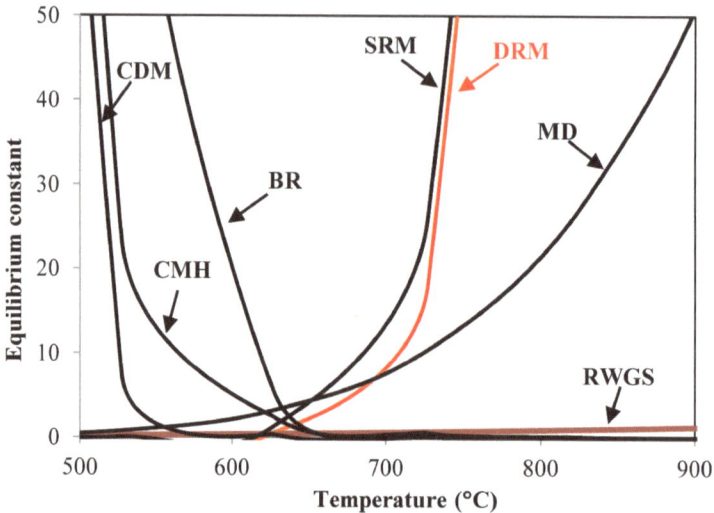

Fig. (3). Evolution of the equilibrium constant *versus* temperature for the dry and steam reforming reactions of methane and some side reactions: DRM (Eq. 1); SRM (Eq. 2); CDM (Eq. 24); RWGS (reversed Eq. 25); MD (Eq. 30); BR (Eq. 31); CMH (Eq. 32).

Carbon Deposition

One of the classifications of carbon deposits is based on the ability of the carbon species to be hydrogenated: C_α is the surface carbide that can be hydrogenated below 50 °C, C_β (or amorphous carbon) can be hydrogenated between 100 and 300 °C, and C_γ (graphitic carbon) can be hydrogenated at above 500 °C [48].

The side reactions represented by equations 30 to 33 (Eqs. 30–33) are mainly responsible for carbon formation during the DRM reaction. Each reaction is favored at a given range of temperature (Fig. **3**). Different carbon species could be deposited on a catalyst surface. The deposited carbon could be amorphous carbon or carbon filaments or C_β, graphitic carbon, carbidic, polymeric, nanotubes or shell–type [26, 49]. The most reactive form of carbon, C_α, is oxidized at lower temperatures than the other forms. It is composed of carbon adsorbed and bound to metallic centers.

Filamentous carbon or whisker carbon is the most deleterious one for nickel based catalysts. This carbon nucleates creating a strong whisker that grows with a nickel crystal eventually pushing it out of the catalyst. Whisker carbon could also cause the catalyst pellets to break [48].

Carbon formation depends on the composition and structure of the catalyst [50]. Furthermore, different studies were devoted to solve this problem by elaborating catalysts that limit the rate of carbon formation.

The use of basic compounds (such as K_2O, MgO, CaO, La_2O_3, *etc…*) decreases carbon formation [47, 51 - 53]. Indeed, basic promoters favor the adsorption and dissociation of CO_2, known to be acidic, leading hence to a decrease of the rate of carbon formation on the surface of catalysts [54]. The addition of oxygen carriers (ZrO_2, CeO_2, La_2O_3,...) oxidizes carbon and reduces its formation [55]. Some researchers suggested the use of bimetallic active phase (and trimetallic) systems instead of monometallic ones. For example, the addition of Co, Fe, Cu, Cr, Sn, Mn, Mo, V, and Rh was studied [26, 47]. The addition of small amounts of noble metals (Rh, Ru) to transition metals (Ni or Co) is also proposed [56]. Furthermore, it was proven that smaller particle size is associated with lower carbon formation [57].

Metal support interaction (MSI) or strong metal support interaction (SMSI) is also responsible for the decrease of the amount of carbon formed. Indeed, metal particle being in strong interaction with the support, are not easily subjected to form agglomerates that increase the particle size. Therefore, high dispersion of metallic nanoparticles and a strong MSI hinder the carbon deposition process. It was reported that highly dispersed nickel with average particles size smaller than a "critical size" can effectively inhibit the coking process compared to big particles over which the kinetics of solid carbon encapsulation and subsequent precipitation into graphitic (deactivating) carbon are much faster [58, 59]. This "critical size" was found to be 10 nm in the work of Zhang *et al.* [60], 15 nm according to Usman *et al.* [61] and 5 nm was reported by Arora and Prasad [62]. Aramouni *et al.* [48] pointed out that not only the amount of carbon could be responsible for the catalyst deactivation but also the type of carbon. In some cases, a low amount of "toxic" carbon could lead to higher deactivation than a high amount of less "toxic" carbon.

Sintering

Sintering is usually referred to the crystalline growth leading to the increase of the particle size and thus the decrease of the active surface area of catalyst. Sintering is favored by high temperatures (thermal agglomeration) and by the composition of the atmosphere that surrounds the catalyst. For example, water presence in the feed stream induces sintering. As already stated, DRM is endothermic and thus more thermodynamically favored at high temperature. Therefore, sintering of the active phase, the support or the bulk catalyst is likely to occur. Sintering is irreversible. To avoid it, it is suggested to use suitable promoters, to increase the MSI, and to choose a suitable catalyst preparation method.

Poisoning

Poisoning is the loss of catalytic activity due to the chemisorption, on the active sites, of impurities originally present in the feed. The most known catalyst poison during reforming process is sulfur present mainly under the form of H_2S. For example, in biogas, H_2S content could range from 0 to 10000 ppm depending on the chemical nature of the waste [63, 64]. H_2S could become chemisorbed on metallic surfaces and regeneration is not all the times feasible.

Oxidation

The active phase in methane reforming reactions is usually a metallic particle. CH_4 molecule can be only activated on a metal surface where it dissociates to produce hydrogen and hydrocarbon–related species [65]. The reduced active phase (supported or not) could be oxidized, by an oxidizing agent, during the reforming process. This oxidation decreases the catalytic activity and can lead to complete deactivation if, in parallel, reduction by a reducing agent (example H_2) does not take place or is not enough to maintain a reduced state [66]. Such deactivation is a main concern facing the development of several reforming catalysts and it depends mainly on the catalyst composition and on the reaction medium.

The deactivation type and/or extent depend mainly on operation conditions and catalyst formulation (chemical and physical properties). Thus, different deactivation mechanisms are associated with different catalytic systems. One type or many types of deactivation could take place on a given catalyst. For example, Hu and Ruckenstein [67] reported that carbon formation rather than sintering is at the origin of the deactivation of platinum–containing catalysts in the DRM reaction. Zhu *et al.* [68] found that the inert carbon is the cause of the deactivation of Ni–Mg/SiO$_2$, whereas sintering of metallic Ni is behind the deactivation of Ni–Co/SiO$_2$.

Catalytic Dry Reforming of Methane (DRM)

An ideal catalyst has high activity, good selectivity, and excellent stability. An affordable price of the catalyst is also an important criterion for its industrial use. Two main types of catalysts exist: bulk catalysts and supported catalysts. Supported catalysts are generally composed of a support and an active phase, to which a promoter is sometimes added to enhance the performance. In the bulk catalyst, the active phase is not dispersed on a support. A promoter could be also added to a bulk catalyst. Different catalyst preparation methods lead to different performances. Hereafter, different supports, active phases, promoters and preparation methods used during the elaboration of catalysts for DRM are given.

Active Phase

Monometallic, bimetallic, and rarely trimetallic active phases were studied in DRM. Noble metals as well as transition metals were studied. Increasing interest for transition metals is noted. The most active phase used and studied is nickel.

Noble Metals (Monometallic)

It has been found that supported precious metals such as Rh, Ru, Pd, Pt, and Ir catalysts are known to exhibit excellent catalytic performances in DRM and less carbon can be formed during the reaction due to the lower solubility of carbon in noble metals, in comparison with non–precious metal catalysts [49, 69].

Safariamin *et al.* [70] studied Ru supported on alumina and ceria. An improvement of the catalytic activity after Ru addition was associated to the high dispersion of Ru species on the support. Ferreira–Aparicio *et al.* [71] studied Ru supported on alumina and silica. Ru/Al_2O_3 presented higher selectivity towards hydrogen and higher thermal stability than Ru/SiO_2. The addition of ruthenium (1wt%) on $Co_xMg_yAl_2$ oxides increases both the CO_2 and CH_4 conversions and decreases the amount of water formed for the supports $CO_2Mg_4Al_2$ and Co_6Al_2 (the number in subscript corresponds to the molar ratio) [56]. It was found that Ru used in small amounts (1wt%) shifted the reducibility of cobalt catalysts to lower temperatures and improved the stability of the catalysts by decreasing carbon formation while produced carbon was not harmful for the catalysts [56]. Nawfal *et al.* [72] added 0.5wt% Ru to mixed NiMgAl oxides and found that Ru enhanced the reactivity and selectivity of the oxides in DRM because of the formation of easily reducible ruthenium and nickel oxides at the catalyst surface. Crisafulli *et al.* [73] showed that the catalytic performances of Ni–based catalysts were enhanced by Ru addition. Bitter *et al.* [74] studied the DRM reaction on Pt/ZrO_2 and suggested a bifunctional mechanism where the adsorbed formates are intermediate, existing on the support surface or at the metal–support interface. They found that the catalysts are not efficient when carbonate formation is not possible.

Apart from powdered catalysts (used in fixed bed reactors), membranes with Pd, Ru, Ag have also been used to study DRM reaction [75 - 78]. Membranes allow permeation of products away from the catalyst bed which drives the reaction towards the formation of CO and H_2, leading to higher conversions at lower temperatures and limiting the occurrence of simultaneous side reactions. Among noble metals, supported Rh showed generally higher activity and stability than Ru, Ir, Pd, and Pt. The low stability of Pd and Pt supported catalysts compared to the other supported noble metals is attributed to the sintering of the metal particles leading to greater particle size and lower dispersion and thus to more carbon deposition [48]. Ternary oxides of perovskite type, $BaZrMO_3$ (M = Ru, Pt, Rh) were investigated in DRM [79]. The Rh based catalyst showed higher activity than the Ru based one. The less active was the Pt based one. The higher reducibility of Rh and Ru was the reason behind their better catalytic performances. A better stability due to lower amount of carbon formation is also noted for Rh and Ru based systems.

Transition Metals (Monometallic)

Despite the reduced carbon formation and the stability of precious metals based catalysts, from an industrial point of view, it is more practical to develop non–precious metal–based catalysts. For this reason, many studies were devoted to the development of transition–metal based catalysts due to their lower cost and wider availability compared to noble metals catalysts. Ni has drawn remarkable attention in this area due to its high activity, wide availability and low cost [80]. Barama *et al.* [81] found comparable activities in the DRM reaction for both 3wt% Rh and 10wt% Ni supported on layered aluminosilicates, montmorillonite, though higher amounts of carbon were deposited on the Ni–based catalyst.

Ruchenstein and Hu [82, 83] reported that NiO/MgO has an excellent stability and a good resistance to carbon formation because of the formation of a solid solution between NiO and MgO, which is believed to suppress CO disproportionation. Ni–MgO solid solution [84], Ni/perovskite [85], and Ni/La_2O_3 [86] were also studied. Asami *et al.* [87] compared Ni, Co, and Fe on CeO_2 and found the following order in activity and selectivity: Ni > Co >> Fe. The DRM reaction was studied over Cu–ferrite/ZrO_2 catalyst [88]. The increase of Cu content led to a decrease in the amount of carbon formed. Arora and Prasad [62] stated that the use of membrane reactors eliminates some difficulties faced with the fixed bed reactors, such as high catalyst amount requirement, poor hydrogen separation, low thermo–chemical and thermo–mechanical durability. Ni–based membrane reactors are economic and advantageous.

Bimetallic

Many authors claimed a possible amelioration for carbon limitation by considering the preparation of bimetallic based–catalysts combining several elements such as Ni–Rh and/or Ni–Ce [81]. Bimetallic catalysts exhibit superior performance in the DRM reaction compared to monometallic catalysts. Indeed, the formation of metal alloy could hinder the metal oxidation process.

Du *et al.* [89] studied the Pt–Ru/Al$_2$O$_3$ catalyst. A higher selectivity towards DRM was observed and more reactive carbon species were formed in comparison with monometallic catalysts. The resistance to agglomeration of the active bimetallic species was mainly responsible for the better catalytic activity and stability.

Combining a noble metal to a non–noble metal to create an alloy was also proposed [49, 90, 91]. This approach could decrease carbon formation, increase the metal dispersion, decrease the metal particle size, decrease the cost, and produce a good activity and stability in the DRM reaction. Rhodium [92] and platinum [93] were added to Ni supported on alumina. Rh increased Ni dispersion leading to higher activity and less carbon formation, Pt enhanced catalyst reducibility leading to higher activities. The addition of Rh to Ni supported on SiO$_2$ also showed an enhancement in catalytic activity and stability [94]. The reviews by Li *et al.* [90] and Bian *et al.* [91] are devoted to Ni catalysts modified with noble metals for application in DRM. Addition of noble metals to Ni based systems is benefic for the activity, the suppression of Ni oxidation, the decrease of carbon formation, and the self–activation of catalyst.

Combining two non–noble metals or transition metals to create an alloy was also proposed. Zhang *et al.* [47] compared the combination of Co, Fe, Cu, and Mn to Ni and found that Ni–Co bimetallic catalyst exhibited the highest catalytic performance on activity and stability because the synergy between Ni and Co had greatly improved the catalytic properties. The high activity, low carbon formation, and excellent stability of Ni–Co catalyst was closely related to its high metal dispersion, strong metal–support interaction, relative easy reducibility, high surface area, small pore diameter, and formation of stable solid solutions. Sharifi *et al.* [95] compared the addition of Co and Cu to Ni/Al$_2$O$_3$–ZrO$_2$. An increase in surface area was observed with both Co and Cu and they both improved the conversions of CH$_4$ and CO$_2$, the product yield, and produced a H$_2$/CO ratio close to 1. However, Co-promoter was better than Cu and its addition improved the particles size uniformity.

Ni–Co bimetallic catalysts are widely studied for the DRM reaction and are associated to much less carbon deposition [62]. The NiCoAlMgO mixed oxides catalyst, prepared using coprecipitation, has shown good stability and activity in DRM [47, 96]. Luisetto *et al.* [97] showed that the DRM reaction is close to thermodynamic equilibrium in the presence of Ni–Co based systems which was not the case with monometallic systems. The higher activity of bimetallic catalysts might be due to an interaction between Ni and Co which promotes the DRM reaction as noted by Al Fatesh [98]. Phongaksorn *et al.* [99] explained the better reactivity of Ni–Co systems compared to Ni or Co by a better dispersion, a stronger metal-support interaction, and the formation of Ni–Co alloy. Takanabe *et al.* [100] also showed that the catalyst activity can be improved through the formation of a homogeneous alloy between Co and Ni. The substitution of a small amount of Ni with Co dramatically improved the catalyst activity and stability. The improvement of catalytic performance was ascribed to the catalyst resistance to metal oxidation. The Co–Ni alloy was reported, to be more reducible than Ni or Co species [101] and to decrease the Ni particle size leading to a better metal dispersion and stronger Ni and Co interaction [102].

It was proposed by Koh *et al.* [103] that cobalt decreases the rate of coke formation by oxidizing the surface carbon to CO or CO_2, since cobalt is an efficient catalyst for soot oxidation. The combination of the two transition metals Ni and Co on various supports (Al_2O_3, ZrO_2, TiO_2, CeO_2, CeO_2–ZrO_2) was considered in different studies where the solids exhibited enhanced catalytic performances in methane reforming due to the synergy between Ni and Co [97, 100, 104 - 108].

Many scientists are still working on the reduction of carbon formation either by using multimetallic catalysts or by modifying the properties of the surface with additives that convey basic character yet they decrease the catalytic activity [109, 110].

Tsoukalou *et al.* [111] studied the $LaNi_{0.8}M_{0.2}O_3$ (M = Ni, Co, and Fe) catalyst from perovskite precursor. Ni–Co was the most active in DRM whereas Ni–Fe showed no activity. Cobalt addition increases the rate of $La_2O_2CO_3$ formation, responsible for the decrease of carbonaceous deposits and thus for a better catalytic activity and stability. The low activity of Ni–Fe was ascribed to the encapsulation of active Ni particles by $LaFeO_3$.

The review of Bian *et al.* [91] is devoted to the study of bimetallic Ni–transition metal catalysts for DRM. It is stated that Co, Fe, and Cu play different roles. Ni–Co alloys adsorb strongly oxygen due to the high affinity of Co to oxygen, and thus less carbon is formed and the catalyst is more stable. Co leads also to small particle size. Small amount of Co is enough to show enhancements whereas a high amount of Co makes the catalyst subject to oxidation. Iron has good redox properties and Ni-Fe shows re–alloying and de–alloying processes that result in a fast removal of carbon deposition. Cu is effective in enhancing carbon resistance but the principle on which its action is based is not yet very clear.

Trimetallic

The addition of Au and Pt to Ni supported on alumina modified by cerium and magnesium oxides was carried out by Wu *et al.* [112]. Bi and trimetallic catalysts were more performing than monometallic ones due to the formation of nickel nanoparticles that interact with Au and Pt. The activity is slightly increased by using a trimetallic catalyst in comparison with the bimetallic one. Synergetic effects are less noticeable in trimetallic systems than in bimetallic ones and the effect of support is less apparent [48, 112]. The use of trimetallic system composed of noble metals does not present an economical interest. Furthermore, interactions in a trimetallic system are more complex than in bimetallic and monometallic ones [48].

Supports

Different supports and supports combinations have been investigated in the DRM reaction. Indeed, the support plays many roles: high and uniform dispersion of the active phase, thermal stability, mechanical resistance, interaction with the active phase (MSI) and so on. From a thermodynamic point of view, a high reaction temperature (750–800 °C) must be chosen to achieve high reactant conversion. Therefore, a thermally resistant high melting point support should be chosen. However, at high temperatures, the active metal would contact each other through thermal motion. Therefore, deactivation by sintering is likely to occur. Thus, the support should present high specific surface area to ensure high dispersion, and it should have enough interaction with the active phase to hinder its motion.

The most common support used in DRM is alumina [70, 107, 109, 113 - 115]. The second support investigated is silica [68, 113]. MgO and CaO were studied due to their high melting points, thus the high thermal stability, and also to their basicity known to decrease carbon formation by favoring the activation of CO_2. TiO_2 [100, 104] and ZrO_2 [108] are reducible oxides on which metals are electron–rich and always show strong metal support interaction (SMSI). As stated before, SMSI plays a crucial role in decreasing carbon formation. Rare–earth elements oxides such as CeO_2 [97, 116] and La_2O_3 [111, 117, 118] are studied because of their ability to release and uptake oxygen. These oxides, used as supports, enhance the catalytic performance by increasing the MSI. Mixed oxides of $Mg_xAl_yO_z$ were used as support, because MgO has a high melting point but a low specific surface area and Al_2O_3 is supposed to compensate the low area of MgO [47, 119, 120]. Combining Al and Mg oxides provides the catalyst with not only relative high surface area but also thermal stability.

Charisiou *et al.* [121] used aluminum oxide (Al_2O_3) modified by CeO_2 and/or La_2O_3 as supports for Ni. Safariamin *et al.* [70] studied CeO_2–Al_2O_3, Chen *et al.* [122] used CeO_2–ZrO_2, Sharifi *et al.* [95] used Al_2O_3–ZrO_2, Muñoz *et al.* [123] used CeO_2–ZrO_2 and CeO_2/ZrO_2–Y_2O_3, and Blanchard *et al.* [124] used Al_2O_3 and ZrO_2–Y_2O_3.

Interests in the research field of mesoporous materials have been expanding since the year 1992 [125, 126] owing to the success of the firstly reported mesoporous silica and due to the potential application of these systems in catalysis [126]. As supports for catalysts, mesoporous oxides have ordered frameworks, high specific surface areas, and narrow pore size distributions. With such special structural features, a supported metal can, in general, be relatively more dispersed and most importantly confined inside a well–defined porous matrix. The confinement (or trapping effect) of small nanoparticles limits their potential growth during DRM process (high temperature…) thus contributing in the preservation of homogenous dispersion of active sites during continuous run.

Ni–loaded mesoporous catalysts based on standard oxides such as silica SBA–15 or mesoporous Al_2O_3 displayed interesting catalytic activity for both dry and steam reforming reactions [127 - 132].

Stabilizing of Ni nanoparticles within well–defined core–shell structures, core–shell structured nanocatalyst, such as Ni coated by SiO_2 shells (Ni@SiO_2), was proven a promising candidate for the DRM reaction [133] since thermal agglomeration due to silica shells and coking of small well–dispersed Ni nanoparticles at high reaction temperature (700°C) were essentially limited.

ZSM5 zeolite was also used as support for Ni [51, 134] or Ni and Co [26] in DRM. Other zeolites were also used as supports such as Y [135].

A peculiar attention was paid to the use of phosphated compounds as supports. Phosphate of nickel–calcium/hydroxyapatite catalysts showed a high activity and good selectivity in DRM [136]. In these catalysts, three types of different nickel species were detected. In addition, the catalyst Ni–Sr/phosphate was studied [137]. It was also found that three different nickel species were present and that the resulting metallic nickel is the active phase.

The use of natural, low cost, and widely available oxides as supports is also reported. Abba *et al.* [138] studied natural phosphates as support for Ni. Their results showed that 15% Ni supported on natural phosphates, prepared by impregnation, has no activity at all. The modification of the support by mechano–chemical and/or acid treatment strongly improved the catalytic performance. The as–obtained catalysts showed good activity (80–85% CO_2 and CH_4 conversion at 750°C) and stability (400 min for 10% decrease in conversion for CH_4 and about 5% for CO_2). The mechano–chemical and/or acid treatment of natural phosphates support mainly modify the interaction between the nickel phase and the support surface. The nickel ions occupy calcium position in the surface of the phosphate phase, which stabilizes and improves the dispersion of nickel species. A much better dispersed metallic phase interacting with the natural phosphates surface was formed. This avoids sintering and results in an improvement in the catalytic activity. Indeed, ball milling led to an increase of surface area [138].

In the work of Barama *et al.* [81], layered aluminosilicates, particularly the montmorillonite family of clay–based substances was chosen as catalytic support for the preparation of several monometallic DRM catalysts with Rh, Ni, Pt, or Ce as active phases. Due to the pillared–layered structure of the support, metallic nanoparticles were homogeneously dispersed and remained stabilized within the clay's layers during reforming.

A great interest among researchers to synthesize catalyst supports from cellulosic materials (plant–based materials) is observed [50]. Indeed, the unique properties of the cellulose which are a well-defined structure and superior mechanical strength could enhance the catalytic activity in the DRM reaction.

In the review of Usman *et al.* [61], it was stated that the support materials ceria–zirconia mixtures, ZrO_2 with alkali metals (Mg^{2+}, Ca^{2+}, Y^{2+}) addition, MgO, SBA–15, ZSM–5, CeO_2, $BaTiO_3$, and $Ca_{0.8}Sr_{0.2}TiO_3$ showed improved catalytic activities and decreased carbon deposition. The modifying effects of cerium (Ce), magnesium (Mg) and yttrium (Y) were significant for DRM.

It is to be noted that no universal support was suggested for all the metals and that the support nature influences the catalyst activity. For example, rhodium promoted catalysts showed the following catalytic activity order: ZrO_2 > Al_2O_3 > TiO_2 > SiO_2 >> MgO [139]; while with nickel based catalysts a different order was obtained: ZrO_2 > CeO_2 > Al_2O_3 > La_2O_3 [110].

A study carried out by Ferreira–Aparicio *et al.* [113] on noble metals (Ru, Rh, Ir, Pt) and transition metals (Fe, Ni, Co) gave the following conclusions. When supported on alumina, the following order of activity was found at 450 °C: Rh > Ni > Ir > Ru ~ Pt > Co. However, when silica is used as support, the order of activity was: Ni > Ru > Rh ~ Ir > Co ~ Pt.

Solymosi *et al.* [140] studied noble metals supported on alumina in DRM. They observed that the catalytic activity and selectivity followed the order: Ru > Pd >

Rh > Pt > Ir; while the amount of deposited carbon corresponded the ranking: Ru > Ir > Rh = Pt

Nakamura *et al.* [141] found for Rh on different supports the following order of activity: Rh/Al_2O_3 > Rh/TiO_2 > Rh/SiO_2. The authors stated that the activation of CH_4 or CO_2 occurs on the metallic oxides used as supports and that the difference between the Rh particle size and the Rh dispersion on the support could lead to different results on different supports.

Promoters

Many promoters were added to reduce carbon formation or sintering. The introduction of promoters is of key importance as it also helps the increase in catalytic activity. Promoters are also capable of modifying the acid–base and redox properties of a catalyst such as to reduce coke formation [142]. Many promoters, such as Sn, Sr, Ca, Ce, K, and Zr, are used to reduce carbon accumulation, and Ce and Zr combined with a transition metal have attracted attention due to their oxygen storage capacity [62]. This gives lattice oxygen in the Ce oxide phase under reducing conditions and generates anionic vacancies that enhance the catalyst activity. Metal oxides, such as MgO, CeO_2, and ZrO_2, also enhance the catalytic activity in the DRM process [62]. Lanthanides have been found to be good promoters as they favor metal dispersion as well as strengthen CO_2 adsorption on the support. Hence, the deposition of carbon through reverse disproportionation [143, 144] is not likely to occur. Furthermore, lanthanides improved the catalytic activity by favoring particularly the DRM among the other side reactions.

The introduction of lanthanum into mixed Ni–Mg–Al oxides affected nickel species distribution in the catalysts and strengthened NiO–MgO interactions inside the solid matrix [142, 145]. The presence of La_2O_3 could favor the formation of superficial carbonates. These carbonates could promote the basicity of the catalysts and help in the rapid transportation of CO_2 molecules to the interface between the support and metal sites thus resulting in an acceleration of the reactions between CH_x species and CO_2 thereby leading to better catalytic performances in DRM [145]. Many scientists evidenced the role of $La_2O_2CO_3$ in the decrease of carbon formation [111, 118].

MgO, CeO_2, and La_2O_3 promoters for metal catalysts supported on mesoporous materials had the highest catalyst stability among all the other promoters [61, 146].

The addition of Cr [147, 148] and Mn [149] to supported Ni catalysts was reported to have significant effect on the inhibition of carbide formation which was believed to be the intermediate for carbon formation [150]. Hassani Rad *et al.* [151] compared the addition of CeO_2 and MgO to Al_2O_3 support in Ni/Al_2O_3 systems. CeO_2 was found to be better than MgO in terms of catalytic activity and stability, mainly because of the smaller and more uniform particles obtained.

Co/SiO_2 systems modified by different promoters were studied in the DRM reaction [152]. The catalytic activity and stability were strongly dependent on the promoter MgO, SrO or BaO. Hence, the addition of alkaline earth oxides to Co/SiO_2 increases remarkably the resistance to carbon formation and contributes to the catalyst stability by accelerating the rate of carbon gasification on the catalyst surface. A group of ternary compounds $NiO–CaO–Al_2O_3$, with a mesoporous structure was also studied in the DRM reaction [131]. A good catalytic activity with a good stability were found and ascribed to the high surface areas and pore volumes and to the confinement of Ni in the mesopores avoiding its agglomeration and allowing its stabilization under the form of nanoparticles.

The addition of MgO to Rh/SiO_2 [141] was also studied in DRM. MgO enhances the catalytic performances by activating the CO_2 dissociation. Osaki and Mori [153] studied potassium as promoter for Ni/Al_2O_3 in DRM. They concluded that this addition enhanced the activity and decreased the carbon formation. The addition of Ce to $Cu–ferrite/ZrO_2$ leads to an enhancement of the catalytic activity in DRM [88]. Zhu *et al.* [68] added La, Co, Mg, and Zn to Ni/SiO_2. La and Mg decreased the contribution of RWGS reaction leading to higher yield of H_2. For a suitable amount of La, La was found to exhibit high activity and stability. The promoter La increased NiO dispersion and the interaction between NiO and SiO_2. The addition of MgO, CaO, and BaO to Ni supported on Al_2O_3 was carried out in the works of Alipour *et al.* [154]. The alkaline earth promoters enhanced the catalytic activity and selectivity (less carbon formation), the effect being more important for MgO. MgO makes the reduction of Ni species easier. It was shown that V_2O_5, used as a promoter, forms an over–layer of VO_x on Rh/SiO_2 which breaks down the larger ensembles into smaller Rh particles [155]. As a result higher Rh dispersion is obtained and therefore less carbon formation.

Bulk Catalysts

Many bulk catalysts were studied in the DRM reaction. Perez–Lopez *et al.* [57] prepared Ni–Mg–Al catalysts by coprecipitation and investigated the influence of different parameters (M^{II}/M^{III} ratio, calcination temperature, reduction temperature, *etc*...) on the DRM reaction. Dahdah *et al.* [145] prepared Ni–Mg–Al–La oxides to examine the effect of lanthanum addition on the DRM reaction.

Different research groups prepared catalysts for DRM by hydrotalcite route. Hydrotalcite structure is composed of positively charged layers, containing bivalent and trivalent cations, separated by interlayers composed of anions and water. It is also called layered double hydroxides (LDH). Calcination of the hydrotalcite under air at temperatures, generally higher than 400–450 °C, leads to the formation of homogeneous oxides with high dispersion and surface area. Different mixed Ni–Mg–Al oxides were prepared by hydrotalcite route for their use in the DRM reaction [72, 156]. Gennequin *et al.* [56] prepared Co–Mg–Al oxides and Long *et al.* [157] as well as Tanios *et al.* [96] prepared Co–Ni–Mg–Al oxides by the same route for applications in the DRM reaction.

The development of novel solid solutions containing Ni with well–defined structures or Ni confined in mesostructured materials is interesting for application in the DRM reaction [158].

It is reported that perovskites (ABO_3) and spinel (AB_2O_4) structures are beneficial for DRM process [62]. Perovskite catalysts have high thermal resistance and metal dispersion with small particle size (thus low carbon formation). Spinel–type catalysts have high sintering resistance and a weakly acidic nature (thus high activity and less carbon formation). In the review of Lavoie [3] a positive impact of metal carbides (Mo_2C, VC, NbC, TaC) in DRM was noted.

Active Phase Content or Metal Loading

Abd Allah and Whitehead [16] demonstrated that lower Ni loading (18%) on alumina gave better catalytic performances than higher one (33%). A smaller particle size for the catalyst is beneficial for hydrogen yield and higher CH_4 and CO_2 conversions. Perez–Lopez *et al.* [57] studied mixed Ni–Mg–Al oxides as DRM catalysts. The catalytic properties (activity and selectivity) of Ni–Mg–Al are rather dependent on the M^{II}/M^{III} ratio than on the Ni/Mg ratio: when M^{II}/M^{III} is maintained constant, the catalyst activity is related to the nickel crystal size and to an adequate Ni/Mg ratio, and is practically independent of the surface area values. The selectivity is little affected by different Ni/Mg ratios. When Ni/Mg is maintained constant and the M^{II}/M^{III} ratio is varied, the catalyst activity is strongly affected; it decreases as M^{II}/M^{III} ratio decreases. The differences in the catalytic properties are due to parameters that affect simultaneously the crystallite size and the acid–base character of surface. The best results for CO_2 reforming of methane were obtained for Ni–Mg–Al samples having Ni/Mg ratio between 1 and 5, Mg/Al ratio at least 1/3 (for M^{II}/M^{III} at around 2).

The catalytic performances of solids $Co_xMg_yAl_2$ mixed oxides in DRM depend on the Co and Mg content. The solid $Co_2Mg_4Al_2$ (calcined at 500 °C) presents interesting catalytic performances [56].

Nawfal *et al.* [72] found that, among the series of catalysts they prepared, those with highest nickel content (Ni_6Al_2 and Ni_5MgAl_2) gave better catalytic activities than those with intermediate nickel loadings ($Ni_4Mg_2Al_2$, $Ni_3Mg_3Al_2$, and $Ni_2Mg_4Al_2$), and the samples with the lower nickel contents ($NiMg_5Al_2$ and

Mg_6Al_2) did not exhibit any activity in DRM at 700–800 °C. These results were correlated to the reduction of the Ni species present. The most–rich nickel samples were more easily reduced than the moderate ones. The samples with lower nickel contents were not reducible.

Different authors investigated different Co and Ni loadings in the DRM reaction. Zhang *et al.* [60] investigated catalyst samples with Ni and Co loadings ranging between 1.83 and 14.5 wt% and 2.76 and 12.9 wt%, respectively. Catalysts with lower Ni–Co content (1.83–3.61 wt% for Ni and 2.76–4.53 wt% for Co) had higher and more stable activity with no deactivation and no detectable carbon formation and catalysts with higher Ni–Co content (5.28–14.5 wt% for Ni and 7.95–12.9 wt% for Co) exhibited apparent deactivation with significant carbon formation in 250 h time–on–stream tests. Catalyst characterizations revealed that the catalyst with lower Ni–Co content has a larger surface area, smaller metal particles, and better metal dispersion and therefore gives rise to better catalytic performance. The absence of large metal particles (>10 nm) is believed essential to the complete suppression of the carbon formation during reaction [60, 158]. Tanios *et al.* [96] found that $Ni_2Co_2Mg_2Al_2$ oxide is a good compromise between metal dispersion, activity in DRM due to nickel, and lower carbon formation due to cobalt. Other researchers also varied the relative ratio Ni/Co and demonstrated that an optimum ratio is required to a better improvement of the catalytic properties. Indeed, Takanabe *et al.* [159] reported a content of 0.5wt% CoNi/TiO$_2$, with a molar ratio Co/Ni = 90/10, Long *et al.* [157] a molar ratio Co/Ni = 2/8 in NiCoMgAl oxides prepared by hydrotalcite route, and Abdollahifar *et al.* [120] 3wt% Co and 10wt% Ni supported on Al_2O_3–MgO.

In the review of Arora and Prasad [62] it was concluded that catalysts with small metal crystallite sizes (~ 5 nm) and optimum metal loading (3–12%) result in high activity for the DRM process.

Catalysts Preparation Methods

A crucial point is the understanding of the effect of the powder synthesis method on the physicochemical properties of the materials. Among the various available synthetic methods, microwave assisted, self–combustion, impregnation, sol–gel, coprecipitation, combustion, and mechanical mixing may be used to prepare catalytic materials for the DRM reaction.

Indeed, even though the same active metals and supports are selected to make a catalyst, the performances of catalysts may be very different since it is easily affected by a variety of factors other than composition, such as preparation methods, thermal treatments, activation procedures, content of active components, precursors of active components, *etc*... This may be the reason why some catalyst systems such as Ni/Al_2O_3 and Ni/MgO have been studied frequently by different research groups resulting in very different results and conclusions.

Most industrial catalysts are manufactured using either precipitation method or impregnation method. Precipitated catalysts are generally prepared by rapid mixing of concentrated solutions of metal salts with a precipitating reagent while impregnated catalysts are made through the impregnation of active components onto a support material.

Bhattacharyya *et al.* [160] compared the activity of mixed Ni–Mg–Al oxides prepared by coprecipitation (bulk catalysts) to that of supported Ni/Al_2O_3 and $Ni/MgAl_2O_4$ prepared by impregnation and found that the bulk catalysts were more active and more stable than the supported ones.

Tsyganok *et al.* [161] prepared NiMgAl oxides by varying the Ni precursor and introducing it at different steps during the synthesis: co–precipitation of Mg^{2+} and Al^{3+} with pre–synthesized $[Ni(EDTA)]^{2-}$ chelate, anion–exchange reaction of NO_3^- of MgAl–NO_3 (prepared by hydrotalcite route) with $[Ni(EDTA)]^{2-}$ chelate in an aqueous solution, reconstruction (memory effect) of hydrotalcite structure of MgAl mixed oxide in an aqueous solution of $[Ni(EDTA)]^{2-}$, co–precipitation of Ni^{2+}, Mg^{2+}, and Al^{3+} with CO_3^{2-} (traditional procedure for hydrotalcite route). All the catalysts exhibited high activity and stability in the DRM reaction. The state of supported nickel and the coke deposition were influenced by the preparation method. The first method led to the lowest nickel particle size and the lowest amount of carbon deposition.

The order of impregnation plays a role in the catalytic performance. Tomishige *et al.* [93] prepared a bimetallic PtNi catalyst supported on alumina. The metal precursors were added by sequential impregnation and by co–impregnation. The sequential impregnation gave better catalytic results than the co–impregnation. The introduction of ruthenium on the Co–Mg–Al oxides, prepared by hydrotalcite route, was performed in two ways of impregnation [56]: impregnation on the dried hydrotalcite and impregnation using the memory effect of hydrotalcite (calcined at 500°C). Ruthenium added using the memory effect allows the generation of both metallic and basic sites which provides an active and stable catalyst. In addition, using the hydrotalcite memory effect provides well dispersed

Ru species on the surface and therefore more accessible sites for oxidizing the carbon deposited on the catalyst.

In the reviews of Arora and Prasad [62] and of Aramouni *et al.* [48], it was reported that, in general, the catalysts synthesized by sol–gel method gave better results in DRM than those prepared by impregnation which in turn are more performant than the catalysts prepared by coprecipitation. The choice of the noble metal precursor for incipient wetness impregnation does not greatly affect the catalyst performances.

Shin *et al.* [162] prepared Ni/ZrO_2–Al_2O_3 catalysts by different methods: modified Pechini sol–gel, urea hydrolysis, and physical mixing. It was evidenced that the interaction between nickel and support depends on the preparation method and plays an important role in resistance to carbon formation. The system Ni/ZrO_2–Al_2O_3 prepared by the sol–gel process was the most stable in DRM due to the interaction between nickel and ZrO_2–Al_2O_3 that favors CO_2 dissociation. These researchers postulated that the catalytic stability in DRM is maintained by the balance between the reaction of deposited carbon and oxygen intermediates derived from the dehydrogenation of methane and the dissociation of CO_2.

Ni/Al_2O_3–CeO_2 and Ni/Al_2O_3–MgO nanocatalysts were prepared by impregnation and sol–gel methods [151]. The sol–gel method gives rise to better NiO dispersion and higher metal–support interaction via the formation of $NiAl_2O_4$. Smaller and more uniform nanoparticles were detected for the sol–gel method and also higher surface areas. The average of particle size was 23.3 nm with a range of 10–40 nm. Higher catalytic performances were also obtained for the sol–gel process. The same research group [95] also compared sol–gel and impregnation method for the elaboration of $Co,Cu,Ni/ Al_2O_3$–ZrO_2. Similar results were obtained and similar conclusions were given.

Matei–Rutkovska *et al.* [163] prepared ceria nanopowders by microwave assisted hydrothermal method. They put into evidence that the preparation under microwave irradiation results in: smaller crystallites size, better stability towards redox treatments, better surface and bulk reducibility by H_2 and CH_4, creation of defects. These properties led to better catalytic performances in DRM compared to the conventionally (without microwave) prepared ceria catalyst.

Atomic layer dispersion resulted in much less carbon accumulation but was slightly complicated due to the need for mass spectroscopy. Some advanced methods, such as co–precipitation with reflux digestion and non–thermal plasma treatment showed also good performances [62].

Thermal Pretreatment Parameters

Effect of Calcination Temperature

The thermal treatment of calcination has major effects on the nature of active sites. For example, high– temperature calcination enhances the metal– support interaction but it also favors the growth of particle size and makes the reduction of catalyst difficult.

Perez–Lopez *et al.* [57] reported that the calcination temperature of the catalyst does not have a significant effect on the catalytic performances in the DRM reaction. The systems studied in their case were NiMgAl mixed oxides prepared by coprecipitation. However, Zhang *et al.* [60] reported that calcination temperature has a great effect on catalysts performances. They found an optimal range for calcination temperature of 700–900 °C for Ni–Co/AlMgO$_x$. In this case, a compromise is reached between particle size, strong metal support interaction, and the easiness of reduction of Co–Ni. In addition, Wang et al. [164] evidenced the phase transformation of Ni catalysts supported on γ–Al$_2$O$_3$ and the formation of spinel nickel aluminate (NiAl$_2$O$_4$) during calcination, the latter makes the reduction of catalyst difficult. Wang *et al.* [165] also reported that the low activity of Ni/MgO in DRM can be ascribed to its low reducibility due to the formation of solid solution during the calcination process.

Effect of Reduction Temperature

Perez–Lopez and al [57]. found that whereas calcination temperature does not significantly influence the catalytic performance, reduction temperature has a noticeable effect on both activity and selectivity in DRM. For example, they stated that for the NiMgAl oxides studied, an optimal reduction temperature is 700 °C.

Catalyst reduction in different environments (H$_2$, He, H$_2$/He, O$_2$/He, H$_2$–N$_2$, and CH$_4$/O$_2$) indicated that probably the mixture of reducing agents will lead to enhanced catalytic activities [61]. Tsyganok *et al.* [161] prepared mixed Ni–Mg–Al oxides and found that these systems did not require a reduction step previous to the DRM reaction. However, an induction period of 0.5 to 1.5 h was necessary to reach maximal activity. Nawfal *et al.* [72] did not reduce their catalytic systems NiMgAl oxides and Ru/NiMgAl oxides before evaluating them in DRM and obtained good catalytic activity and selectivity. Such a result is very interesting since fewer steps are required for the DRM reaction with less reducing gas consumption. Indeed, it is stated in literature [82, 166 - 168] that the reductive pretreatment of Ni–containing solid solutions and spinels at 790–900 °C should be applied before catalytic DRM. Without this operation, the reaction will not be catalytically accelerated or a much long induction time will be necessary [82].

Dry Reforming of Methane in the Presence of Impurities

In most of the studies, the DRM reaction was conducted in the presence of a mixture of model gas without taking into account the contaminants present in the biogas. However, these contaminants, even in trace amounts, could lead to catalyst poisoning. It is to be reminded that the impurities or contaminants present in biogas are generally classified into four categories: nitrogen–based compounds, sulfur–based compounds, terpenes, and Volatile Organic Compounds (VOCs). Many works were carried out in presence of sulfur compounds evidencing deactivation of catalysts even with very low amounts of contaminants.

Chattanathan *et al.* [169] studied the effect of addition of different concentrations of H_2S (0.5 ; 1 and 1.5 mol%) during the DRM reaction and found that even the introduction of small amount of H_2S decreases the conversions of CH_4 and of CO_2 by 20% directly after 10 min on–stream. Chiodo *et al.* [170] obtained a total deactivation of Ni–based catalysts after addition of 1 to 2 ppm of H_2S, with different deactivation extents. Appari *et al.* [171] suggested a detailed kinetic model to simulate the biogas reforming in the presence of H_2S on Ni–based catalysts. They observed, for temperatures close to 700°C, a total deactivation of the catalyst. Jablonski *et al.* [172] showed a complete deactivation of Ni–based catalysts with contents of 1, 3, and 5 ppm of different sulfur species. They found that deactivation by sulfur is so high that it is necessary to eliminate sulfur, to contents lower than ppm, from hydrocarbons before their use.

Kinetics

The CH_4 cracking and the surface reaction between carbon species and oxygen species are generally accepted by different research groups as rate–determining steps over various Ni based catalysts [86, 167, 173, 174].

Bradfrod and Vannice [175] investigated the kinetics of CO_2 reforming of CH_4 over various Ni catalysts and developed a kinetic model based on CH_4 dissociative adsorption to form CH_x species and CH_xO decomposition as rate–determining steps. Also, Tsipouriari *et al.* [117] reported another kinetic model by assuming that the CH_4 cracking and surface reaction between carbon and oxy–carbonate species are rate–determining steps over Ni/La_2O_3 catalyst. Even though different rate–determining steps are suggested in mechanism studies, a model of Langmuir–Hinshelwood type has been commonly used in the kinetic studies of CO_2 reforming of CH_4 [69, 86, 117, 164, 175 - 177]. However, the Langmuir–Hinshelwood model has different formula depending on catalyst systems [49].

The kinetics of DRM was studied by Zhang *et al.* [178] over Ni–Co/AlMgO$_x$ bimetallic catalyst. It was found that the reforming rate in terms of CH_4 consumption was less sensitive to CO_2 partial pressures than to CH_4 partial pressures. At a constant CH_4 partial pressure, the increase in CO_2 partial pressure did not cause significant change in the reforming rate, whereas at a constant CO_2

partial pressure the reforming rate increased with the increase in CH_4 partial pressure. The increase in CO_2 pressure at a constant CH_4 pressure led to decreases in hydrogen (H_2) formation but increase in carbon monoxide (CO) formation due to the simultaneous occurrence of the reverse water–gas shift reaction. A Langmuir–Hinshelwood model was developed assuming that the dissociation of CH_4 and the reaction between the carbon species and the activated carbon dioxide are the rate determining steps over the $Ni–Co/AlMgO_x$. This assumption satisfactorily fits the experimental data obtained by the authors.

RECENT ADVANCES IN CO_2 REFORMING OF ALCOHOLS

Dry Reforming of Methanol (DRMe)

Methanol is the most common industrial solvent. It is found in the waste gaseous emissions of various industries such as paper and pulp industries [179]. Methanol is a volatile organic compound causing harmful effects to human health and to the environment [180]. Since methanol is produced nowadays from syngas, it defeats the purpose to carry out its DR reaction to obtain syngas. Utilization of methanol will make sense if it originates from industrial waste emissions or from biomass derived sources. It can be then considered a valuable source to generate H_2 rich stream for feeding fuel cell systems *via* DR reaction. Few studies were reported in the literature on DRMe due to its inefficiency to produce net gained energy. Most of the studies dealt with methanol steam reforming, because it is more interesting than DR in terms of H_2 yields [181, 182]. According to Shah and Gardner [2], the most common reactions that occur during DRMe are those represented by equations 4 to 7 (Eqs. 4-7).

Recently, Zhang *et al.* [183] studied the DRMe reaction using plasma technology in a rotating gliding arc reactor to produce clean syngas with high CO_2 conversion. They demonstrated that this technology is promising for clean syngas production and high efficiency CO_2 conversion.

Dry Reforming of Ethanol (DRE)

Ethanol is the most produced biofuel and there is an increasing interest in producing it from biomass derived materials such as ligno–cellulosic and municipal wastes [184]. The steam reforming of ethanol is a possible pathway for ethanol valorization [185], but using ethanol and CO_2 as reactants provides many environmental and social benefits such as the reduction of greenhouse gases (CO_2) and valorization of agro, forest or municipal wastes to generate useful energy for fuel cell devices [19, 184]. Dry reforming of ethanol process is used to produce H_2 rich stream, syngas and also carbon nanofilaments [21]. Optimized conditions in terms of CO_2 to ethanol ratio (C/E), pressure, and temperature influence the products concentrations. Thermodynamically, above 800°C a maximum amount of H_2 and CO can be produced with a C/E ratio of 2 to 5 at ambient pressure [21]. Below 800°C, producing H_2 rich stream is possible but the carbon formation is high, nevertheless it can be controlled through operating parameters and catalysts formulations [186, 187]. Wang and Wang [188] studied the thermodynamics of DRE and concluded that the optimum conditions for H_2 production were a temperature around 900°C, a pressure of 1 bar and a C/E ratio of 1.2 to 1.3. Under these conditions, the H_2 yield was 97% and the ethanol conversion was complete. However, most of the DRE catalytic studies were performed in the 700–800°C temperature range with different C/E ratios aiming to reduce carbon formation (Table **3**). An experimental study performed at 550°C with a C/E ratio of 3, showed that almost ~100% ethanol conversion and ~26% CO_2 conversion were achieved [189]. Equations 1–3; 8–14; 25–27 and 28–31 correspond to the main desirable and undesirable reactions that take place during DRE [17].

Table 3. List of catalysts and its performance in recent DRE reaction studies.

Catalyst	Preparation Method	Reaction Conditions	Conversions (X%) Yields (Y%) or Molar Ratio	Sailent Features	Reference
nano–NiO/SiO$_2$ (10%Ni)	Sol–gel	T = 750°C C/E = 1.4	X_{C2H5OH} = 100 XCO_2 = 76 Y_{H2} = 100 Y_{CO} = 35	Optimal Ni loading and favorable operating conditions for H_2 rich stream Smaller Ni crystallite sizes are active, selective and more stable than commercial catalyst	[17]
Ni/CeO$_2$ (5%Ni)	Impregnation	T = 750°C C/E = 1.5	X_{C2H5OH} = 99.7 XCO_2 = 40 H_2/CO = 1	Influence of oxide support type on activity CeO$_2$ is the most suitable support due to redox properties and better Ni–support interactions Very low filamentous carbon formation	[192]

(Table 3) cont.....

Catalyst	Preparation Method	Reaction Conditions	Conversions (X%) Yields (Y%) or Molar Ratio	Sailent Features	Reference
Rh/NiO–Al$_2$O$_3$ (1%Rh–20% Ni)	Sol–gel Rh wet impregnation	T = 800°C C/E = 1.0	X$_{C2H5OH}$ = 100 H$_2$/CO >>> 1 H$_2$/C$_2$H$_5$OH = 2.3	Presence of the NiAl$_2$O$_4$ spinel phase Rh migration into Al support avoids deactivation High Rh dispersion favored over Ni	[186]
Stainless steel (SS) (Fe<61%, Ni 10–14%, Cr 16–19% and others)	Commercial SS316	T = 600°C C/E = 0.32	X$_{C2H5OH}$ = 86 Y$_{H2}$ = 98	Pre–treated SS316 at 800°C is active and at optimal E/C = 3 highest H$_2$ yield at low T was achieved. SS catalyst is cheap and easily recyclable	[190]
Co–Ni/Al$_2$O$_3$ (3%Co–10% Ni)	Co–impregnation	T = 700°C C/E = 2.5 P$_{C2H5OH}$ = 0.3 bar TOS = 1h	X$_{C2H5OH}$ = 60 XCO$_2$ = 40 Y$_{H2}$ = 33 Y$_{CO}$ = 22	Cobalt promoted produced spinel phases and high performance Partial pressure of ethanol had influence on H$_2$ and CO yields	[193]
Cu/Ce$_{0.8}$Zr$_{02}$O$_2$ (15%Cu)	Facile co–precipitation	T = 700°C C/E = 1 P$_{C2H5OH}$ = 0.3 bar TOS = 1h	X$_{C2H5OH}$ = 100 Y$_{H2}$ = 42 Y$_{CO}$ = 39	Highly stable and active for 90 h Strong metal–support interactions High redox capacity	[194]
Carbonsteel	Not mentioned	T = 550°C C/E = 3 TOS = 4h	X$_{C2H5OH}$ = 100 XCO$_2$ = 26 Y$_{H2}$ = 73 Y$_{CO}$ = 29	Carbon nano fibers are formed at low C/E High H$_2$ yield at low temperature	[189]
Rh/CeO$_2$	Not mentioned	T=800°C C/E = 1 TOS = 25h	X$_{C2H5OH}$ = 100 XCO$_2$ = 89 Y$_{H2}$ = 33 Y$_{CO}$ = 50	High surface area catalyst showed better results High stability at high T due to reverse Boudouard reaction	[191]

*TOS: Time on stream

During the dry reforming of alcohols, the operating conditions and catalysts composition affects significantly the produced H$_2$/CO molar ratio [17, 25]. A highly H$_2$ rich stream is possible *via* DR but depends on the CO$_2$ to ethanol ratio. De Oliveira–Vigier *et al.* [190] showed that a higher ethanol to CO$_2$ molar ratio is suitable for the production of high H_2 yield with relatively low CO formation. Moreover, ethanol conversion at low temperatures is more favorable than CO_2 conversion. Therefore, ethanol conversion to H_2 is thermodynamically possible at low temperatures, thus producing H_2 rich stream but unwanted side reactions such as carbon formation and decomposition are also favored (Eq. 28-31). In DR reactions, first CO_2 converts to CO and then CO leads to carbon formation *via* the Boudouard reaction (Eq. 31).

Catalyst development for DRE is quite new in the scientific community and few studies are published on this topic. Nickel based catalysts are the most active catalysts for DRE [19, 186] because it presents the advantages of being less expensive and more abundant than noble metal catalysts. However, the major drawback of nickel based catalysts is its deactivation *via* metal sintering, poisoning and fouling [27]. The nature of support, the preparation methods and the active phase composition have a significant effect on the overall reaction performance. Drif *et al.* [186] reported that a rhodium doped $NiO–Al_2O_3$ catalyst was more active and stable compared to other rhodium doped mixed oxides supports. Avoiding carbon formation (Eqs. 28–31) is the biggest challenge in DR of alcohols. Previous studies discussed the strategies to avoid or minimize the carbon formation during DR [21, 27, 188]. A high surface area of the catalyst can significantly improve its redox properties as shown by Da Silva *et al.* [191] who successfully developed a stable, high surface area Rh/CeO_2 containing a high amount of oxygen vacancies in CeO_2. These properties led to efficient carbon removal *via* the reverse Boudouard reaction at 800°C. During the DRE reaction over various Ni–based catalysts, ethanol is initially either dehydrogenated or decomposed and then subsequent steam and dry reforming reactions take place according to the reactions presented with equations 8 to 14 (Eqs. 8-14). A list of catalysts studied in the DRE reaction along with its performance and best features are presented in Table **3**.

Cu based catalysts were studied in the DRE reaction and the Ce–Zr support played a significant role in avoiding the catalyst deactivation, *i.e.* inhibiting the encapsulated graphite (carbon type) formation on the proximity of the catalyst surface [194]. Henceforth, the support played a vital role in the stability of the catalyst. Therefore, the influence of support composition, reaction conditions and chemical state of metal catalyst are important considerations during the design of an efficient catalyst [19, 194].

Dry Reforming of Glycerol (DRG)

Glycerol is a biomass molecule and a byproduct during biodiesel production (Table **2**). Recently, the market for glycerol has flourished due to its increased production, and future prospectus for biodiesel market and soap manufacture [195]. Glycerol reforming is used in the field of renewable H_2 production for fuel cell applications. Steam reforming of glycerol is a widely studied reaction over different heterogeneous catalysts for fuel cell applications [196 - 198]. However, few studies on dry reforming of glycerol (DRG) are available in the scientific literature. Wang *et al.* [199] reported the main possible reactions involved in the DRG reaction (Eqs. 1; 15; 23–27; 30–31). They performed a thermodynamic analysis of DRG reaction, and found that optimized conditions for H_2 production are temperatures above 700°C and CO_2 to glycerol (C/G) ratios of 0 to 1. For a C/G = 1 and at atmospheric pressure, they obtained 6.4 moles of syngas (H_2/CO = 1) while complete glycerol conversion and 33% CO_2 conversion were achieved. Table **4** summarizes the experimental results of few available studies on the DRG reaction using different heterogeneous catalysts. Tables **3** and **4** show that ethanol is more converted than glycerol at similar conditions, thus, ethanol is more suitable as raw material for H_2–rich stream production which explains the presence of more DRE reaction studies in the literature.

Table 4. List of catalysts and their performance in recent DRG reaction studies.

Catalyst	Preparation Method	Reaction Conditions	Conversions (X%) Yields (Y%) or Molar Ratio	Sailent Features	Reference
La–Ni/Al$_2$O$_3$ (3%La–20% Ni)	Wet impregnation	T = 750°C, C/G = 1.4, GHSV = 36000 ml g^{-1} h^{-1}	X_{C3H8O3} = 14 H_2/CO = 1.7 Y_{H2} = 13 Y_{CO} = 9	La promoted catalyst had higher carbon resistance due to redox property 3wt% La is the optimal loading	[200]
La–Ni/Al$_2$O$_3$ (3%La–20% Ni)	Wet impregnation	T = 600°C, C/G = 1.4, GHSV = 36000 ml g^{-1} h^{-1}	X_{C3H8O3} = 24.5 H_2/CO = 1.7 Y_{H2} = 10 Y_{CO} = 10	La promotion enhances the Ni dispersion	[6]
Ni/cement clinker (20%Ni, 62%CaO, 17%SiO$_2$)	Wet impregnation	T=750°C C/G =1.4 GHSV = 36000 ml g^{-1} h^{-1}	X_{C3H8O3} = 77 H_2/CO = 1.5 Y_{H2} = 69 Y_{CO} = 55	Cement clinker is highly stable and inexpensive compared to many commercial catalysts	[201]

*GHSV: gas hourly space velocity

Dry Reforming of Butanol (DRB)

Butanol is another interesting and promising alcohol feedstock due to the aforementioned advantages (Table **2**). The dry reforming of butanol (DRB) requires higher amount of energy compared to the other alcohols considered in this review. As the C/H ratio increases the reaction network becomes more complex and also carbon deposition will be pronounced. A complete thermodynamic analysis was done by Wang [23] to determine the optimum conditions for DRB. DRB is a series of multistep reaction pathways, *i.e.* butanol dehydrogenation, butanol dehydration, butanol decomposition and CO_2 reforming (Eqs. 17-25; 29-33). The optimal conditions for DRB reaction to achieve 100% butanol conversion and to produce ~38% H_2 and 58% CO yields, suitable for Solid Oxide Fuel Cells (SOFC) would be a temperature in the range 877°C-927°C and a CO_2 to butanol of 3.5-4 [23]. To the best of our knowledge, there are no studies on catalyst development for DRB.

RECENT ADVANCES IN CO_2 REFORMING OF TAR

Catalytic CO_2 Reforming of Tar

The elimination of tar *via* catalytic reforming can be performed under different gaseous atmospheres such as air, steam or carbon dioxide [42, 202, 203]. Because many gasification processes use steam as the gasifying agent and for the sake of favoring the water–gas shift reaction and increase the H_2 yield, steam reforming of tar is widely considered [180, 204 - 207]. However, the CO_2 volume percentage in gases produced from biomass gasification ranges between 10 and 30 vol.% [208, 209]. Therefore, the CO_2 reforming or dry reforming (DR) of tar is a scientifically sound solution for tar removal and may even be considered superior to steam reforming for several reasons. For instance, using CO_2 as the reforming agent decreases its emissions to the atmosphere (environmental protection), increases the energetic value of the produced gases by converting CO_2 into CO (more effective biomass conversion) and avoids the extra energy consumption accompanying steam injection in steam reforming processes [39, 210]. Furthermore, the dry reforming of tar leads to a higher CO/H_2 ratio facilitating the eventual Fischer–Tropsch synthesis of long–chain hydrocarbons [211]. The dry reforming of methane into syngas is widely considered in the literature [180, 212] and recent developments are also detailed in this chapter. Nevertheless, a relatively small number of studies have considered the dry reforming of tar. Usually, individual tar components are used in laboratory experiments, and therefore most of the literature studies consider the dry reforming of toluene [210, 213 - 215], naphthalene [216], benzene and some heavy alkanes [39, 217] and in few cases real tar or some type of biomass [39, 218]. According to literature [219, 220], the general chemical reactions pertaining to the dry reforming of tar are:

Dry reforming of tar

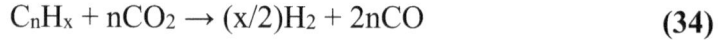

$$C_nH_x + nCO_2 \rightarrow (x/2)H_2 + 2nCO \qquad (34)$$

Tar cracking reaction

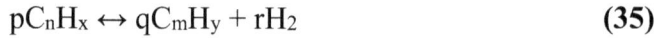

$$pC_nH_x \leftrightarrow qC_mH_y + rH_2 \qquad (35)$$

along with the methanation reactions (Eqs. 23–24), the carbon formation reactions (Eqs. 29–30) and the Boudouard reaction (Eq.31).

The non-catalytic CO_2 reforming of tar has been discussed in some papers where it was systematically concluded that it enhances tar conversion but remains less interesting compared to the use of catalysts that drastically improve the reaction outcome [218, 221, 222]. Therefore, in the following paragraph, a review of the findings and developments in the field of catalytic dry reforming of tar is presented.

Catalysts for CO_2 Reforming of Tar

The catalytic dry reforming of tar can be used as a primary method where the catalyst operates inside the gasifier. In this case, the gasifying agent is either the CO_2 product from biomass gasification process or additional CO_2 injected from an external source. However, it is difficult to determine the contribution of the dry reforming reaction to tar conversion in primary methods as moisture is usually present and the steam reforming reaction occurs simultaneously. On the other hand, a secondary method consists of an independent catalytic reformer where reaction parameters are better monitored [42]. Table **5** lists recent references that considered the DRT reaction along with the used catalysts, their preparation method and conditioning, the reaction parameters and achieved conversions.

Table 5. List of catalysts and its performance in recent DRT reaction studies.

Catalyst	Preparation/ Conditioning	Reaction Conditions	Tar Surrogate Conversion (%)	Reference
5%Ni/α–Al$_2$O$_3$		T=600°C/7h	63	
5%Ni/γ–Al$_2$O$_3$	Impregnation/	87.7% Ar	32	
5%Ni/SiO$_2$	700°C–6h–air/	11.6% CO$_2$	9	[213]
5%Ni/ZrO$_2$	700°C–1h–H$_2$	0.7% Toluene	12	
5%Ni/MgO		0.5mL catalyst	86	

(Table 5) cont.....

Catalyst	Preparation/ Conditioning	Reaction Conditions	Tar Surrogate Conversion (%)	Reference
5%Ni/MgO	Impregnation/ (550–600–700–800°C)–6h–air/ 700°C–1h–H_2		59 76 63 6	[223]
5%Ni/MgO	Impregnation/ 700°C–6h–air/ (500–600–700–800°C)–1h–H_2	T=570°C/7h 285mL/min Ar 14.5mL/min CO_2 0.3mL/h Toluene 0.45g catalyst	44 51 63 48	[224]
5% Co/MgO 9% Co/MgO 12% Co/MgO 15% Co/MgO	Impregnation/ 700°C–6h–air/ 700°C–1h–H_2		0 40 76 93	[210]
2% Ni/Palygorskite 5% Ni/Palygorskite 8% Ni/Palygorskite	Impregnation/ 500°C–3h–N_2/ 500°C–2h–H_2	T=650 to 800°C/20min 200 to 250mL/min Ar 0 to 50mL/min CO_2 5.92mg/min Toluene 2g catalyst	–	[214]
$xCeO_2.9\%$ Ni/MCM–22 (CeO_2: 0; 1; 2; 4; 8; 12%)	Sol–gel method/ 550°C–6h–air	T=700°C/until total conversion of corncob 0.2g catalyst	100	[225]
Ni:Al (1:2 ratio)	Coprecipitation/ 750°C–3h–air	T=700°C/40–85min 1457.75mL/min N_2 257.25mL/min CO_2 21.2–34.1g/h Pine sawdust 2–16g catalyst	90.4 to 100	[226]
13.3%Ni/α–Al_2O_3	Impregnation 900°C–1h–(H_2/N_2)	T=900°C/2MPa Different inlet gas compositions 720ppmv Toluene 0.3g catalyst	100	[227]
Biomass char	Cotton stalk/Microwave heating 800°C/30min	T=700°C/20min 80 to 200mL/min N_2 0 to 120mL/min CO_2 35.7mg/min Toluene 6g catalyst	> 90% for 80 mL/min CO_2	[215]

(Table 5) cont.....

Catalyst	Preparation/ Conditioning	Reaction Conditions	Tar Surrogate Conversion (%)	Reference
Ni–Co–Ce/γ–Al$_2$O$_3$ (10%–3.33%–3.33%)	Co–impregnation 700°C–2h–air 750°C–2h–(H$_2$/N$_2$)	T=800°C 300mL/min N$_2$ 200mL/min CO$_2$ 0.1g/min Benzene (B) 0.15g/min n–heptane (H) 2g catalyst	65 (B) 78 (H)	[39]
Ni–Co–Ce/Activated Carbon (10%–3.33%–3.33%)	Co–impregnation 700°C–2h– N$_2$ 750°C–2h–(H$_2$/N$_2$)		28 (B) 36 (H)	
10%Ni/Palygorskite *Ni–Fe/ Palygorskite *Ni–Fe/MgO–γ–Al$_2$O$_3$ *Ni–Fe/La$_{0.8}$Ca$_{0.2}$CrO$_3$ *Ni–Fe/[MgO–γ–Al$_2$O$_3$+La$_{0.8}$Ca$_{0.2}$CrO$_3$] *(5%Ni–5%Fe)	Co–impregnation 900°C–air 400°C–6h–H$_2$	T=700°C Total flow rate 100mL/min CO$_2$/Naphtalene=1 0.1g catalyst	90%	[228]
3.7%NiO/(3.8% γ–Al$_2$O$_3$–cordierite)	Wet impregnation of NiO 800°C–air 750°C–2h–(H$_2$/N$_2$)	T=750°C 278mL/min mixture (CH$_4$, CO, H$_2$, N$_2$,C$_2$H$_4$) 22mL/min CO$_2$ 4.26g/m^3 Naphtalene	> 99%	[216]
14%Ni2%La$_2$O$_3$/ γ–Al$_2$O$_3$	Ultrasound impregnation 450°C–3h–air 550°C–3h–(H$_2$/N$_2$)	T=550°C 40mL/min CO$_2$ 1.5mL/h Benzene 1.5g catalyst	–	[217]

Simell *et al.* [227] studied the catalytic reforming of toluene under different atmospheres. They concluded that Ni/α–Al$_2$O$_3$ was very active almost completely converting toluene at 900°C. Moreover, in their testing conditions, the dry reforming reaction was faster than the steam reforming reaction. Kong *et al.* [213] tested different oxide supports (α–Al$_2$O$_3$, γ–Al$_2$O$_3$, SiO$_2$, ZrO$_2$, MgO) impregnated with 5 wt% nickel in the dry reforming of toluene reaction. They discovered that the Ni/MgO catalyst was the most active with 91% toluene conversion after 0.5 h that slightly dropped to 86% after 7 h on stream. Nickel forms a solid solution with MgO which creates a strong interaction between the promoter and the support. This has led to the formation of smaller and well dispersed Ni particles that favor the considered reaction. Moreover, the basic environment in the Ni/MgO catalyst led to lower coke deposition and therefore less deactivation compared to the other catalysts. In an additional study [223], the authors found a positive correlation between the surface Ni amount and the conversion of toluene.

Therefore, they concluded that it is the surface Ni amount (Ni surface area) in Ni/MgO catalysts and not the Ni particles size that is the key parameter affecting catalytic activity. The catalyst calcination temperature played a major role in the formation of surface Ni. In a subsequent work [224], the authors showed that a reduction temperature of 700°C before test is optimal to obtain the maximum amount of surface Ni while avoiding extensive sintering of Ni particles. Moreover, they showed that increasing the CO_2/Toluene ratio in the feed led to a better catalytic performance, overweighing the effect of increased reaction temperature. In the same reaction setup [210], it was found that the initial catalytic activity of Co/MgO catalysts increases with cobalt content while the stability on stream depended on the amount of reduced cobalt species present at the reaction temperature. The deactivation of Co/MgO catalysts was attributed to the oxidation of the metallic cobalt species with CO_2 at the reaction temperature. In 2013, Chen *et al.* [214] studied the dry reforming of toluene reaction over Ni/Palygorskite catalysts. They showed that the CO_2 adsorption is favored with increasing nickel content and that maximum H_2 yield is obtained for the 5%Ni/Palygorskite catalyst. Increasing the temperature and/or the CO_2 feed decreases slightly the H_2 yield while inhibiting carbon deposition on the catalyst.

Wang *et al.* [216] showed that naphthalene conversion is complete during the dry reforming reaction at 750°C over a Ni/γ–Al_2O_3–cordierite catalyst. Carbon deposition on the catalyst surface was effectively reduced when a small amount of O_2 was introduced in the feed gas and the catalytic activity remained stable for 108 hours under these partial oxidation dry reforming conditions. Laosiripojana *et al.* [228] prepared Ni and Ni–Fe supported on Palygorskite, MgO–Al_2O_3 or $La_{0.8}Ca_{0.2}CrO_3$ and tested it in the dry reforming of naphthalene. The H_2 yield at 700°C was increased from 50% (catalytic cracking) to 75% when CO_2 (CO_2/C=1 molar ratio) was introduced in the feed in the presence of the Ni–Fe/$La_{0.8}Ca_{0.2}CrO_3$/MgO/Al_2O_3 catalyst. Moreover, carbon deposition was cut by around 40% during dry reforming compared to catalytic cracking.

In 2016, Amin *et al.* [225] concluded that a 9%Ni/MCM–22 catalyst doped with 4 wt% CeO_2 is very catalytically active in the dry reforming of corncob powders. The authors claimed that the presence of an optimum amount of CeO_2 contributes to a better dispersion of reduced Ni particles, and decreases carbon deposition and sintering of Ni particles at high temperature. Garcia *et al.* studied the CO_2 gasification of pine sawdust using Ni/Al catalysts prepared using the coprecipitation method [226]. The Ni/Al catalyst was very efficient, especially for a "catalyst weight (W)/biomass flow rate (bm)" $\geq$ 0.3h, for which the thermodynamic equilibrium values were reached. The W/bm ratio had also a major impact on the reaction outcome. For instance, when W/bm ratio decreased, the syngas yield decreased. The authors also concluded that CO_2 gasification produces lower H_2/CO ratios compared to steam reforming of the same biomass.

Li *et al.* [217] studied the effect of ultrasound treatment during Ni/γ–Al$_2$O$_3$ catalyst preparation on the dry reforming of benzene. It appears that ultrasound treatment can affect the pore size distribution in the catalyst leading to enhanced resistance to coke deposition. The authors also described a reforming mechanism during which CO_2 reacts with coke precursors and deposited coke leading to its elimination. De Caprariis *et al.* [39] investigated the catalytic dry reforming of benzene, n–heptane and real tar. They concluded that a Ni–Co bimetallic catalyst promoted with Ce and supported on γ–Al$_2$O$_3$ is efficient in reforming biomass tar and reduces its content by almost 80% at 800°C in the presence of excess CO_2.

Several studies considered the use of biomass–derived materials, especially biochars, as catalysts to enhance tar conversion. Li *et al.* [215] investigated the microwave assisted dry reforming of toluene over biochars produced from cotton stalk. They discovered that the microwave heating is superior to electrical heating. In addition, they concluded that dry reforming yields better results compared to cracking because the presence of CO_2 contributed to the partial removal of deposited carbon but also regenerated pore structure and catalytic sites. An optimum value for CO_2 in the feed was determined (80 mL/min) that corresponded to maximum toluene conversions as higher values led to substantial biochar (catalyst) gasification. Feng *et al.* compared the effects of H_2O and CO_2 on the reforming of tar over biochar. In one study [229] they concluded that the effect of 15 vol% H_2O is significantly higher than that of 29 vol% CO_2 in the homogeneous reforming of tar from rice husk. However, in a different study but using the same setup [230], the same team came to the conclusion that the presence of CO_2 maintained or even improved the reactivity of biochar in the presence of tar while H_2O could not. Shen *et al.* [231] discussed the usage of biochar catalysts for tar reforming and mentioned that biochar has the advantage of being low cost and can still be directly combusted to recover its chemical energy avoiding traditional catalysts reprocessing costs. The authors stated that most of the studies that considered biochars as catalysts pointed out the importance of the alkali/alkaline earth metals (AAEM) content on the tar reforming reaction. These AAEM are the key in catalyzing the tar reforming reaction where they are activated by free radicals such as oxygen radicals that are generated by the reaction "$CO_2 + e^* \rightarrow CO + O^{\cdot} + e$".

RECENT DEVELOPMENTS IN PROCESS TECHNOLOGIES

The previous part of this chapter reviewed in details the alternative catalysts for dry reforming processes. This section considers technical aspects of the dry reforming including using of plasma, microwaves, electrical current and solar energy to improve the energy efficiency of the process. Some examples of reactor technologies are also presented including chemical looping reforming, membrane reactors and ceramic counterflow reactor developed reactor technologies are also presented including chemical looping reforming, membrane reactors and ceramic counterflow reactor.

Use of Plasma

Plasma has been observed to improve the conversion of methane [232, 233]. It has been proposed, that plasma technologies offer certain advantages over catalytic dry reforming, namely, no requirement for high temperature, lower capital costs, smaller equipment footprint and fewer catalyst deactivation problems [183]. Plasma is a state where ionized gaseous compounds become electrically conductive. At that state the electric and magnetic fields give the material its special character. Plasma can be created for example by heating with a strong electromagnetic field. Electrically generated plasma can be divided further into thermal and non–thermal plasma. The temperature of the thermal plasma is around 10000°C while non–thermal plasma temperature is closer to 1000°C. Non–thermal plasma has an advantage of being chemically more selective. Furthermore, the generation of non–thermal plasma requires less energy [234]. Plasma contains electrons, ions, excited atoms and molecules, radicals and meta–stable ions that induce and accelerate the chemical reactions. The involvement of catalyst in the plasma reactor has been observed to improve the efficiency [235]. In DR applications, the use of non–thermal plasma is more common, and it seems that dielectric barrier discharge plasma (DBD) is the most reported in the literature. Yap *et al.* [233] used 10%La_2O_3/alumina spheres as a catalyst at the temperature range, where the catalyst alone is inactive, and obtained a 3–fold higher methane conversion in the presence of DBD plasma. The plasma discharges were generated using a high voltage probe with a power of 8W by varying voltage of 22 kV (at 300°C) with a frequency of 800 Hz. In another study [235], a 10–30%NiO/Al_2O_3 catalyst was tested in combination with a DBD plasma. The catalyst was introduced in the reactor as a packed bed, and it was observed that the syngas selectivity increased in the presence of the catalyst. Zhang *et al.* [183] reported results obtained during the plasma assisted dry reforming of methanol by using novel rotating gliding arc plasma reactor. They aimed at the portability of the technology, and found out that the resulting CO_2 conversion was significantly higher compared with other non–thermal plasma technologies. In addition, an improved capacity and a relatively clean syngas were produced. In the rotating gliding arc plasma reactor the reactants are fed in the reactor tangentially from the bottom of the reactor. These gases create a swirling flow field inside the reactor. Together with the Lorentz force in the magnetic field, the swirling gas flow induces and intensifies the rotation of the arc. This increases the residence time and improves the contact between the reactants and plasma [234]. The plasma technologies seem interesting from the energetic point of view, but it did not reach the commercial phase yet. Wang *et al.* [236] made an environmental assessment of the plasma technologies using gliding arc discharge reactor as a reference. They concluded that the largest obstacle of the technology is the electricity consumption. Since the process is more compact compared with the traditional reforming process, the plasma reforming units would have an advantage (in terms of energy) when combined with solid oxide fuel cells. The scalability of the plasma process has also risen certain criticism [3].

Use of Microwaves

The use of microwaves has been studied in connection of dry reforming. In many cases, carbon material was used as the microwave receptor and catalyst [237 - 239]. Fidalgo *et al.* [237] studied the methane dry reforming reaction in the presence of activated carbon under microwave heating. They found that at optimal operation temperature of 700–800°C the methane and CO_2 conversions of 100% were maintained for long periods. They concluded that the improved efficiency compared with the convective heating is due to micro–plasmas that appear inside activated carbon. The activated carbon had a dual role: microwave receptor and a catalyst. In their experiments they used a single mode microwave device with a quartz reactor where the activated carbon was packed [237]. In another study [238], the energy consumption in a pilot scale microwave assisted dry reforming of methane process was evaluated. The experiments were done using activated carbon, and activated carbon mixed with Ni/Al_2O_3 catalysts. In both cases the H_2 yields were high, but in the presence of the mixed catalyst the energy consumption was significantly smaller (4.6 kWh per m^3 of H_2 produced). They concluded that the process is promising for future application, since the energy consumption for methane steam reforming is about 1.2 kWh per m^3 of H_2 produced [238]. Tsodikov *et al.* [9] used lignin as a waste–based raw material for the production of syngas. Their approach was very interesting, since they deposited nanosized (6 nm) Ni–particles directly on the lignin after which the lignin was mixed with carbon, which absorbs microwaves better than many inorganic materials. In the experimental set–up, the flow reactor was packed with this mechanically mixed material through which CO_2 flow was passed. The microwave facility contained a 540 W magnetron giving radiation with ~2.45 GHz frequency and 100-150 mA current densities. The excess radiation was absorbed by water at the reactor outlet. Temperature was measured inside the reactor and it was observed that about 700-750°C where reached during the reaction. Automated magnetron controller was used to keep the temperature in the required range. The authors noticed that at these conditions plasma was generated. and that microwaves enhance significantly the syngas yield compared to the convective heating [9, 240].

Use of Electric Current

An electrical current in combination with steel wool in a fixed bed has shown to result in high methane conversion [3]. Banville *et al.* [241] concluded that the steel wool works best under specific temperature and feed composition. A 300W electricity input at 950°C gave almost full conversion of methane and CO_2. The researchers also speculated that even though the energy consumption in their test reactor was considerably very high, adding the steam could decrease significantly. Yabe *et al.* [242] used La–doped Ni/ZrO_2 catalyst in an electro–dry reforming of methane reaction. They obtained high methane conversions even at surprisingly low temperature of 150°C over 1wt%Ni/10mol%La/ZrO_2 catalyst. In their reactor set–up, they used two high voltage electrodes that were placed before and after the catalytic bed inside a tubular reactor. They used a tubular furnace to heat the reactor system up to 150°C, but due to the Joule heating induced by the electrical conduction the temperature of the catalyst bed was actually higher. A direct current high voltage power supply was used in the production of the electric heat. The used current was 3.0–12.0 mA and the applied voltage was monitored with an oscilloscope. In the tests, a good selectivity, high H_2/CO ratio and low carbon deposition were observed. The authors speculated that the electric field could activate methane, formate species and promote surface reactions as well [242].

Use of Solar Energy

A lot of studies have considered the use of solar energy to promote the dry reforming of methane reaction, since sunlight is a clean and renewable energy source. Moreover, in this technology, very high temperatures are achievable [243 - 245]. In 1998, Wörner and Tamme [245] studied the reforming of methane in solar chemical receiver reactor. In this type of reactor the solar light receiver and the reforming reactor are combined in the same equipment. The solar light was received through a quartz window behind which is located a specific material that is able to absorb the solar radiation. In Wörner's case, the absorber material was a ceramic foam made either from α–Al_2O_3 or SiC. During the tests, Rh/γ–Al_2O_3 was used as the catalytically active material. Using the ceramic foam as an absorber material gives an advantage of high gas permeability, good turbulence of flow, easy shaping and good solar absorption. In practice, using darker colored materials will improve the absorption of the solar light. During the operation of the reactor, the reactant gases were passed between the window and the absorber and the product gases were collected at the opposite side of the catalytic absorber. The system was designed for power input of maximum 300 kW and pressure of 3.5 bars that gave about 700–860°C operation temperature which allowed 80%

methane conversion. The authors concluded that the system is promising for methane reforming but recommended more development in terms of temperature resistant materials and higher temperature operation [245]. Kodama *et al.* [243] studied the effect of direct light irradiation on Ni–catalyzed methane dry reforming. They produced the solar irradiation by Xe–arc lamp light. As a comparison, they also used a UV–cut Xe light and infrared irradiation. The methane reforming experiments were carried out in a typical flow–through tubular quartz reactor. Based on the experiments they concluded that the activity of $Ni/\alpha–Al_2O_3$ catalyst was only very slightly affected by the photochemical effect. The main improvement was due to the increase in the catalyst bed temperature. Dahl *et al.* [246] studied the possibility of carrying out methane dry reforming in a solar–thermal aerosol flow reactor. In this type of reactor a high flux solar furnace facility was used as a source of solar energy. The reactor consisted of three vertical graphite tubes placed inside one another. The innermost tube, made of solid graphite, was heated by the concentrated sunlight. The heat from that tube was then radiated to the porous graphite tube located in between the other two tubes. The outer tube was made of quartz. Ar flow was fed between the inner and porous graphite tube and forced through the porous tube to minimize the possible carbon deposition on the central tube. Methane and CO_2 were fed from the top of the porous tube and the products were collected from the bottom of the reactor. The temperature of the solid graphite tube was followed by an optical pyrometer. Dahl *et al.* [246] were able to achieve 70% conversion of methane and 65% conversion of CO_2 without using a catalyst. The residence time in the reactor was 10 ms and the temperature achieved by solar irradiation was ~1700°C. During the experiments they observed the formation of amorphous carbon black particles that were able to enhance the heat transfer to the gas phase.

Chemical Looping Reforming

The first example of the processes presented here for the dry reforming is a chemical looping technology. Chemical looping reforming is a process that can be applied in connection of steam reforming, dry reforming or in the combination of the two. The basic principle for chemical looping was already introduced in 1954, but started to attract lot of interests in early 2000 due to the possibility to avoid CO_2 formation especially in the combustion processes [247, 248]. The principle of this technology is to capture CO_2 formed by using oxygen carriers that prevent the contact between fuel and air [8].

In practice, the chemical–looping process is similar to circulating–fluidized bed boiler. It uses two separate reactors, one for air and another one for fuel. A solid oxygen carrier (reducible metal oxide) transports the oxygen between the two reactors. The solid oxygen carrier releases oxygen for the fuel combustion and then is regenerated in the other reactor unit by using air (or steam) [247, 249]. This reactor concept was first used for combustion purposes and later applied also in reforming processes. The autothermal chemical looping reforming process was first proposed in 2001 by Mattison and Lyngfelt [250]. The difference between chemical–looping combustion and reforming is that in the latter the oxygen/fuel ratio is kept low to maintain the partial oxidation of the fuel. Autothermal chemical looping reforming process can be run either in atmospheric pressure or in a pressurized reactor. With the pressurized reactor one can get about 5% higher H_2 yield [251]. In the autothermal chemical looping reforming the heat balance is important. The regeneration of metal oxide is exothermic while the reforming reaction is endothermic. The solid oxygen carrier circulating between the two reactors supplies the heat required for the reforming reactions. In this type of operation, it must be ensured that the exothermic heat from oxidation reactions is enough to maintain the reforming without external heat supply [247]. It has been observed, that in methane reforming oxygen/fuel molar ratio should be more than 1.25 for autothermal operation [252]. The oxygen–carriers need to be highly reactive and keep their activity up to temperatures of $1100°C$ for long duration of time. In practice, active oxygen carrier materials are supported on reasonably inert particles, such as Al_2O_3, SiO_2, TiO_2, bentonite, *etc…* [247]. These particles need to be sufficiently resistant mechanically so that they can be used in the continuous fluidized bed reactors. According to Adanez *et al.* [247], by 2012 more than 700 different materials based on Ni, Cu, Fe, Mn, Co oxides, mixed oxides and low–cost materials have been studied for the oxygen carrying in chemical looping combustion. Of these metal oxides, Fe, Ni, Cu and Mn–based materials are also studied in chemical looping reforming applications. Ni–based oxygen carriers have been studied most extensively as they have shown to be active at high working temperatures ($900–1100°C$). The drawback of Ni–containing particles is their toxicity and ability to form inactive spinel ($NiAl_2O_4$) when supported on Al_2O_3. The strategies to overcome the spinel formation in addition of using high loading of nickel oxide include utilization of low surface area $α–Al_2O_3$ and chemical deactivation of the support with *e.g.* addition of CaO or MgO causing the spinel formation with Ca and Mg instead of Ni. The use of alternatives to alumina–based materials have shown problems with activity, mechanical stability, deactivation by carbon and defluidization [247, 253]. The CuO–based oxygen carrier have shown high reactivity in chemical looping combustion, however, some precautions should be made due to low melting temperature of metallic Cu that may cause agglomeration in the fluidized bed [254]. Furthermore, in the case of chemical looping reforming, the activity of Cu–oxygen carrier is not as good as in the case of Ni–oxygen carrier [255]. Iron oxide has attracted interests in chemical looping combustion especially due to its low cost. The problem of its application is its poor redox behavior [256]. The good features of Fe–oxide carriers are its low tendency for carbon formation [257] and good tolerance to sulfur [258] but its activity is not sufficient in the methane reforming reaction [251, 255]. In addition to synthetic materials some naturally occurring materials have been tested, such as iron ore, Norwegian ilmenite, manganese ore and wastes of steel and aluminum oxide production (red mud) [247]. When using different oxygen carrier materials special attention should be paid to the support material. Very often the oxygen carriers are not active alone and supporting is needed to improve the mechanical and chemical stability of the particles. Furthermore, different support materials are suitable for different oxygen carriers. It has been shown that the preparation of the oxygen carrier material is an important factor affecting the activity of the material in the chemical looping process. A good review on the development of chemical looping combustion and chemical looping reforming technologies including the development of oxygen carriers was published by Adánez *et al.* [247].

Membrane Reactor Technology

The utilization of membrane reactors allows online separation of hydrogen that shifts the equilibrium of the reaction towards synthesis gas production according to Le Chatelier's principle. This is especially interesting for hydrogen production since hydrogen selective membranes allow collection of high purity product [259]. Kim *et al.* [260] have made a techno–economic analysis by using process simulation (Aspen HYSYS) and economic analysis where they compare a membrane reactor with a packed–bed reactor in DRM reaction. They concluded that H_2 production costs in membrane reactor are about 42% lower than in packed–bed reactor. The sensitivity analysis showed also that reactant price and labor were main factors affecting the H_2 final price in both cases of packed bed reactor and membrane reactor. There are certain advantages when working with membrane reactors. Using membrane reactors presents the advantage of working at significantly lower temperatures compared to fixed–bed reactors. In addition, the catalyst can be incorporated in the membrane allowing simultaneous reaction and separation. In this case smaller amount of catalytically active materials are required. Thus, catalytic membrane reactor is one of the very promising technologies for the thermodynamically limited dry reforming reaction [77]. The membrane permeability and selectivity are two important factors that need to be considered. Dense membranes are very selective, and those made from palladium or palladium–alloys such as PdAgCu [261] and PdCu [262] are selective for hydrogen production. The transport through the membrane occurs in this case *via* a solution–diffusion phenomenon. However, the flux through the membrane is typically quite low and the membrane suffers from deactivation due to impurities, hydrogen embrittlement and carbon formation [259]. Raybold and Huff demonstrated in 2002 [263] that using palladium foil membranes coupled to

Pt/γ–Al$_2$O$_3$ enhances the dry reforming significantly. At 550°C, conversions were two times higher compared to the conventional reactors and H$_2$O formation was suppressed. They also tested Pd/ZrO$_2$ catalyst in the same system, which showed to be slightly more active and more stable than alumina–supported catalyst. Pd–membrane showed some deactivation in certain operation conditions, but when larger hydrogen amounts were passed through the membrane the deactivation was suppressed. As an alternative to Pd–membranes, Haag *et al.* [264] developed a nickel composite membrane by electrodeless plating of asymmetric porous alumina. The major benefit of the membrane was the limitation of the reverse water gas shift reaction. Furthermore, when they used Ni/Al$_2$O$_3$ catalyst that suffers typically from carbon deactivation, the deactivation was suppressed due to restriction of the Boudouard reaction. They also used Ni–Co/Al$_2$O$_3$ catalyst, which is inherently more resistant to coking and did not observe any beneficial effect when using the membrane in terms of the carbon deposition. The latter catalyst showed a significant increase in methane conversion (+67% at 450°C) and hydrogen production (+42% at 450°C) compared to packed–bed reactor. Haag *et al.* [264] showed also a good thermal resistance of the Ni membrane at the reaction conditions (up to 550°C), which makes it a potential alternative to palladium–based membranes. Less selective membranes with higher fluxes can be made from different microporous and mesoporous materials, such as ceramic materials. The transport through these types of membranes follows viscous flow, Knudsen diffusion, surface diffusion, capillary condensation and molecular sieving mechanisms. These mechanisms may also occur simultaneously, which makes the material more versatile [259]. Fereira–Aparicio *et al.* [259] tested a system consisting of a Pt/Al$_2$O$_3$ catalyst packed inside a tubular membrane made of mullite (3Al$_2$O$_3$–2SiO$_2$), which inner surface was coated with SiO$_2$ *via* a sol–gel technique. Without using a sweep gas, only a small conversion enhancement was observed and the concentrations of the species on both sides of the membrane were practically in equilibrium. When helium was introduced as a sweep gas, the lighter gases (H$_2$ and CH$_4$) were concentrated on the sweep gas side of the membrane while CO$_2$ and CO remained at the reaction side of the membrane. The selectivity of H$_2$ was increased due to the inhibition of reverse water gas shift reaction. In this case, only moderate conversion increase was observed, since also the reactants are removed through this type of membrane.

More recently, the tubular porous membranes were used as a support for palladium foil membranes. This kind of configuration allows higher fluxes per unit volume compared to the flat palladium foil membranes. Garcia–Garcia *et al.* [77] compared the performance of fixed–bed reactor, Pd–based tubular membrane reactor and a hollow fiber membrane reactor. The Pd–based tubular membrane reactor was using stainless–steel substrate that has very good thermal, mechanical and chemical resistance. The drawback of this system is the inter–diffusion of metal between the Pd–foil and the stainless steel membrane even at rather low temperature (275–350°C). Tubular ceramic composite substrates are more economical, and they have narrower pore size distribution, smaller pore size and a smooth surface that facilitates the deposition of Pd thin layer. The authors also foresaw that the costs of the hollow fiber ceramic membrane could be significantly decreased when using one step phase inversion technique. In their tests, they used $Ru/ZrO_2–La_2O_3$ catalyst. In the case of tubular membrane reactor, the catalyst pellets were packed inside the tubular Pd–membrane supported on stainless–steel substrate. The reactants flow inside the catalyst bed and the products are recovered on the other side of the membrane with the help of the He sweep gas. In the case of hollow fiber membrane reactor (made of Al_2O_3 and coated with Pd), the membrane was placed in the middle of the reactor and the catalyst was packed in the outer shell. The reactants were fed to the outer shell of the reactor and the products were collected from the middle of the reactor with the help of the He sweep gas. The diameter of the used hollow fiber membranes was about 1 to 2 mm, while the stainless steel tubular membrane had a diameter of about 1 cm. Pd coated Al_2O_3 hollow fiber membrane was found to give conversions equal to the stainless steel tubular membrane with fifteen times less Pd– coating, which significantly decreases the costs of the membrane. When using hollow fiber membrane, a high purity of H_2 was observed at wide range of temperature. These results are very promising, since the efficiency of this type of reactor can be easily improved by increasing the number of the hollow fibers [77].

Tsodikov *et al.* [265] studied the use of nickel and cobalt containing porous ceramic membrane–catalytic reactor in dry reforming of methane, ethanol and model mixture of fermentation products (19.1% propanol, 17.5% isobutyl alcohol, 63.4% isopentyl alcohol). The membranes were prepared by using self–propagating high–temperature synthesis realized by thermal explosion of the precursors (Ni powder, aluminum and cobalt oxide powders) in a vacuum. In this case, the active catalytic species were inside the porous membrane material, making the membrane as a hybrid material. The catalysts were nano–sized (1–50 nm) spherical Ni–Co alloy particles distributed in the massive clusters of Al_2O_3. As a result, the authors reported somehow better performance of the system compared with the packed–bed. When it comes to the selectivity, H_2, CO and CH_4 were observed as the products of ethanol and fermentation products reforming.

Ceramic Counterflow Technology

Successful application of the dry reforming in an industrial scale could be also realized *via* developing completely new reactor configurations. As an example, a ceramic counterflow reactor concept was proposed by Kelling *et al.* [1] for autothermal dry reforming. Since the highly endothermic dry reforming reaction has to be carried out at high temperature, ceramic construction materials are a good choice due to their temperature stability. In the proposed concept, partial oxidation of methane is used in production of enough heat for the dry reforming reaction. The reactor contains several ceramic tubes, which are placed inside a ceramic reactor with high temperature insulation. The tubes are coated at one end with the catalytic material, which makes it work as both heat exchangers and reforming catalyst carriers. The reactants (CH_4 and CO_2) flow inside the ceramic reactor outside the tubes and then reach the end of the reactor and flow back inside the tubes to exit the reactor. Oxygen needed in partial oxidation is injected *via* separate tubes close to the catalytically coated tube zone. The catalytic and homogeneous methane partial oxidation occurs close to the area where oxygen is fed and provides the heat for the catalytic dry reforming reactions. When the reforming products flow out of the reactor through the tubes, they still release heat through the ceramic tubes and heat up the cold reactants [1]. The authors developed also a reactor model for their concept and tested it experimentally. As a result they were able to demonstrate the successful operation of the developed autothermal ceramic counterflow reactor. The developed kinetic model also gives indications of successful operation at the industrial scale [1].

CONCLUDING REMARKS

There is a large debate on whether to use dry reforming or steam reforming to produce H_2–rich stream from different feedstocks. Theoretically, a high H_2 selectivity and yield are produced by steam reforming reactions (SR) while a lower H_2 yield is obtained during dry reforming reactions (DR). On the other hand, the DR reactions show significant environmental benefits by consuming the CO_2 (greenhouse gas) as an oxidant. In DR reactions, the CO_2 to feedstock molar ratio is the very important parameter to produce H_2 rich stream or syngas mixture with different H_2/CO ratios. Unfortunately, carbon deposition is more pronounced in DR than in SR.

The use of catalysts for the DR reactions leads to the decrease of the energy required for the process and the increase of the rate and selectivity towards the desired products. However, despite of all its advantages, catalytic DR reactions still face several difficulties making it not yet used at an industrial scale. The DR reactions are highly endothermic and are therefore carried out at high temperatures that can rapidly deactivate the catalyst by sintering of the active phase as well as by carbon deposition. The latter can also possibly lead to internal reactor blockage, bed clogging, and high pressure drop. Furthermore, metal oxidation could occur which decreases the catalytic activity. Many strategies were adopted to enhance the catalytic option viability: the choice of the support, the metal content, the preparation method, and the introduction of promoters into the catalyst composition. Preparation methods played an important role in the synthesis of smaller particle size and higher dispersion of active metals. Calcination temperature and treatment duration imparted significant changes to the morphology of the catalysts. For example, particle size smaller than 10–15 nm had a significant influence on the suppression of carbon deposition and catalytic activity. The challenge is to combine a good support and active phase with an affordable price. It would be ideal if a natural low cost support could be combined to a transition metal and a suitable promoter. Till now, transition metals still exhibit deactivation by carbon and/or sintering and natural supports investigations remain shy. Furthermore, poisoning by impurities present in the feed is also a barrier that cannot be circumvented however it is not yet well developed in the literature.

There is a promising scope and lot of room to improve overall reaction performance in DR processes. A possibility would be to combine DR with SR, *i.e.* addition of steam that will deliberately reduce the carbon deposition. Another possibility is to use tri–reforming, *i.e.* $H_2O+CO_2+O_2$ (combination of dry, steam and autothermal reforming reactions), which is a quite new research topic. Other process technologies can be utilized in order to improve the efficiency of DR technology such as plasma technology, chemical looping and membrane reactors.

Finally, it seems that the research related to the development of dry reforming processes is progressively evolving towards commercialization. In the development of catalytic materials, a vast amount of research is done. Furthermore, new possibilities to induce chemical reactions with and without a catalyst are sought. In addition, new reactor configurations are developed to make the autothermal dry reforming possible. The future will show us which one of these technologies will become the market leader; however, membrane processes seem very promising when high purity hydrogen production is required.

CONSENT FOR PUBLICATION

Not applicable.

CONFLICT OF INTEREST

The authors declare no conflict of interest, financial or otherwise.

ACKNOWLEDGEMENT

The BRG 5/2016, the ARCUS E2D2 project, the French Ministry of Foreign Affairs, and the « Région Nord–Pas de Calais » are gratefully acknowledged.

REFERENCES

[1] Kelling, R.; Eigenberger, G.; Nieken, U. Ceramic counterflow reactor for autothermal dry reforming at high temperatures. *Catal. Today,* **2016**, *273*, 196-204.
[http://dx.doi.org/10.1016/j.cattod.2016.02.056]

[2] Shah, Y.T.; Gardner, T.H. Dry reforming of hydrocarbon feedstocks. *Catal. Rev., Sci. Eng.,* **2014**, *56*, 476-536.
[http://dx.doi.org/10.1080/01614940.2014.946848]

[3] Lavoie, J-M. Review on dry reforming of methane, a potentially more environmentally-friendly approach to the increasing natural gas exploitation. *Front Chem.,* **2014**, *2*, 81.
[http://dx.doi.org/10.3389/fchem.2014.00081] [PMID: 25426488]

[4] Kumar, A.; Bhosale, R.R.; Malik, S.S.; Abusrafa, A.E.; Saleh, M.A.H.; Ghosh, U.K.; Al-Marri, M.J.; Almomani, F.A.; Khader, M.M.; Abu-Reesh, I.M. Thermodynamic investigation of hydrogen enrichment and carbon suppression using chemical additives in ethanol dry reforming. *Int. J. Hydrogen Energy,* **2016**, *41*, 15149-15157.
[http://dx.doi.org/10.1016/j.ijhydene.2016.06.157]

[5] Mattos, L.V.; Jacobs, G.; Davis, B.H.; Noronha, F.B. Production of hydrogen from ethanol: review of reaction mechanism and catalyst deactivation. *Chem. Rev.,* **2012**, *112*(7), 4094-4123.
[http://dx.doi.org/10.1021/cr2000114] [PMID: 22617111]

[6] Siew, K.W.; Lee, H.C.; Gimbun, J.; Cheng, C.K. Production of CO-rich hydrogen gas from glycerol dry reforming over La-promoted Ni/Al_2O_3 catalyst. *Int. J. Hydrogen Energy,* **2014**, *39*, 6927-6936.
[http://dx.doi.org/10.1016/j.ijhydene.2014.02.059]

[7] Wang, S.; Wang, Q.; Song, X.; Chen, J. Dry autothermal reforming of glycerol with in situ hydrogen separation *via* thermodynamic evaluation. *Int. J. Hydrogen Energy,* **2017**, *42*, 838-847.
[http://dx.doi.org/10.1016/j.ijhydene.2016.09.103]

[8] García-Díez, E.; García-Labiano, F.; de Diego, L.F.; Abad, A.; Gayán, P.; Adánez, J.; Ruíz, J.A.C. Steam, dry, and steam-dry chemical looping reforming of diesel fuel in a 1 kWth unit. *Chem. Eng. J.,* **2017**, *325*, 369-377.
[http://dx.doi.org/10.1016/j.cej.2017.05.042]

[9] Tsodikov, M.V.; Ellert, O.G.; Nikolaev, S.A.; Arapova, O.V.; Konstantinov, G.I.; Bukhtenko, O.V.; Vasil'kov, A.Y. The role of nanosized nickel particles in microwave-assisted dry reforming of lignin. *Chem. Eng. J.,* **2017**, *309*, 628-637.
[http://dx.doi.org/10.1016/j.cej.2016.10.031]

[10] Saad, J.M.; Williams, P.T. Catalytic dry reforming of waste plastics from different waste treatment plants for production of synthesis gases. *Waste Manag.,* **2016**, *58*, 214-220.
[http://dx.doi.org/10.1016/j.wasman.2016.09.011] [PMID: 27650631]

[11] Homsi, D.; Aouad, S.; Gennequin, C.; Nakat, J.E.; Aboukaïs, A.; Abi-Aad, E. The effect of copper content on the reactivity of Cu/Co_6Al_2 solids in the catalytic steam reforming of methane reaction. *C. R. Chim.,* **2014**, *17*, 454-458.
[http://dx.doi.org/10.1016/j.crci.2013.07.004]

[12] Homsi, D.; Aouad, S.; Gennequin, C.; Aboukaïs, A.; Abi-Aad, E. Hydrogen production by methane steam reforming over Ru and Cu supported on hydrotalcite precursors. *Adv. Mat. Res.,* **2011**, *324*, 453-456.
[http://dx.doi.org/10.4028/www.scientific.net/AMR.324.453]

[13] Homsi, D.; Aouad, S.; Gennequin, C.; Aboukaïs, A.; Abi-Aad, E. A highly reactive and stable $Ru/Co_{6-x}Mg_xAl_2$ catalyst for hydrogen production *via* methane steam reforming. *Int. J. Hydrogen Energy,* **2014**, *39*, 10101-10107.
[http://dx.doi.org/10.1016/j.ijhydene.2014.04.151]

[14] Horn, R.; Schloegl, R. Methane activation by heterogeneous catalysis. *Catal. Lett.,* **2015**, *145*, 23-39.

[http://dx.doi.org/10.1007/s10562-014-1417-z]

[15]   Ross, J.R.H. Natural gas reforming and CO_2 mitigation. *Catal. Today*, **2005**, *100*, 151-158.
[http://dx.doi.org/10.1016/j.cattod.2005.03.044]

[16]   Abd Allah, Z.; Whitehead, J.C. Plasma-catalytic dry reforming of methane in an atmospheric pressure AC gliding arc discharge. *Catal. Today*, **2015**, *256*, 76-79.
[http://dx.doi.org/10.1016/j.cattod.2015.03.040]

[17]   Bej, B.; Bepari, S.; Pradhan, N.C.; Neogi, S. Production of hydrogen by dry reforming of ethanol over alumina supported nano-NiO/SiO_2 catalyst. *Catal. Today*, **2017**, *291*, 58-66.
[http://dx.doi.org/10.1016/j.cattod.2016.12.010]

[18]   Kale, G.R.; Kulkarni, B.D. Thermodynamic analysis of dry autothermal reforming of glycerol. *Fuel Process. Technol.*, **2010**, *91*, 520-530.
[http://dx.doi.org/10.1016/j.fuproc.2009.12.015]

[19]   Kawi, S.; Kathiraser, Y. CO_2 as an oxidant for high-temperature reactions. *Front. Energy Res.*, **2015**, *3* article 13

[20]   Seelam, P.K.; Huuhtanen, M.; Sápi, A.; Szabó, M.; Kordás, K.; Turpeinen, E.; Tóth, G.; Keiski, R.L. CNT-based catalysts for H_2 production by ethanol reforming. *Int. J. Hydrogen Energy*, **2010**, *35*, 12588-12595.
[http://dx.doi.org/10.1016/j.ijhydene.2010.07.160]

[21]   Kale, G.R.; Gaikwad, T.M. Thermodynamic analysis of ethanol dry reforming: effect of combined parameters. *ISRN Thermodyn.*, **2014**, *2014*, 1-10.
[http://dx.doi.org/10.1155/2014/929676]

[22]   Nahar, G.A.; Madhani, S.S. Thermodynamics of hydrogen production by the steam reforming of butanol: Analysis of inorganic gases and light hydrocarbons. *Int. J. Hydrogen Energy*, **2010**, *35*, 98-109.
[http://dx.doi.org/10.1016/j.ijhydene.2009.10.013]

[23]   Wang, W. Hydrogen production via dry reforming of butanol: Thermodynamic analysis. *Fuel*, **2011**, *90*, 1681-1688.
[http://dx.doi.org/10.1016/j.fuel.2010.11.001]

[24]   Guo, Y.; Wan, T.; Zhu, A.; Shi, T.; Zhang, G.; Wang, C.; Yu, H.; Shao, Z. Performance and durability of a layered proton conducting solid oxide fuel cell fueled by the dry reforming of methane. *RSC Advances*, **2017**, *7*, 44319-44325.
[http://dx.doi.org/10.1039/C7RA07710F]

[25]   Saad, J.M.; Williams, P.T. Manipulating the H_2/CO ratio from dry reforming of simulated mixed waste plastics by the addition of steam. *Fuel Process. Technol.*, **2017**, *156*, 331-338.
[http://dx.doi.org/10.1016/j.fuproc.2016.09.016]

[26]   Estephane, J.; Aouad, S.; Hany, S.; El Khoury, B.; Gennequin, C.; El Zakhem, H.; El Nakat, J.; Aboukaïs, A.; Abi Aad, E. CO_2 reforming of methane over Ni-Co/ZSM5 catalysts. Aging and carbon deposition study. *Int. J. Hydrogen Energy*, **2015**, *40*, 9201-9208.
[http://dx.doi.org/10.1016/j.ijhydene.2015.05.147]

[27]   Argyle, M.; Bartholomew, C. Heterogeneous catalyst deactivation and regeneration: a review. *Catalysts*, **2015**, *5*, 145-269.
[http://dx.doi.org/10.3390/catal5010145]

[28]   Kummamuru, B. WBA Global Bioenergy Statistics **2017**.

[29]   Shen, Y.; Yoshikawa, K. Recent progresses in catalytic tar elimination during biomass gasification or pyrolysis—A review. *Renew. Sustain. Energy Rev.*, **2013**, *21*, 371-392.
[http://dx.doi.org/10.1016/j.rser.2012.12.062]

[30]   Göransson, K.; Söderlind, U.; He, J.; Zhang, W. Review of syngas production *via* biomass DFBGs.

Renew. Sustain. Energy Rev., **2011**, *15*, 482-492.
[http://dx.doi.org/10.1016/j.rser.2010.09.032]

[31]   Siedlecki, M.; De Jong, W.; Verkooijen, A.H.M. Fluidized bed gasification as a mature and reliable technology for the production of bio-syngas and applied in the production of liquid transportation fuels—a review. *Energies,* **2011**, *4*, 389-434.
[http://dx.doi.org/10.3390/en4030389]

[32]   Cohce, M.K.; Dincer, I.; Rosen, M.A. Thermodynamic analysis of hydrogen production from biomass gasification. *Int. J. Hydrogen Energy,* **2010**, *35*, 4970-4980.
[http://dx.doi.org/10.1016/j.ijhydene.2009.08.066]

[33]   Corella, J.; Toledo, J.M.; Molina, G. A review on dual fluidized-bed biomass gasifiers. *Ind. Eng. Chem. Res.,* **2007**, *46*, 6831-6839.
[http://dx.doi.org/10.1021/ie0705507]

[34]   Abu El-Rub, Z.; Bramer, E.A.; Brem, G. Review of catalysts for tar elimination in biomass gasification processes. *Ind. Eng. Chem. Res.,* **2004**, *43*, 6911-6919.
[http://dx.doi.org/10.1021/ie0498403]

[35]   Sutton, D.; Kelleher, B.; Ross, J.R.H. Review of literature on catalysts for biomass gasification. *Fuel Process. Technol.,* **2001**, *73*, 155-173.
[http://dx.doi.org/10.1016/S0378-3820(01)00208-9]

[36]   Coll, R.; Salvadó, J.; Farriol, X.; Montané, D. Steam reforming model compounds of biomass gasification tars: conversion at different operating conditions and tendency towards coke formation. *Fuel Process. Technol.,* **2001**, *74*, 19-31.
[http://dx.doi.org/10.1016/S0378-3820(01)00214-4]

[37]   Sarioğlan, A. Tar removal on dolomite and steam reforming catalyst: benzene, toluene and xylene reforming. *Int. J. Hydrogen Energy,* **2012**, *37*, 8133-8142.
[http://dx.doi.org/10.1016/j.ijhydene.2012.02.045]

[38]   Han, J.; Kim, H. The reduction and control technology of tar during biomass gasification/pyrolysis: an overview. *Renew. Sustain. Energy Rev.,* **2008**, *12*, 397-416.
[http://dx.doi.org/10.1016/j.rser.2006.07.015]

[39]   De Caprariis, B.; Bassano, C.; Deiana, P.; Palma, V.; Petrullo, A.; Scarsella, M.; De Filippis, P. Carbon dioxide reforming of tar during biomass gasification. *Chem. Eng. Trans.,* **2014**, *37*, 97-102.

[40]   Li, C.; Suzuki, K. Tar property, analysis, reforming mechanism and model for biomass gasification-an overview. *Renew. Sustain. Energy Rev.,* **2009**, *13*, 594-604.
[http://dx.doi.org/10.1016/j.rser.2008.01.009]

[41]   Ni, M.; Leung, D.Y.C.; Leung, M.K.H.; Sumathy, K. An overview of hydrogen production from biomass. *Fuel Process. Technol.,* **2006**, *87*, 461-472.
[http://dx.doi.org/10.1016/j.fuproc.2005.11.003]

[42]   Devi, L.; Ptasinski, K.J.; Janssen, F.J.J. A review of the primary measures for tar elimination in biomass gasification processes. *Biomass Bioenergy,* **2003**, *24*, 125-140.
[http://dx.doi.org/10.1016/S0961-9534(02)00102-2]

[43]   Zhang, R.; Brown, R.C.; Suby, A.; Cummer, K. Catalytic destruction of tar in biomass derived producer gas. *Energy Convers. Manage.,* **2004**, *45*, 995-1014.
[http://dx.doi.org/10.1016/j.enconman.2003.08.016]

[44]   Anis, S.; Zainal, Z.A. Tar reduction in biomass producer gas *via* mechanical, catalytic and thermal methods: A review. *Renew. Sustain. Energy Rev.,* **2011**, *15*, 2355-2377.
[http://dx.doi.org/10.1016/j.rser.2011.02.018]

[45]   Paula Peres, A.G.; Lunelli, B.H. Maciel FIlho, R. Application of biomass to hydrogen and syngas production. *Chem. Eng. Trans.,* **2013**, *32*, 589-594.
[http://dx.doi.org/10.3303/CET1332099]

[46]   Valderrama Rios, M.L.; González, A.M.; Lora, E.E.S.; Almazán del Olmo, O.A. Reduction of tar generated during biomass gasification: A review. *Biomass Bioenergy,* **2018**, *108*, 345-370.
       [http://dx.doi.org/10.1016/j.biombioe.2017.12.002]

[47]   Zhang, J.; Wang, H.; Dalai, A.K. Development of stable bimetallic catalysts for carbon dioxide reforming of methane. *J. Catal.,* **2007**, *249*, 300-310.
       [http://dx.doi.org/10.1016/j.jcat.2007.05.004]

[48]   Aramouni, N.A.K.; Touma, J.G.; Tarboush, B.A.; Zeaiter, J.; Ahmad, M.N. Catalyst design for dry reforming of methane: analysis review. *Renew. Sustain. Energy Rev.,* **2018**, *82*, 2570-2585.
       [http://dx.doi.org/10.1016/j.rser.2017.09.076]

[49]   Pakhare, D.; Spivey, J. A review of dry (CO$_2$) reforming of methane over noble metal catalysts. *Chem. Soc. Rev.,* **2014**, *43*(22), 7813-7837.
       [http://dx.doi.org/10.1039/C3CS60395D] [PMID: 24504089]

[50]   Abdullah, B.; Abd Ghani, N.A.; Vo, D.V.N. Recent advances in dry reforming of methane over Ni-based catalysts. *J. Clean. Prod.,* **2017**, *162*, 170-185.
       [http://dx.doi.org/10.1016/j.jclepro.2017.05.176]

[51]   Chang, J-S.; Park, S-E.; Chon, H. Catalytic activity and coke resistance in the carbon dioxide reforming of methane to synthesis gas over zeolite-supported Ni catalysts. *Appl. Catal. A Gen.,* **1996**, *145*, 111-124.
       [http://dx.doi.org/10.1016/0926-860X(96)00150-0]

[52]   Debecker, D.P.; Gaigneaux, E.M.; Busca, G. Exploring, tuning, and exploiting the basicity of hydrotalcites for applications in heterogeneous catalysis. *Chemistry,* **2009**, *15*(16), 3920-3935.
       [http://dx.doi.org/10.1002/chem.200900060] [PMID: 19301329]

[53]   Özkara-Aydinoğlu, Ş.; Aksoylu, A.E. Carbon dioxide reforming of methane over Co-X/ZrO$_2$ catalysts (X = La, Ce, Mn, Mg, K). *Catal. Commun.,* **2010**, *11*, 1165-1170.
       [http://dx.doi.org/10.1016/j.catcom.2010.07.001]

[54]   Lucrédio, A.F.; Assaf, J.M.; Assaf, E.M. Reforming of a model biogas on Ni and Rh-Ni catalysts: Effect of adding La. *Fuel Process. Technol.,* **2012**, *102*, 124-131.
       [http://dx.doi.org/10.1016/j.fuproc.2012.04.020]

[55]   Soria, M.A.; Mateos-Pedrero, C.; Guerrero-Ruiz, A.; Rodríguez-Ramos, I. Thermodynamic and experimental study of combined dry and steam reforming of methane on Ru/ ZrO$_2$-La$_2$O$_3$ catalyst at low temperature. *Int. J. Hydrogen Energy,* **2011**, *36*, 15212-15220.
       [http://dx.doi.org/10.1016/j.ijhydene.2011.08.117]

[56]   Gennequin, C.; Hany, S.; Tidahy, H.L.; Aouad, S.; Estephane, J.; Aboukaïs, A.; Abi-Aad, E. Influence of the presence of ruthenium on the activity and stability of Co-Mg-Al-based catalysts in CO$_2$ reforming of methane for syngas production. *Environ. Sci. Pollut. Res. Int.,* **2016**, *23*(22), 22744-22760.
       [http://dx.doi.org/10.1007/s11356-016-7453-z] [PMID: 27562810]

[57]   Perez-Lopez, O.W.; Senger, A.; Marcilio, N.R.; Lansarin, M.A. Effect of composition and thermal pretreatment on properties of Ni-Mg-Al catalysts for CO$_2$ reforming of methane. *Appl. Catal. A Gen.,* **2006**, *303*, 234-244.
       [http://dx.doi.org/10.1016/j.apcata.2006.02.024]

[58]   Liu, C.J.; Ye, J.; Jiang, J.; Pan, Y. Progresses in the preparation of coke resistant Ni-based catalyst for steam and CO$_2$ reforming of methane. *ChemCatChem,* **2011**, *3*, 529-541.
       [http://dx.doi.org/10.1002/cctc.201000358]

[59]   Zhou, L.; Li, L.; Wei, N.; Li, J.; Basset, J.M. Effect of NiAl$_2$O$_4$ Formation on Ni/Al$_2$O$_3$ Stability during Dry Reforming of Methane. *ChemCatChem,* **2015**, *7*, 2508-2516.
       [http://dx.doi.org/10.1002/cctc.201500379]

[60]   Zhang, J.; Wang, H.; Dalai, A.K. Effects of metal content on activity and stability of Ni-Co bimetallic

catalysts for CO_2 reforming of CH_4. *Appl. Catal. A Gen.*, **2008**, *339*, 121-129.
[http://dx.doi.org/10.1016/j.apcata.2008.01.027]

[61]   Usman, M.; Wan Daud, W.M.A.; Abbas, H.F. Dry reforming of methane: Influence of process parameters—A review. *Renew. Sustain. Energy Rev.*, **2015**, *45*, 710-744.
[http://dx.doi.org/10.1016/j.rser.2015.02.026]

[62]   Arora, S.; Prasad, R. An overview on dry reforming of methane: strategies to reduce carbonaceous deactivation of catalysts. *RSC Advances*, **2016**, *6*, 108668-108688.
[http://dx.doi.org/10.1039/C6RA20450C]

[63]   Rasi, S.; Veijanen, A.; Rintala, J. Trace compounds of biogas from different biogas production plants. *Energy*, **2007**, *32*, 1375-1380.
[http://dx.doi.org/10.1016/j.energy.2006.10.018]

[64]   Jingura, R.M.; Matengaifa, R. Optimization of biogas production by anaerobic digestion for sustainable energy development in Zimbabwe. *Renew. Sustain. Energy Rev.*, **2009**, *13*, 1116-1120.
[http://dx.doi.org/10.1016/j.rser.2007.06.015]

[65]   Xu, Y.; Bao, X.; Lin, L. Direct conversion of methane under nonoxidative conditions. *J. Catal.*, **2003**, *216*, 386-395.
[http://dx.doi.org/10.1016/S0021-9517(02)00124-0]

[66]   Moulijn, J.A.; Van Diepen, A.E.; Kapteijn, F. Catalyst deactivation: Is it predictable? What to do? *Appl. Catal. A Gen.*, **2001**, *212*, 3-16.
[http://dx.doi.org/10.1016/S0926-860X(00)00842-5]

[67]   Hu, Y.H.; Ruckenstein, E. Catalytic Conversion of Methane to Synthesis Gas by Partial Oxidation and CO_2 Reforming. *Adv. Catal.*, **2004**, *48*, 297-345.
[http://dx.doi.org/10.1016/S0360-0564(04)48004-3]

[68]   Zhu, J.; Peng, X.; Yao, L.; Shen, J.; Tong, D.; Hu, C. The promoting effect of La, Mg, Co and Zn on the activity and stability of Ni/SiO$_2$ catalyst for CO_2 reforming of methane. *Int. J. Hydrogen Energy*, **2011**, *36*, 7094-7104.
[http://dx.doi.org/10.1016/j.ijhydene.2011.02.133]

[69]   Bradford, M.C.J.; Vannice, M.A. CO_2 reforming of CH_4. *Catal. Rev.*, **1999**, *41*, 1-42.
[http://dx.doi.org/10.1081/CR-100101948]

[70]   Safariamin, M.; Tidahy, L.H.; Abi-Aad, E.; Siffert, S.; Aboukaïs, A. Dry reforming of methane in the presence of ruthenium-based catalysts. *C. R. Chim.*, **2009**, *12*, 748-753.
[http://dx.doi.org/10.1016/j.crci.2008.10.021]

[71]   Ferreira-Aparicio, P.; Rodríguez-Ramos, I.; Anderson, J.; Guerrero-Ruiz, A. Mechanistic aspects of the dry reforming of methane over ruthenium catalysts. *Appl. Catal. A Gen.*, **2000**, *202*, 183-196.
[http://dx.doi.org/10.1016/S0926-860X(00)00525-1]

[72]   Nawfal, M.; Gennequin, C.; Labaki, M.; Nsouli, B.; Aboukaïs, A.; Abi-Aad, E. Des catalyseurs Ni-Mg-Al préparés par voie hydrotalcite et imprégnés par le ruthénium pour le reformage à sec du méthane. *Leban. Sci. J.*, **2015**, *16*, 117-125.

[73]   Crisafulli, C.; Scirè, S.; Maggiore, R.; Minicò, S.; Galvagno, S. CO_2 reforming of methane over Ni-Ru and Ni-Pd bimetallic catalysts. *Catal. Lett.*, **1999**, *59*, 21-26.
[http://dx.doi.org/10.1023/A:1019031412713]

[74]   Bitter, J.H.; Seshan, K.; Lercher, J.A. The state of zirconia supported platinum catalysts for CO_2/CH_4 reforming. *J. Catal.*, **1997**, *171*, 279-286.
[http://dx.doi.org/10.1006/jcat.1997.1792]

[75]   Faroldi, B.; Carrara, C.; Lombardo, E.A.; Cornaglia, L.M. Production of ultrapure hydrogen in a Pd-Ag membrane reactor using Ru/La$_2$O$_3$ catalysts. *Appl. Catal. A Gen.*, **2007**, *319*, 38-46.
[http://dx.doi.org/10.1016/j.apcata.2006.11.015]

[76] Faroldi, B.; Bosko, M.L.; Múnera, J.; Lombardo, E.; Cornaglia, L. Comparison of Ru/La$_2$O$_2$CO$_3$ performance in two different membrane reactors for hydrogen production. *Catal. Today,* **2013**, *213*, 135-144.
[http://dx.doi.org/10.1016/j.cattod.2013.02.024]

[77] Garcia-Garcia, F.R.; Soria, M.A.; Mateos-Pedrero, C.; Guerrero-Ruiz, A.; Rodriguez-Ramos, I.; Li, K. Dry reforming of methane using Pd-based membrane reactors fabricated from different substrates. *J. Membr. Sci.,* **2013**, *435*, 218-225.
[http://dx.doi.org/10.1016/j.memsci.2013.02.029]

[78] Múnera, J.F.; Coronel, L.; Faroldi, B.; Carrara, C.; Lombardo, E.A.; Cornaglia, L.M. Production of ultrapure hydrogen in a Pd-Ag membrane reactor using noble metal supported on La-based oxides. Modeling for the dry reforming of methane reaction. *Asia-Pac. J. Chem. Eng.,* **2010**, *5*, 35-47.
[http://dx.doi.org/10.1002/apj.380]

[79] De Caprariis, B.; De Filippis, P.; Palma, V.; Petrullo, A.; Ricca, A.; Ruocco, C.; Scarsella, M. Rh, Ru and Pt ternary perovskites type oxides BaZr$_{(1-x)}$Me$_x$O$_3$ for methane dry reforming. *Appl. Catal. A Gen.,* **2016**, *517*, 47-55.
[http://dx.doi.org/10.1016/j.apcata.2016.02.029]

[80] Rostrup-Nielsen, J.R. Industrial relevance of coking. *Catal. Today,* **1997**, *37*, 225-232.
[http://dx.doi.org/10.1016/S0920-5861(97)00016-3]

[81] Barama, S.; Dupeyrat-Batiot, C.; Capron, M.; Bordes-Richard, E.; Bakhti-Mohammedi, O. Catalytic properties of Rh, Ni, Pd and Ce supported on Al-pillared montmorillonites in dry reforming of methane. *Catal. Today,* **2009**, *141*, 385-392.
[http://dx.doi.org/10.1016/j.cattod.2008.06.025]

[82] Ruckenstein, E.; Hu, Y.H. Carbon dioxide reforming of methane over nickel/alkaline earth metal oxide catalysts. *Appl. Catal. A Gen.,* **1995**, *133*, 149-161.
[http://dx.doi.org/10.1016/0926-860X(95)00201-4]

[83] Ruckenstein, E.; Hu, Y.H. Role of support in CO$_2$ reforming of CH$_4$ to syngas over Ni catalysts. *J. Catal.,* **1996**, *162*, 230-238.
[http://dx.doi.org/10.1006/jcat.1996.0280]

[84] Yamazaki, O.; Nozaki, T.; Omata, K.; Fujimoto, K. Reduction of carbon-dioxide by methane with ni-on-mgo-cao containing catalysts. *Chem. Lett.,* **1992**, *21*, 1953-1954.
[http://dx.doi.org/10.1246/cl.1992.1953]

[85] Hayakawa, T.; Suzuki, S.; Nakamura, J.; Uchijima, T.; Hamakawa, S.; Suzuki, K.; Shishido, T.; Takehira, K. CO$_2$ reforming of CH$_4$ over Ni/perovskite catalysts prepared by solid phase crystallization method. *Appl. Catal. A Gen.,* **1999**, *183*, 273-285.
[http://dx.doi.org/10.1016/S0926-860X(99)00071-X]

[86] Zhang, Z.; Verykios, X.E. Carbon dioxide reforming of methane to synthesis gas over Ni/La$_2$O$_3$ catalysts. *Appl. Catal. A Gen.,* **1996**, *138*, 109-133.
[http://dx.doi.org/10.1016/0926-860X(95)00238-3]

[87] Asami, K.; Li, X.; Fujimoto, K.; Koyama, Y.; Sakurama, A.; Kometani, N.; Yonezawa, Y. CO$_2$ reforming of CH$_4$ over ceria-supported metal catalysts. *Catal. Today,* **2003**, *84*, 27-31.
[http://dx.doi.org/10.1016/S0920-5861(03)00297-9]

[88] Cha, K.S.; Kim, H.S.; Yoo, B.K.; Lee, Y.S.; Kang, K.S.; Park, C.S.; Kim, Y.H. Reaction characteristics of two-step methane reforming over a Cu-ferrite/Ce-ZrO$_2$ medium. *Int. J. Hydrogen Energy,* **2009**, *34*, 1801-1808.
[http://dx.doi.org/10.1016/j.ijhydene.2008.12.063]

[89] Du, J.; Yang, X-X.; Ding, J.; Wei, X-L.; Yang, J-P.; Wang, W-L.; Yang, M-L. Carbon dioxide reforming of methane over bimetallic catalysts of Pt-Ru/γ-Al$_2$O$_3$ for thermochemical energy storage. *J. Cent. South Univ.,* **2013**, *20*, 1307-1313.

[http://dx.doi.org/10.1007/s11771-013-1616-6]

[90]   Li, D.; Nakagawa, Y.; Tomishige, K. Methane reforming to synthesis gas over Ni catalysts modified with noble metals. *Appl. Catal. A Gen.,* **2011**, *408*, 1-24.
[http://dx.doi.org/10.1016/j.apcata.2011.09.018]

[91]   Bian, Z.; Das, S.; Wai, M.H.; Hongmanorom, P.; Kawi, S. A review on bimetallic nickel-based catalysts for CO_2 reforming of methane. *ChemPhysChem,* **2017**, *18*(22), 3117-3134.
[http://dx.doi.org/10.1002/cphc.201700529] [PMID: 28710875]

[92]   Hou, Z.; Chen, P.; Fang, H.; Zheng, X.; Yashima, T. Production of synthesis gas *via* methane reforming with CO_2 on noble metals and small amount of noble-(Rh-) promoted Ni catalysts. *Int. J. Hydrogen Energy,* **2006**, *31*, 555-561.
[http://dx.doi.org/10.1016/j.ijhydene.2005.06.010]

[93]   Tomishige, K.; Kanazawa, S.; Sato, M.; Ikushima, K.; Kunimori, K. Catalyst design of Pt-modified Ni/Al_2O_3 catalyst with flat temperature profile in methane reforming with CO_2 and O_2. *Catal. Lett.,* **2002**, *84*, 69-74.
[http://dx.doi.org/10.1023/A:1021076601653]

[94]   Jóźwiak, W.K.; Nowosielska, M.; Rynkowski, J. Reforming of methane with carbon dioxide over supported bimetallic catalysts containing Ni and noble metal I. Characterization and activity of SiO_2 supported Ni-Rh catalysts. *Appl. Catal. A Gen.,* **2005**, *280*, 233-244.
[http://dx.doi.org/10.1016/j.apcata.2004.11.003]

[95]   Sharifi, M.; Haghighi, M.; Rahmani, F.; Karimipour, S. Syngas production *via* dry reforming of CH_4 over Co- and Cu-promoted Ni/Al_2O_3-ZrO_2 nanocatalysts synthesized *via* sequential impregnation and sol-gel methods. *J. Nat. Gas Sci. Eng.,* **2014**, *21*, 993-1004.
[http://dx.doi.org/10.1016/j.jngse.2014.10.030]

[96]   Tanios, C.; Bsaibes, S.; Gennequin, C.; Labaki, M.; Cazier, F.; Billet, S.; Tidahy, H.L.; Nsouli, B.; Aboukaïs, A.; Abi-Aad, E. Syngas production by the CO_2 reforming of CH_4 over Ni–Co–Mg–Al catalysts obtained from hydrotalcite precursors. *Int. J. Hydrogen Energy,* **2017**, *42*, 12818-12828.
[http://dx.doi.org/10.1016/j.ijhydene.2017.01.120]

[97]   Luisetto, I.; Tuti, S.; Di Bartolomeo, E. Co and Ni supported on CeO_2 as selective bimetallic catalyst for dry reforming of methane. *Int. J. Hydrogen Energy,* **2012**, *37*, 15992-15999.
[http://dx.doi.org/10.1016/j.ijhydene.2012.08.006]

[98]   Al-Fatesh, A. Suppression of carbon formation in CH_4–CO_2 reforming by addition of Sr into bimetallic Ni–Co/γ-Al_2O_3 catalyst. *J. King Saud Univ. Eng. Sci.,* **2015**, *27*, 101-107.
[http://dx.doi.org/10.1016/j.jksues.2013.09.006]

[99]   Phongaksorn, M.; Tungkamani, S.; Dharmasaroja, N.; Sornchamni, T.; Chuvaree, R.; Kanjanabat, N.; Siri-Nguan, N. Elucidation of the influence of Ni-Co catalytic properties on dry methane reforming performance. *Chem. Eng. Trans.,* **2015**, *43*, 925-930.
[http://dx.doi.org/10.3303/CET1543155]

[100]   Takanabe, K.; Nagaoka, K.; Nariai, K.; Aika, K.I. Titania-supported cobalt and nickel bimetallic catalysts for carbon dioxide reforming of methane. *J. Catal.,* **2005**, *232*, 268-275.
[http://dx.doi.org/10.1016/j.jcat.2005.03.011]

[101]   Arbag, H.; Yasyerli, S.; Yasyerli, N.; Dogu, G.; Dogu, T. Enhancement of catalytic performance of Ni based mesoporous alumina by Co incorporation in conversion of biogas to synthesis gas. *Appl. Catal. B,* **2016**, *198*, 254-265.
[http://dx.doi.org/10.1016/j.apcatb.2016.05.064]

[102]   Sidik, S.M.; Triwahyono, S.; Jalil, A.A.; Majid, Z.A.; Salamun, N.; Talib, N.B.; Abdullah, T.A.T. CO_2 reforming of CH_4 over Ni-Co/MSN for syngas production: Role of Co as a binder and optimization using RSM. *Chem. Eng. J.,* **2016**, *295*, 1-10.
[http://dx.doi.org/10.1016/j.cej.2016.03.041]

[103] Koh, A.C.W.; Chen, L.; Kee Leong, W.; Johnson, B.F.G.; Khimyak, T.; Lin, J. Hydrogen or synthesis gas production *via* the partial oxidation of methane over supported nickel-cobalt catalysts. *Int. J. Hydrogen Energy,* **2007**, *32*, 725-730.
[http://dx.doi.org/10.1016/j.ijhydene.2006.08.002]

[104] Nagaoka, K.; Takanabe, K.; Aika, K.I. Modification of Co/TiO$_2$ for dry reforming of methane at 2 MPa by Pt, Ru or Ni. *Appl. Catal. A Gen.,* **2004**, *268*, 151-158.
[http://dx.doi.org/10.1016/j.apcata.2004.03.029]

[105] Chen, L.; Zhu, Q.; Wu, R. Effect of Co-Ni ratio on the activity and stability of Co-Ni bimetallic aerogel catalyst for methane oxy-CO$_2$ reforming. *Int. J. Hydrogen Energy,* **2011**, *36*, 2128-2136.
[http://dx.doi.org/10.1016/j.ijhydene.2010.11.042]

[106] Djinović, P. Osojnik črnivec, I.G.; Erjavec, B.; Pintar, A. Influence of active metal loading and oxygen mobility on coke-free dry reforming of Ni-Co bimetallic catalysts. *Appl. Catal. B,* **2012**, *125*, 259-270.
[http://dx.doi.org/10.1016/j.apcatb.2012.05.049]

[107] Foo, S.Y.; Cheng, C.K.; Nguyen, T-H.; Adesina, A. Kinetic study of methane CO$_2$ reforming on Co–Ni/Al$_2$O$_3$ and Ce–Co–Ni/Al$_2$O$_3$ catalysts. *Catal. Today,* **2011**, *164*, 221-226.
[http://dx.doi.org/10.1016/j.cattod.2010.10.092]

[108] Gonzalez-Delacruz, V.M.; Pereñiguez, R.; Ternero, F.; Holgado, J.P.; Caballero, A. *In situ* XAS study of synergic effects on Ni-Co/ZrO$_2$ methane reforming catalysts. *J. Phys. Chem. C,* **2012**, *116*, 2919-2926.
[http://dx.doi.org/10.1021/jp2092048]

[109] San José-Alonso, D.; Illán-Gómez, M.J.; Román-Martínez, M.C. K and Sr promoted Co alumina supported catalysts for the CO$_2$ reforming of methane. *Catal. Today,* **2011**, *176*, 187-190.
[http://dx.doi.org/10.1016/j.cattod.2010.11.093]

[110] Barroso-Quiroga, M.M.; Castro-Luna, A.E. Catalytic activity and effect of modifiers on Ni-based catalysts for the dry reforming of methane. *Int. J. Hydrogen Energy,* **2010**, *35*, 6052-6056.
[http://dx.doi.org/10.1016/j.ijhydene.2009.12.073]

[111] Tsoukalou, A.; Imtiaz, Q.; Kim, S.M.; Abdala, P.M.; Yoon, S.; Muller, C.R. Dry-reforming of methane over bimetallic Ni-M/La$_2$O$_3$ (M=Co, Fe): The effect of the rate of La$_2$O$_2$CO$_3$ formation and phase stability on the catalytic activity and stability. *J. Catal.,* **2016**, *343*, 208-214.
[http://dx.doi.org/10.1016/j.jcat.2016.03.018]

[112] Wu, H.; Pantaleo, G.; La Parola, V.; Venezia, A.M.; Collard, X.; Aprile, C.; Liotta, L.F. Bi- and trimetallic Ni catalysts over Al$_2$O$_3$ and Al$_2$O$_3$-MO$_x$ (M=Ce or Mg) oxides for methane dry reforming: Au and Pt additive effects. *Appl. Catal. B,* **2014**, *156–157*, 350-361.
[http://dx.doi.org/10.1016/j.apcatb.2014.03.018]

[113] Ferreira-Aparicio, P.; Guerrero-Ruiz, A.; Rodríguez-Ramos, I. Comparative study at low and medium reaction temperatures of syngas production by methane reforming with carbon dioxide over silica and alumina supported catalysts. *Appl. Catal. A Gen.,* **1998**, *170*, 177-187.
[http://dx.doi.org/10.1016/S0926-860X(98)00048-9]

[114] Nagai, M.; Nakahira, K.; Ozawa, Y.; Namiki, Y.; Suzuki, Y. CO$_2$ reforming of methane on Rh/Al$_2$O$_3$ catalyst. *Chem. Eng. Sci.,* **2007**, *62*, 4998-5000.
[http://dx.doi.org/10.1016/j.ces.2006.12.017]

[115] Althenayan, F.M.; Yei Foo, S.; Kennedy, E.M.; Dlugogorski, B.Z.; Adesina, A.A. Bimetallic Co-Ni/Al$_2$O$_3$ catalyst for propane dry reforming: estimation of reaction metrics from longevity runs. *Chem. Eng. Sci.,* **2010**, *65*, 66-73.
[http://dx.doi.org/10.1016/j.ces.2009.03.037]

[116] Löfberg, A.; Guerrero-Caballero, J.; Kane, T.; Rubbens, A.; Jalowiecki-Duhamel, L. Ni/CeO$_2$ based catalysts as oxygen vectors for the chemical looping dry reforming of methane for syngas production. *Appl. Catal. B,* **2017**, *212*, 159-174.

[http://dx.doi.org/10.1016/j.apcatb.2017.04.048]

[117] Tsipouriari, V.A.; Verykios, X.E.; Zhang, Z.; Xenophon, V.; MacDonald, S.; Affrossman, S. Kinetic study of the catalytic reforming of methane with carbon dioxide to synthesis gas over Ni/La_2O_3 catalyst. *Catal. Today,* **2001**, *64*, 83-90.
[http://dx.doi.org/10.1016/S0920-5861(00)00511-3]

[118] Li, X.; Li, D.; Tian, H.; Zeng, L.; Zhao, Z.J.; Gong, J. Dry reforming of methane over Ni/La_2O_3 nanorod catalysts with stabilized Ni nanoparticles. *Appl. Catal. B,* **2017**, *202*, 683-694.
[http://dx.doi.org/10.1016/j.apcatb.2016.09.071]

[119] Guo, J.; Lou, H.; Zhao, H.; Chai, D.; Zheng, X. Dry reforming of methane over nickel catalysts supported on magnesium aluminate spinels. *Appl. Catal. A Gen.,* **2004**, *273*, 75-82.
[http://dx.doi.org/10.1016/j.apcata.2004.06.014]

[120] Abdollahifar, M.; Haghighi, M.; Babaluo, A.A.; Talkhoncheh, S.K. Sono-synthesis and characterization of bimetallic $Ni-Co/Al_2O_3-MgO$ nanocatalyst: Effects of metal content on catalytic properties and activity for hydrogen production *via* CO_2 reforming of CH_4. *Ultrason. Sonochem.,* **2016**, *31*, 173-183.
[http://dx.doi.org/10.1016/j.ultsonch.2015.12.010] [PMID: 26964938]

[121] Charisiou, N.D.; Siakavelas, G.; Papageridis, K.N.; Baklavaridis, A.; Tzounis, L.; Avraam, D.G.; Goula, M.A. Syngas production *via* the biogas dry reforming reaction over nickel supported on modified with CeO_2 and/or La_2O_3 alumina catalysts. *J. Nat. Gas Sci. Eng.,* **2016**, *31*, 164-183.
[http://dx.doi.org/10.1016/j.jngse.2016.02.021]

[122] Chen, J.; Yao, C.; Zhao, Y.; Jia, P. Synthesis gas production from dry reforming of methane over $Ce_{0.75}Zr_{0.25}O_2$-supported Ru catalysts. *Int. J. Hydrogen Energy,* **2010**, *35*, 1630-1642.
[http://dx.doi.org/10.1016/j.ijhydene.2009.12.043]

[123] Muñoz, M.A.; Calvino, J.J.; Rodríguez-Izquierdo, J.M.; Blanco, G.; Arias, D.C.; Pérez-Omil, J.A.; Hernández-Garrido, J.C.; González-Leal, J.M.; Cauqui, M.A.; Yeste, M.P. Highly stable ceria-zirconia-yttria supported Ni catalysts for syngas production by CO_2 reforming of methane. *Appl. Surf. Sci.,* **2017**, *426*, 864-873.
[http://dx.doi.org/10.1016/j.apsusc.2017.07.210]

[124] Blanchard, J.; Nsungui, A.J.; Abatzoglou, N.; Gitzhofer, F. Dry reforming of methane with a Ni/Al_2O_3-YSZ catalyst: the role of the catalyst preparation protocol. *Can. J. Chem. Eng.,* **2008**, *85*, 889-899.
[http://dx.doi.org/10.1002/cjce.5450850610]

[125] Kresge, C.T.; Leonowicz, M.E.; Roth, W.J.; Vartuli, J.C.; Beck, J.S. Ordered mesoporous molecular sieves synthesized by a liquid-crystal template mechanism. *Nature,* **1992**, *359*, 710-712.
[http://dx.doi.org/10.1038/359710a0]

[126] Vartuli, J.C.; Shih, S.S.; Kresge, C.T.; Beck, J.S. Potential applications for M41S type mesoporous molecular sieves. *Stud. Surf. Sci. Catal.,* **1998**, *117*, 13-21.
[http://dx.doi.org/10.1016/S0167-2991(98)80973-7]

[127] Wan, H.; Li, X.; Ji, S.; Huang, B.; Wang, K.; Li, C. Effect of Ni loading and $Ce_xZr_{1-x}O_2$ promoter on Ni-based SBA-15 catalysts for steam reforming of methane. *J. Nat. Gas Chem.,* **2007**, *16*, 139-147.
[http://dx.doi.org/10.1016/S1003-9953(07)60039-5]

[128] Wang, N.; Xu, Z.; Deng, J.; Shen, K.; Yu, X.; Qian, W.; Chu, W.; Wei, F. One-pot synthesis of ordered mesoporous NiCeAl oxide catalysts and a study of their performance in methane dry reforming. *ChemCatChem,* **2014**, *6*, 1470-1480.
[http://dx.doi.org/10.1002/cctc.201300720]

[129] Shen, W.; Komatsubara, K.; Hagiyama, T.; Yoshida, A.; Naito, S. Steam reforming of methane over ordered mesoporous Ni-Mg-Al oxides. *Chem. Commun. (Camb.),* **2009**, *0*(42), 6490-6492.
[http://dx.doi.org/10.1039/b910071g] [PMID: 19841819]

[130] Xu, L.; Song, H.; Chou, L. Carbon dioxide reforming of methane over ordered mesoporous NiO–MgO–Al$_2$O$_3$ composite oxides. *Appl. Catal. B,* **2011**, *108–109*, 177-190.
[http://dx.doi.org/10.1016/j.apcatb.2011.08.028]

[131] Xu, L.; Song, H.; Chou, L. One-pot synthesis of ordered mesoporous NiO-CaO-Al$_2$O$_3$ composite oxides for catalyzing CO$_2$ reforming of CH$_4$. *ACS Catal.,* **2012**, *2*, 1331-1342.
[http://dx.doi.org/10.1021/cs3001072]

[132] Fang, X.; Peng, C.; Peng, H.; Liu, W.; Xu, X.; Wang, X.; Li, C.; Zhou, W. Methane dry reforming over coke-resistant mesoporous Ni-Al$_2$O$_3$ catalysts prepared by evaporation-induced self-assembly method. *ChemCatChem,* **2015**, *7*, 3753-3762.
[http://dx.doi.org/10.1002/cctc.201500538]

[133] Li, Z.; Mo, L.; Kathiraser, Y.; Kawi, S. Yolk–satellite–shell structured Ni–Yolk@Ni@SiO$_2$ nanocomposite: superb catalyst toward methane CO$_2$ reforming reaction. *ACS Catal.,* **2014**, *4*, 1526-1536.
[http://dx.doi.org/10.1021/cs401027p]

[134] Moradi, G.; Khezeli, F.; Hemmati, H. Syngas production with dry reforming of methane over Ni/ZSM-5 catalysts. *J. Nat. Gas Sci. Eng.,* **2016**, *33*, 657-665.
[http://dx.doi.org/10.1016/j.jngse.2016.06.004]

[135] Abdollahifar, M.; Haghighi, M.; Sharifi, M. Dry reforming of methane over nanostructured Co/Y catalyst for hydrogen production: Effect of ultrasound irradiation and Co-loading on catalyst properties and performance. *Energy Convers. Manage.,* **2015**, *103*, 1101-1112.
[http://dx.doi.org/10.1016/j.enconman.2015.04.053]

[136] Jun, J.H.; Lee, T-J.; Lim, T.H.; Nam, S-W.; Hong, S-A.; Yoon, K.J. Nickel–calcium phosphate/hydroxyapatite catalysts for partial oxidation of methane to syngas: characterization and activation. *J. Catal.,* **2004**, *221*, 178-190.
[http://dx.doi.org/10.1016/j.jcat.2003.07.004]

[137] Lee, S.J.; Jun, J.H.; Lee, S-H.; Yoon, K.J.; Lim, T.H.; Nam, S-W.; Hong, S-A. Partial oxidation of methane over nickel-added strontium phosphate. *Appl. Catal. A Gen.,* **2002**, *230*, 61-71.
[http://dx.doi.org/10.1016/S0926-860X(01)00995-4]

[138] Abba, M.O.; Gonzalez-DelaCruz, V.M.; Colón, G.; Sebti, S.; Caballero, A. *In situ* XAS study of an improved natural phosphate catalyst for hydrogen production by reforming of methane. *Appl. Catal. B,* **2014**, *150–151*, 459-465.
[http://dx.doi.org/10.1016/j.apcatb.2013.12.031]

[139] Tsipouriari, V.A.; Efstathiou, A.M.; Zhang, Z.L.; Verykios, X.E. Reforming of methane with carbon dioxide to synthesis gas over supported Rh catalysts. *Catal. Today,* **1994**, *21*, 579-587.
[http://dx.doi.org/10.1016/0920-5861(94)80182-7]

[140] Solymosi, F.; Kutsan, G.; Erdohelyi, A. Catalytic reaction of CH$_4$ with CO$_2$ over alumina-supported Pt metals. *Catal. Lett.,* **1991**, *11*, 149-156.
[http://dx.doi.org/10.1007/BF00764080]

[141] Nakamura, J.; Aikawa, K.; Sato, K.; Uchijima, T. Role of support in reforming of CH$_4$ with CO$_2$ over Rh catalysts. *Catal. Lett.,* **1994**, *25*, 265-270.
[http://dx.doi.org/10.1007/BF00816306]

[142] Yu, X.; Wang, N.; Chu, W.; Liu, M. Carbon dioxide reforming of methane for syngas production over La-promoted NiMgAl catalysts derived from hydrotalcites. *Chem. Eng. J.,* **2012**, *209*, 623-632.
[http://dx.doi.org/10.1016/j.cej.2012.08.037]

[143] Serrano-Lotina, A.; Martin, A.J.; Folgado, M.A.; Daza, L. Dry reforming of methane to syngas over La-promoted hydrotalcite clay-derived catalysts. *Int. J. Hydrogen Energy,* **2012**, *37*, 12342-12350.
[http://dx.doi.org/10.1016/j.ijhydene.2012.06.041]

[144] Serrano-Lotina, A.; Rodríguez, L.; Muñoz, G.; Daza, L. Biogas reforming on La-promoted NiMgAl

catalysts derived from hydrotalcite-like precursors. *J. Power Sources,* **2011**, *196*, 4404-4410.
[http://dx.doi.org/10.1016/j.jpowsour.2010.10.107]

[145] Dahdah, E.; Abou Rached, J.; Aouad, S.; Gennequin, C.; Tidahy, H.L.; Estephane, J.; Aboukaïs, A.; Abi Aad, E. CO_2 reforming of methane over $Ni_xMg_{6-x}Al_2$ catalysts: effect of lanthanum doping on catalytic activity and stability. *Int. J. Hydrogen Energy,* **2017**, *42*, 12808-12817.
[http://dx.doi.org/10.1016/j.ijhydene.2017.01.197]

[146] Zhuang, Q.; Qin, Y.; Chang, L. Promoting effect of cerium oxide in supported nickel catalyst for hydrocarbon steam-reforming. *Appl. Catal.,* **1991**, *70*, 1-8.
[http://dx.doi.org/10.1016/S0166-9834(00)84149-4]

[147] Wang, J.B.; Kuo, L.E.; Huang, T.J. Study of carbon dioxide reforming of methane over bimetallic Ni-Cr/yttria-doped ceria catalysts. *Appl. Catal. A Gen.,* **2003**, *249*, 93-105.
[http://dx.doi.org/10.1016/S0926-860X(03)00180-7]

[148] Chen, P.; Zhang, H. Bin; Lin, G.D.; Tsai, K.R. Development of coking-resistant Ni-based catalyst for partial oxidation and CO_2-reforming of methane to syngas. *Appl. Catal. A Gen.,* **1998**, *166*, 343-350.
[http://dx.doi.org/10.1016/S0926-860X(97)00291-3]

[149] Choi, J.S.; Moon, K.I.; Kim, Y.G.; Lee, J.S.; Kim, C.H.; Trimm, D.L. Stable carbon dioxide reforming of methane over modified Ni/Al_2O_3 catalysts. *Catal. Lett.,* **1998**, *52*, 43-47.
[http://dx.doi.org/10.1023/A:1019002932509]

[150] Trimm, D.L. Coke formation and minimisation during steam reforming reactions. *Catal. Today,* **1997**, *37*, 233-238.
[http://dx.doi.org/10.1016/S0920-5861(97)00014-X]

[151] Hassani Rad, S.J.; Haghighi, M.; Alizadeh Eslami, A.; Rahmani, F.; Rahemi, N. Sol–gel vs. impregnation preparation of MgO and CeO_2 doped Ni/Al_2O_3 nanocatalysts used in dry reforming of methane: Effect of process conditions, synthesis method and support composition. *Int. J. Hydrogen Energy,* **2016**, *41*, 5335-5350.
[http://dx.doi.org/10.1016/j.ijhydene.2016.02.002]

[152] Bouarab, R.; Cherifi, O. Effet de l'ajout alcalino-terreux sur les propriétés catalytiques de Co/SiO_2 en réaction $CH_4 + CO_2$. *Nat. Technol.,* **2013**, *9*, 9-12.

[153] Osaki, T.; Mori, T. Role of potassium in carbon-free CO_2 reforming of methane on K-promoted Ni/Al_2O_3 catalysts. *J. Catal.,* **2001**, *204*, 89-97.
[http://dx.doi.org/10.1006/jcat.2001.3382]

[154] Alipour, Z.; Rezaei, M.; Meshkani, F. Effect of alkaline earth promoters (MgO, CaO, and BaO) on the activity and coke formation of Ni catalysts supported on nanocrystalline Al_2O_3 in dry reforming of methane. *J. Ind. Eng. Chem.,* **2014**, *20*, 2858-2863.
[http://dx.doi.org/10.1016/j.jiec.2013.11.018]

[155] Sigl, M.; Bradford, M.C.J.; Knözinger, H.; Vannice, M.A. CO_2 reforming of methane over vanadia-promoted Rh/SiO_2 catalysts. *Top. Catal.,* **1999**, *8*, 211-222.
[http://dx.doi.org/10.1023/A:1019121429843]

[156] Wang, Q.; Ren, W.; Yuan, X.; Mu, R.; Song, Z.; Wang, X. Ni catalysts derived from Mg-Al layered double hydroxides for hydrogen production from landfill gas conversion. *Int. J. Hydrogen Energy,* **2012**, *37*, 11488-11494.
[http://dx.doi.org/10.1016/j.ijhydene.2012.05.010]

[157] Long, H.; Xu, Y.; Zhang, X.; Hu, S.; Shang, S.; Yin, Y.; Dai, X. Ni-Co/Mg-Al catalyst derived from hydrotalcite-like compound prepared by plasma for dry reforming of methane. *J. Energy Chem.,* **2013**, *22*, 733-739.
[http://dx.doi.org/10.1016/S2095-4956(13)60097-2]

[158] Nair, M.M.; Kaliaguine, S. Structured catalysts for dry reforming of methane. *New J. Chem.,* **2016**, *40*, 4049-4060.

[http://dx.doi.org/10.1039/C5NJ03268G]

[159] Takanabe, K.; Nagaoka, K.; Aika, K. Improved resistance against coke deposition of titania supported cobalt and nickel bimetallic catalysts for carbon dioxide reforming of methane. *Catal. Lett.,* **2005**, *102*, 153-157.
[http://dx.doi.org/10.1007/s10562-005-5848-4]

[160] Bhattacharyya, A.; Chang, V.W.; Schumacher, D.J. CO_2 reforming of methane to syngas I: evaluation of hydrotalcite clay-derived catalysts. *Appl. Clay Sci.,* **1998**, *13*, 317-328.
[http://dx.doi.org/10.1016/S0169-1317(98)00030-1]

[161] Tsyganok, A.I.; Tsunoda, T.; Hamakawa, S.; Suzuki, K.; Takehira, K.; Hayakawa, T. Dry reforming of methane over catalysts derived from nickel-containing Mg-Al layered double hydroxides. *J. Catal.,* **2003**, *213*, 191-203.
[http://dx.doi.org/10.1016/S0021-9517(02)00047-7]

[162] Shin, S.A.; Noh, Y.S.; Hong, G.H.; Park, J.I.; Song, H.T.; Lee, K.-Y.; Moon, D.J. Dry reforming of methane over Ni/ZrO_2 -Al_2O_3 catalysts: Effect of preparation methods *J. Taiwan Inst. Chem. Eng,* **2017**.
[http://dx.doi.org/10.1016/j.jtice.2017.11.032]

[163] Matei-Rutkovska, F.; Postole, G.; Rotaru, C.G.; Florea, M.; Pârvulescu, V.I.; Gelin, P. Synthesis of ceria nanopowders by microwave-assisted hydrothermal method for dry reforming of methane. *Int. J. Hydrogen Energy,* **2016**, *41*, 2512-2525.
[http://dx.doi.org/10.1016/j.ijhydene.2015.12.097]

[164] Wang, S.; Lu, G.Q. A comprehensive study on carbon dioxide reforming of methane over Ni/γ-Al_2O_3 catalysts. *Ind. Eng. Chem. Res.,* **1999**, *38*, 2615-2625.
[http://dx.doi.org/10.1021/ie980489t]

[165] Wang, S.; Lu, G.Q.M. CO_2 reforming of methane on Ni catalysts: Effects of the support phase and preparation technique. *Appl. Catal. B,* **1998**, *16*, 269-277.
[http://dx.doi.org/10.1016/S0926-3373(97)00083-0]

[166] Gadalla, A.M.; Bower, B. The role of catalyst support on the activity of nickel for reforming methane with CO_2. *Chem. Eng. Sci.,* **1988**, *43*, 3049-3062.
[http://dx.doi.org/10.1016/0009-2509(88)80058-7]

[167] Hu, Y.H.; Ruckenstein, E. Transient response analysis *via* a broadened pulse combined with a step change or an isotopic pulse. Application to CO_2 reforming of methane over NiO/SiO_2. *J. Phys. Chem. B,* **1997**, *101*, 7563-7565.
[http://dx.doi.org/10.1021/jp971711v]

[168] Tomishige, K.; Chen, Y.; Fujimoto, K. Studies on carbon deposition in CO_2 reforming of CH_4 over nickel–magnesia solid solution catalysts. *J. Catal.,* **1999**, *181*, 91-103.
[http://dx.doi.org/10.1006/jcat.1998.2286]

[169] Chattanathan, S.A.; Adhikari, S.; McVey, M.; Fasina, O. Hydrogen production from biogas reforming and the effect of H2S on CH_4 conversion. *Int. J. Hydrogen Energy,* **2014**, *39*, 19905-19911.
[http://dx.doi.org/10.1016/j.ijhydene.2014.09.162]

[170] Chiodo, V.; Maisano, S.; Zafarana, G.; Urbani, F. Effect of pollutants on biogas steam reforming. *Int. J. Hydrogen Energy,* **2017**, *42*, 1622-1628.
[http://dx.doi.org/10.1016/j.ijhydene.2016.07.251]

[171] Appari, S.; Janardhanan, V.M.; Bauri, R.; Jayanti, S.; Deutschmann, O. A detailed kinetic model for biogas steam reforming on Ni and catalyst deactivation due to sulfur poisoning. *Appl. Catal. A Gen.,* **2014**, *471*, 118-125.
[http://dx.doi.org/10.1016/j.apcata.2013.12.002]

[172] Jablonski, W.S.; Villano, S.M.; Dean, A.M. A comparison of H_2S, SO_2, and COS poisoning on Ni/YSZ and Ni/K_2O-$CaAl_2O_4$ during methane steam and dry reforming. *Appl. Catal. A Gen.,* **2015**,

502, 399-409.
[http://dx.doi.org/10.1016/j.apcata.2015.06.009]

[173] Wang, H.Y.; Au, C.T. CH_4/CD_4 isotope effects in the carbon dioxide reforming of methane to syngas over SiO_2-supported nickel catalysts. *Catal. Lett.,* **1996**, *38*, 77-79.
[http://dx.doi.org/10.1007/BF00806903]

[174] Slagtern, Å.; Olsbye, U.; Blom, R.; Dahl, I.M.; Fjellvåg, H. Characterization of Ni on La modified Al_2O_3 catalysts during CO_2 reforming of methane. *Appl. Catal. A Gen.,* **1997**, *165*, 379-390.
[http://dx.doi.org/10.1016/S0926-860X(97)00220-2]

[175] Bradford, M.C.J.; Vannice, M.A. Catalytic reforming of methane with carbon dioxide over nickel catalysts II. Reaction kinetics. *Appl. Catal. A Gen.,* **1996**, *142*, 97-122.
[http://dx.doi.org/10.1016/0926-860X(96)00066-X]

[176] Nandini, A.; Pant, K.K.; Dhingra, S.C. Kinetic study of the catalytic carbon dioxide reforming of methane to synthesis gas over Ni-K/CeO_2-Al_2O_3 catalyst. *Appl. Catal. A Gen.,* **2006**, *308*, 119-127.
[http://dx.doi.org/10.1016/j.apcata.2006.04.014]

[177] Olsbye, U.; Wurzel, T.; Mleczko, L. Kinetic and reaction engineering studies of dry reforming of methane over a Ni/La/Al_2O_3 catalyst. *Ind. Eng. Chem. Res.,* **1997**, *36*, 5180-5188.
[http://dx.doi.org/10.1021/ie970246l]

[178] Zhang, J.; Wang, H.; Dalai, A.K. Kinetic studies of carbon dioxide reforming of methane over Ni−Co/Al−Mg−O bimetallic catalyst. *Ind. Eng. Chem. Res.,* **2009**, *48*, 677-684.
[http://dx.doi.org/10.1021/ie801078p]

[179] Ojala, S.; Lassi, U.; Ylonen, R.; Keiski, R.; Laakso, I.; Maunula, T.; Silvonen, R. Abatement of malodorous pulp mill emissions by catalytic oxidation - pilot experiments in Stora Enso Pulp Mill, Oulu, Finland. *Tappi J.,* **2005**, *4*, 9-14.

[180] Ojala, S.; Koivikko, N.; Laitinen, T.; Mouammine, A.; Seelam, P.K.; Laassiri, S.; Ainassaari, K.; Brahmi, R.; Keiski, R.L. Utilization of volatile organic compounds as an alternative for destructive abatement. *Catalysts,* **2015**, *5*, 1092-1151.
[http://dx.doi.org/10.3390/catal5031092]

[181] Iulianelli, A.; Ribeirinha, P.; Mendes, A.; Basile, A. Methanol steam reforming for hydrogen generation *via* conventional and membrane reactors: a review. *Renew. Sustain. Energy Rev.,* **2014**, *29*, 355-368.
[http://dx.doi.org/10.1016/j.rser.2013.08.032]

[182] Aouad, S.; Gennequin, C.; Mrad, M.; Tidahy, H.L.; Estephane, J.; Aboukaïs, A.; Abi-Aad, E. Steam reforming of methanol over ruthenium impregnated ceria, alumina and ceria–alumina catalysts. *Int. J. Energy Res.,* **2016**, *40*, 1287-1292.
[http://dx.doi.org/10.1002/er.3531]

[183] Zhang, H.; Li, X.; Zhu, F.; Cen, K.; Du, C.; Tu, X. Plasma assisted dry reforming of methanol for clean syngas production and high-efficiency CO_2 conversion. *Chem. Eng. J.,* **2017**, *310*, 114-119.
[http://dx.doi.org/10.1016/j.cej.2016.10.104]

[184] Iqbal, M.S.; Ahmed, M.S.; Ogras, T.T.; Li, C.; Arshad, M.; Khattak, J.Z.K.; Asif, J.H.; Ullah, S.; Bashir, R. Ethanol production from waste materials. *J. Biochem. Technol.,* **2012**, *4*, 285-288.

[185] Homsi, D.; Rached, J.A.; Aouad, S.; Gennequin, C.; Dahdah, E.; Estephane, J.; Tidahy, H.L.; Aboukaïs, A.; Abi-Aad, E. Steam reforming of ethanol for hydrogen production over Cu/Co-Mg-Al based catalysts prepared by hydrotalcite route. *Environ. Sci. Pollut. Res. Int.,* **2017**, *24*(11), 9907-9913.
[http://dx.doi.org/10.1007/s11356-016-7480-9] [PMID: 27552997]

[186] Drif, A.; Bion, N.; Brahmi, R.; Ojala, S.; Pirault-Roy, L.; Turpeinen, E.; Seelam, P.K.; Keiski, R.L.; Epron, F. Study of the dry reforming of methane and ethanol using Rh catalysts supported on doped alumina. *Appl. Catal. A Gen.,* **2015**, *504*, 576-584.
[http://dx.doi.org/10.1016/j.apcata.2015.02.019]

[187] Li, D.; Li, X.; Gong, J. Catalytic reforming of oxygenates: state of the art and future prospects. *Chem. Rev.*, **2016**, *116*(19), 11529-11653.
[http://dx.doi.org/10.1021/acs.chemrev.6b00099] [PMID: 27527927]

[188] Wang, W.; Wang, Y. Dry reforming of ethanol for hydrogen production: Thermodynamic investigation. *Int. J. Hydrogen Energy*, **2009**, *34*, 5382-5389.
[http://dx.doi.org/10.1016/j.ijhydene.2009.04.054]

[189] Jankhah, S.; Abatzoglou, N.; Gitzhofer, F. Thermal and catalytic dry reforming and cracking of ethanol for hydrogen and carbon nanofilaments' production. *Int. J. Hydrogen Energy*, **2008**, *33*, 4769-4779.
[http://dx.doi.org/10.1016/j.ijhydene.2008.06.058]

[190] De Oliveira-Vigier, K.; Abatzoglou, N.; Gitzhofer, F. Dry-reforming of ethanol in the presence of a 316 stainless steel catalyst. *Can. J. Chem. Eng.*, **2005**, *83*, 978-984.
[http://dx.doi.org/10.1002/cjce.5450830607]

[191] Da Silva, A.M.; De Souza, K.R.; Jacobs, G.; Graham, U.M.; Davis, B.H.; Mattos, L.V.; Noronha, F.B. Steam and CO_2 reforming of ethanol over Rh/CeO_2 catalyst. *Appl. Catal. B*, **2011**, *102*, 94-109.
[http://dx.doi.org/10.1016/j.apcatb.2010.11.030]

[192] Zawadzki, A.; Bellido, J.D.A.; Lucrédio, A.F.; Assaf, E.M. Dry reforming of ethanol over supported Ni catalysts prepared by impregnation with methanolic solution. *Fuel Process. Technol.*, **2014**, *128*, 432-440.
[http://dx.doi.org/10.1016/j.fuproc.2014.08.006]

[193] Bahari, M.B.; Fayaz, F.; Ainirazali, N.; Phuc, N.H.H.; Dai Viet, N. V. Evaluation of Co-promoted Ni/Al_2O_3 catalyst for CO_2 reforming of ethanol. *ARPN J. Eng. Appl. Sci.*, **2016**, *11*, 7249-7253.

[194] Cao, D.; Cai, W.; Li, Y.; Li, C.; Yu, H.; Zhang, S.; Qu, F. Syngas production from ethanol dry reforming over $Cu/Ce_{0.8}Zr_{0.2}O_2$ catalyst. *Catal. Lett.*, **2017**, *147*, 2929-2939.
[http://dx.doi.org/10.1007/s10562-017-2216-0]

[195] Zhou, C.H.; Beltramini, J.N.; Fan, Y.X.; Lu, G.Q. Chemoselective catalytic conversion of glycerol as a biorenewable source to valuable commodity chemicals. *Chem. Soc. Rev.*, **2008**, *37*(3), 527-549.
[http://dx.doi.org/10.1039/B707343G] [PMID: 18224262]

[196] Vaidya, P.D.; Rodrigues, A.E. Glycerol reforming for hydrogen production: a review. *Chem. Eng. Technol.*, **2009**, *32*, 1463-1469.
[http://dx.doi.org/10.1002/ceat.200900120]

[197] Kousi, K.; Chourdakis, N.; Matralis, H.; Kontarides, D.; Papadopoulou, C.; Verykios, X. Glycerol steam reforming over modified Ni-based catalysts. *Appl. Catal. A Gen.*, **2016**, *518*, 129-141.
[http://dx.doi.org/10.1016/j.apcata.2015.11.047]

[198] Iulianelli, A.; Seelam, P.K.; Liguori, S.; Longo, T.; Keiski, R.; Calabr, V.; Basile, A. Hydrogen production for PEM fuel cell by gas phase reforming of glycerol as byproduct of bio-diesel. The use of a Pd-Ag membrane reactor at middle reaction temperature. *Int. J. Hydrogen Energy*, **2011**, *36*, 3827-3834.
[http://dx.doi.org/10.1016/j.ijhydene.2010.02.079]

[199] Wang, X.; Li, M.; Wang, M.; Wang, H.; Li, S.; Wang, S.; Ma, X. Thermodynamic analysis of glycerol dry reforming for hydrogen and synthesis gas production. *Fuel*, **2009**, *88*, 2148-2153.
[http://dx.doi.org/10.1016/j.fuel.2009.01.015]

[200] Siew, K.W.; Lee, H.C.; Gimbun, J.; Cheng, C.K. Characterization of La-promoted Ni/Al_2O_3 catalysts for hydrogen production from glycerol dry reforming. *J. Energy Chem.*, **2014**, *23*, 15-21.
[http://dx.doi.org/10.1016/S2095-4956(14)60112-1]

[201] Lee, H.C.; Siew, K.W.; Khan, M.R.; Chin, S.Y.; Gimbun, J.; Cheng, C.K. Catalytic performance of cement clinker supported nickel catalyst in glycerol dry reforming. *J. Energy Chem.*, **2014**, *23*, 645-656.

[http://dx.doi.org/10.1016/S2095-4956(14)60196-0]

[202] Kumar, A.; Eskridge, K.; Jones, D.D.; Hanna, M.A. Steam-air fluidized bed gasification of distillers grains: Effects of steam to biomass ratio, equivalence ratio and gasification temperature. *Bioresour. Technol.,* **2009**, *100*(6), 2062-2068.
[http://dx.doi.org/10.1016/j.biortech.2008.10.011] [PMID: 19028089]

[203] Li, C.; Hirabayashi, D.; Suzuki, K. Steam reforming of biomass tar producing H_2-rich gases over Ni/MgO$_x$/CaO$_{1-x}$ catalyst. *Bioresour. Technol.,* **2010**, *101* Suppl. 1, S97-S100.
[http://dx.doi.org/10.1016/j.biortech.2009.03.043] [PMID: 19369062]

[204] Cao, L.; Jia, Z.; Ji, S.; Hu, J. Catalytic steam reforming of biomass over Ni-based catalysts: conversion from poplar leaves to hydrogen-rich syngas. *J. Nat. Gas Chem.,* **2011**, *20*, 377-383.
[http://dx.doi.org/10.1016/S1003-9953(10)60195-8]

[205] Guan, G.; Kaewpanha, M.; Hao, X.; Abudula, A. Catalytic steam reforming of biomass tar: Prospects and challenges. *Renew. Sustain. Energy Rev.,* **2016**, *58*, 450-461.
[http://dx.doi.org/10.1016/j.rser.2015.12.316]

[206] Abou Rached, J.; El Hayek, C.; Dahdah, E.; Gennequin, C.; Aouad, S.; Tidahy, H.L.; Estephane, J.; Nsouli, B.; Aboukaïs, A.; Abi-Aad, E. Ni based catalysts promoted with cerium used in the steam reforming of toluene for hydrogen production. *Int. J. Hydrogen Energy,* **2017**, *42*, 12829-12840.
[http://dx.doi.org/10.1016/j.ijhydene.2016.10.053]

[207] Abou Rached, J.; Dahdah, E.; Estephane, J.; Aouad, S.; Gennequin, C.; Tidahy, H.L.; Nsouli, B.; Aboukaïs, A.; Abi-Aad, E. Steam reforming of toluene for hydrogen production over NiMgAlCe catalysts prepared via hydrotalcite route in IREC 2016 - 7th International Renewable Energy Congress **2016**, pp. 1-6.
[http://dx.doi.org/10.1109/IREC.2016.7507629]

[208] Weerachanchai, P.; Horio, M.; Tangsathitkulchai, C. Effects of gasifying conditions and bed materials on fluidized bed steam gasification of wood biomass. *Bioresour. Technol.,* **2009**, *100*(3), 1419-1427.
[http://dx.doi.org/10.1016/j.biortech.2008.08.002] [PMID: 18793834]

[209] Alauddin, Z.A.B.Z.; Lahijani, P.; Mohammadi, M.; Mohamed, A.R. Gasification of lignocellulosic biomass in fluidized beds for renewable energy development: a review. *Renew. Sustain. Energy Rev.,* **2010**, *14*, 2852-2862.
[http://dx.doi.org/10.1016/j.rser.2010.07.026]

[210] Bao, X.; Kong, M.; Lu, W.; Fei, J.; Zheng, X. Performance of Co/MgO catalyst for CO_2 reforming of toluene as a model compound of tar derived from biomass gasification. *J. Energy Chem.,* **2014**, *23*, 795-800.
[http://dx.doi.org/10.1016/S2095-4956(14)60214-X]

[211] Hu, Y.H.; Ruckenstein, E. Binary MgO-based solid solution catalysts for methane conversion to syngas. *Catal. Rev.,* **2002**, *44*, 423-453.
[http://dx.doi.org/10.1081/CR-120005742]

[212] Fan, M-S.; Abdullah, A.Z.; Bhatia, S. Catalytic technology for carbon dioxide reforming of methane to synthesis gas. *ChemCatChem,* **2009**, *1*, 192-208.
[http://dx.doi.org/10.1002/cctc.200900025]

[213] Kong, M.; Fei, J.; Wang, S.; Lu, W.; Zheng, X. Influence of supports on catalytic behavior of nickel catalysts in carbon dioxide reforming of toluene as a model compound of tar from biomass gasification. *Bioresour. Technol.,* **2011**, *102*(2), 2004-2008.
[http://dx.doi.org/10.1016/j.biortech.2010.09.054] [PMID: 20943380]

[214] Chen, T.; Liu, H.; Shi, P.; Chen, D.; Song, L.; He, H.; Frost, R.L. CO_2 reforming of toluene as model compound of biomass tar on Ni/Palygorskite. *Fuel,* **2013**, *107*, 699-705.
[http://dx.doi.org/10.1016/j.fuel.2012.12.036]

[215] Li, L.; Song, Z.; Zhao, X.; Ma, C.; Kong, X.; Wang, F. Microwave-induced cracking and CO_2

reforming of toluene on biomass derived char. *Chem. Eng. J.,* **2016**, *284*, 1308-1316.
[http://dx.doi.org/10.1016/j.cej.2015.09.040]

[216]   Wang, C.G.; Wang, T.J.; Ma, L.L.; Gao, Y.; Wu, C.Z. Partial oxidation reforming of biomass fuel gas over nickel-based monolithic catalyst with naphthalene as model compound. *Korean J. Chem. Eng.,* **2008**, *25*, 738-743.
[http://dx.doi.org/10.1007/s11814-008-0121-3]

[217]   Li, B.; Chen, H.P.; Yang, H.P.; Yang, G.L.; Wang, X.H.; Zhang, S.H. Ni/γ-Al$_2$O$_3$ catalyst for CO$_2$ reforming of benzene as a model compound of biomass gasification tar: Promotional effect of ultrasonic treatment on catalytic performance. *Proceedings of the 20th International Conference on Fluidized Bed Combustion,* **2009**, pp. 576-582.
[http://dx.doi.org/10.1007/978-3-642-02682-9_87]

[218]   Hoang, T.M.C.; van Eck, E.R.H.; Bula, W.P.; Gardeniers, J.G.E.; Lefferts, L.; Seshan, K. Humin based by-products from biomass processing as a potential carbonaceous source for synthesis gas production. *Green Chem.,* **2015**, *17*, 959-972.
[http://dx.doi.org/10.1039/C4GC01324G]

[219]   Devi, L.; Craje, M.; Thüne, P.; Ptasinski, K.J.; Janssen, F.J.J.G. Olivine as tar removal catalyst for biomass gasifiers: catalyst characterization. *Appl. Catal. A Gen.,* **2005**, *294*, 68-79.
[http://dx.doi.org/10.1016/j.apcata.2005.07.044]

[220]   Devi, L.; Ptasinski, K.J.; Janssen, F.J.J.G. Pretreated olivine as tar removal catalyst for biomass gasifiers: investigation using naphthalene as model biomass tar. *Fuel Process. Technol.,* **2005**, *86*, 707-730.
[http://dx.doi.org/10.1016/j.fuproc.2004.07.001]

[221]   Tursunov, O.; Zubek, K.; Dobrowolski, J.; Czerski, G.; Grzywacz, P. Effect of Ni/Al$_2$O$_3$-SiO$_2$ and Ni/Al$_2$O$_3$-SiO$_2$ with K$_2$O promoter catalysts on H$_2$, CO and CH$_4$ concentration by CO$_2$ gasification of Rosa Multiflora biomass. *Oil Gas Sci. Technol.,* **2017**, *72* article 37.
[http://dx.doi.org/10.2516/ogst/2017037]

[222]   Saad, J.M.; Williams, P.T. Pyrolysis-catalytic-dry reforming of waste plastics and mixed waste plastics for syngas production. *Energy Fuels,* **2016**, *30*, 3198-3204.
[http://dx.doi.org/10.1021/acs.energyfuels.5b02508]

[223]   Kong, M.; Yang, Q.; Lu, W.; Fan, Z.; Fei, J.; Zheng, X.; Wheelock, T.D. Effect of calcination temperature on characteristics and performance of Ni/MgO catalyst for CO$_2$ reforming of toluene. *Chin. J. Catal.,* **2012**, *33*, 1508-1516.
[http://dx.doi.org/10.1016/S1872-2067(11)60424-5]

[224]   Kong, M.; Yang, Q.; Fei, J.; Zheng, X. Experimental study of Ni/MgO catalyst in carbon dioxide reforming of toluene, a model compound of tar from biomass gasification. *Int. J. Hydrogen Energy,* **2012**, *37*, 13355-13364.
[http://dx.doi.org/10.1016/j.ijhydene.2012.06.108]

[225]   Amin, R.; Liu, B.S.; Zhao, Y.C.; Huang, Z.B. Hydrogen production by corncob/CO$_2$ dry reforming over CeO$_2$ modified Ni-based MCM-22 catalysts. *Int. J. Hydrogen Energy,* **2016**, *41*, 12869-12879.
[http://dx.doi.org/10.1016/j.ijhydene.2016.05.233]

[226]   Garcia, L.; Salvador, M.L.; Arauzo, J.; Bilbao, R. CO$_2$ as a gasifying agent for gas production from pine sawdust at low temperatures using a Ni/Al coprecipitated catalyst. *Fuel Process. Technol.,* **2001**, *69*, 157-174.
[http://dx.doi.org/10.1016/S0378-3820(00)00138-7]

[227]   Simell, P.; Hepola, J.; Krause, O. Effects of gasification gas components on tar and ammonia decomposition over hot gas cleanup catalysts. *Fuel,* **1997**, *76*, 1117-1127.
[http://dx.doi.org/10.1016/S0016-2361(97)00109-9]

[228]   Laosiripojana, N.; Sutthisripok, W.; Charojrochkul, S.; Assabumrungrat, S. Development of Ni–Fe bimetallic based catalysts for biomass tar cracking/reforming: effects of catalyst support and co-fed

reactants on tar conversion characteristics. *Fuel Process. Technol.,* **2014**, *127*, 26-32.
[http://dx.doi.org/10.1016/j.fuproc.2014.06.015]

[229] Feng, D.; Zhao, Y.; Zhang, Y.; Sun, S. Effects of H_2O and CO_2 on the homogeneous conversion and heterogeneous reforming of biomass tar over biochar. *Int. J. Hydrogen Energy,* **2017**, *42*, 13070-13084.
[http://dx.doi.org/10.1016/j.ijhydene.2017.04.018]

[230] Feng, D.; Zhao, Y.; Zhang, Y.; Zhang, Z.; Xu, H.; Zhang, L.; Sun, S. Synergies and progressive effects of $H_2\backslash O/CO_2$ and nascent tar on biochar structure and reactivity during gasification. *Fuel Process. Technol.,* **2017**, *168*, 1-10.
[http://dx.doi.org/10.1016/j.fuproc.2017.08.030]

[231] Shen, Y.; Ma, D.; Ge, X. CO_2 looping in biomass pyrolysis or gasification. *Sustain. Energy Fuels,* **2017**, *1*, 1700-1729.
[http://dx.doi.org/10.1039/C7SE00279C]

[232] Tu, X.; Whitehead, J.C. Plasma-catalytic dry reforming of methane in an atmospheric dielectric barrier discharge: Understanding the synergistic effect at low temperature. *Appl. Catal. B,* **2012**, *125*, 439-448.
[http://dx.doi.org/10.1016/j.apcatb.2012.06.006]

[233] Yap, D.; Tatibouët, J.M.; Batiot-Dupeyrat, C. Catalyst assisted by non-thermal plasma in dry reforming of methane at low temperature. *Catal. Today,* **2018**, *299*, 263-271.
[http://dx.doi.org/10.1016/j.cattod.2017.07.020]

[234] Zhang, H.; Li, X.D.; Zhang, Y.Q.; Chen, T.; Yan, J.H.; Du, C.M. Rotating gliding arc codriven by magnetic field and tangential flow. *IEEE Trans. Plasma Sci.,* **2012**, *40*, 3493-3498.
[http://dx.doi.org/10.1109/TPS.2012.2220984]

[235] Mahammadunnisa, S.; Manoj Kumar Reddy, P.; Ramaraju, B.; Subrahmanyam, C. Catalytic nonthermal plasma reactor for dry reforming of methane. *Energy Fuels,* **2013**, *27*, 4441-4447.
[http://dx.doi.org/10.1021/ef302193e]

[236] Wang, Q.; Spasova, B.; Hessel, V.; Kolb, G. Methane reforming in a small-scaled plasma reactor - Industrial application of a plasma process from the viewpoint of the environmental profile. *Chem. Eng. J.,* **2015**, *262*, 766-774.
[http://dx.doi.org/10.1016/j.cej.2014.09.091]

[237] Fidalgo, B.; Domínguez, A.; Pis, J.J.; Menéndez, J.A. Microwave-assisted dry reforming of methane. *Int. J. Hydrogen Energy,* **2008**, *33*, 4337-4344.
[http://dx.doi.org/10.1016/j.ijhydene.2008.05.056]

[238] Fidalgo, B.; Menéndez, J.A. Study of energy consumption in a laboratory pilot plant for the microwave-assisted CO_2 reforming of CH_4. *Fuel Process. Technol.,* **2012**, *95*, 55-61.
[http://dx.doi.org/10.1016/j.fuproc.2011.11.012]

[239] Li, L.; Jiang, X.; Wang, H.; Wang, J.; Song, Z.; Zhao, X.; Ma, C. Methane dry and mixed reforming on the mixture of bio-char and nickel-based catalyst with microwave assistance. *J. Anal. Appl. Pyrolysis,* **2017**, *125*, 318-327.
[http://dx.doi.org/10.1016/j.jaap.2017.03.009]

[240] Tsodikov, M.V.; Konstantinov, G.I.; Chistyakov, A.V.; Arapova, O.V.; Perederii, M.A. Utilization of petroleum residues under microwave irradiation. *Chem. Eng. J.,* **2016**, *292*, 315-320.
[http://dx.doi.org/10.1016/j.cej.2016.02.028]

[241] Banville, M.; Labrecque, R.; Lavoie, J.M. Dry reforming of methane under an electro-catalytic bed: effect of electrical current and catalyst composition. *WIT Trans. Ecol. Environ.,* **2014**, *186*, 603-611.
[http://dx.doi.org/10.2495/ESUS140531]

[242] Yabe, T.; Mitarai, K.; Oshima, K.; Ogo, S.; Sekine, Y. Low-temperature dry reforming of methane to produce syngas in an electric field over La-doped Ni/ZrO_2 catalysts. *Fuel Process. Technol.,* **2017**,

158, 96-103.
[http://dx.doi.org/10.1016/j.fuproc.2016.11.013]

[243] Kodama, T.; Ohtake, H.; Shimizu, K.I.; Kitayama, Y. Nickel catalyst driven by direct light irradiation for solar CO_2-reforming of methane. *Energy Fuels,* **2002**, *16*, 1016-1023.
[http://dx.doi.org/10.1021/ef000226n]

[244] Welte, M.; Warren, K.; Scheffe, J.R.; Steinfeld, A. Combined ceria reduction and methane reforming in a solar-driven particle-transport reactor. *Ind. Eng. Chem. Res.,* **2017**, *56*(37), 10300-10308.
[http://dx.doi.org/10.1021/acs.iecr.7b02738] [PMID: 28966440]

[245] Wörner, A.; Tamme, R. CO_2 reforming of methane in a solar driven volumetric receiver–reactor. *Catal. Today,* **1998**, *46*, 165-174.
[http://dx.doi.org/10.1016/S0920-5861(98)00338-1]

[246] Dahl, J.K.; Weimer, A.W.; Lewandowski, A.; Bingham, C.; Bruetsch, F.; Steinfeld, A. Dry reforming of methane using a solar-thermal aerosol flow reactor. *Ind. Eng. Chem. Res.,* **2004**, *43*, 5489-5495.
[http://dx.doi.org/10.1021/ie030307h]

[247] Adanez, J.; Abad, A.; Garcia-Labiano, F.; Gayan, P.; De Diego, L.F. Progress in chemical-looping combustion and reforming technologies. *Pror. Energy Combust. Sci.,* **2012**, *38*, 215-282.
[http://dx.doi.org/10.1016/j.pecs.2011.09.001]

[248] Lewis, W.K.; Gilliland, E.R.; Sweeney, M.P. Gasification of carbon. Metal oxides in a fluidized powder bed. *Chem. Eng. Prog.,* **1951**, *47*, 251-256.

[249] Rydén, M.; Moldenhauer, P.; Mattisson, T.; Lyngfelt, A.; Younes, M.; Niass, T.; Fadhel, B.; Ballaguet, J.P. Chemical-looping combustion with liquid fuels. *Energy Procedia,* **2013**, *37*, 654-661.
[http://dx.doi.org/10.1016/j.egypro.2013.05.153]

[250] Mattisson, T.; Lyngfelt, A. Applications of chemical-looping combustion with capture of CO_2. *Control,* **2001**, 46-51.

[251] Rydén, M.; Lyngfelt, A.; Mattisson, T.; Chen, D.; Holmen, A.; Bjorgum, E. Novel oxygen-carrier materials for chemical-looping combustion and chemical-looping reforming $La_xSr_{1-x}FeyCo_{1-y}O_{3-\delta}$ perovskites and mixed-metal oxides of NiO, Fe_2O_3 and Mn_3O_4. *Int. J. Greenh. Gas Control,* **2008**, *2*, 21-36.
[http://dx.doi.org/10.1016/S1750-5836(07)00107-7]

[252] Ortiz, M.; Abad, A.; De Diego, L.F.; Gayán, P.; García-Labiano, F.; Adánez, J. Optimization of a chemical-looping auto-thermal reforming system working with a Ni-based oxygen-carrier. *Energy Procedia,* **2011**, *4*, 425-432.
[http://dx.doi.org/10.1016/j.egypro.2011.01.071]

[253] Gayán, P.; de Diego, L.F.; García-Labiano, F.; Adánez, J.; Abad, A.; Dueso, C. Effect of support on reactivity and selectivity of Ni-based oxygen carriers for chemical-looping combustion. *Fuel,* **2008**, *87*, 2641-2650.
[http://dx.doi.org/10.1016/j.fuel.2008.02.016]

[254] Cho, P.; Mattisson, T.; Lyngfelt, A. Comparison of iron-, nickel-, copper- and manganese-based oxygen carriers for chemical-looping combustion. *Fuel,* **2004**, *83*, 1215-1225.
[http://dx.doi.org/10.1016/j.fuel.2003.11.013]

[255] Zafar, Q.; Mattisson, T.; Gevert, B. Integrated hydrogen and power production with CO_2 capture using chemical-looping reforming-redox reactivity of particles of CuO, Mn_2O_3, NiO, and Fe_2O_3 using SiO_2 as a support. *Ind. Eng. Chem. Res.,* **2005**, *44*, 3485-3496.
[http://dx.doi.org/10.1021/ie048978i]

[256] Abad, A.; Adánez, J.; García-Labiano, F.; de Diego, L.F.; Gayán, P.; Celaya, J. Mapping of the range of operational conditions for Cu, Fe, and Ni-based oxygen carriers in chemical-looping combustion. *Chem. Eng. Sci.,* **2007**, *62*, 533-549.
[http://dx.doi.org/10.1016/j.ces.2006.09.019]

[257] Cho, P.; Mattisson, T.; Lyngfelt, A. Carbon formation on nickel and iron oxide-containing oxygen carriers for chemical-looping combustion. *Ind. Eng. Chem. Res.,* **2005**, *44*, 668-676.
[http://dx.doi.org/10.1021/ie049420d]

[258] Jerndal, E.; Mattisson, T.; Lyngfelt, A. Thermal analysis of chemical-looping combustion. *Chem. Eng. Res. Des.,* **2006**, *84*, 795-806.
[http://dx.doi.org/10.1205/cherd05020]

[259] Ferreira-Aparicio, P.; Rodríguez-Ramos, I.; Guerrero-Ruiz, A. On the applicability of membrane technology to the catalysed dry reforming of methane. *Appl. Catal. A Gen.,* **2002**, *237*, 239-252.
[http://dx.doi.org/10.1016/S0926-860X(02)00337-X]

[260] Kim, S.; Ryi, S-K.; Lim, H. Techno-economic analysis (TEA) for CO_2 reforming of methane in a membrane reactor for simultaneous CO_2 utilization and ultra-pure H_2 production. *Int. J. Hydrogen Energy,* **2018**, *43*, 5881-5893.
[http://dx.doi.org/10.1016/j.ijhydene.2017.09.084]

[261] Sumrunronnasak, S.; Tantayanon, S.; Kiatgamolchai, S.; Sukonket, T. Improved hydrogen production from dry reforming reaction using a catalytic packed-bed membrane reactor with Ni-based catalyst and dense PdAgCu alloy membrane. *Int. J. Hydrogen Energy,* **2016**, *41*, 2621-2630.
[http://dx.doi.org/10.1016/j.ijhydene.2015.10.129]

[262] Acha, E.; van Delft, Y.C.; Cambra, J.F.; Arias, P.L. Thin PdCu membrane for hydrogen purification from in-situ produced methane reforming complex mixtures containing H_2S. *Chem. Eng. Sci.,* **2018**, *176*, 429-438.
[http://dx.doi.org/10.1016/j.ces.2017.11.019]

[263] Raybold, T.M.; Huff, M.C. Analyzing enhancement of CO_2 reforming of CH_4 in Pd membrane reactors. *AIChE J.,* **2002**, *48*, 1051-1061.
[http://dx.doi.org/10.1002/aic.690480514]

[264] Haag, S.; Burgard, M.; Ernst, B. Beneficial effects of the use of a nickel membrane reactor for the dry reforming of methane: comparison with thermodynamic predictions. *J. Catal.,* **2007**, *252*, 190-204.
[http://dx.doi.org/10.1016/j.jcat.2007.09.022]

[265] Tsodikov, M.V.; Fedotov, A.S.; Antonov, D.O.; Uvarov, V.I.; Bychkov, V.Y.; Luck, F.C. Hydrogen and syngas production by dry reforming of fermentation products on porous ceramic membrane-catalytic converters. *Int. J. Hydrogen Energy,* **2016**, *41*, 2424-2431.
[http://dx.doi.org/10.1016/j.ijhydene.2015.11.113]

CHAPTER 4

Nico and NiCu Based-Materials for Hydrogen Production and Electro-oxidation Reactions

Moisés R. Cesario[1,*], Daniel A. Macedo[2,*], Glageane S. Souza[2], Francisco J.A. Loureiro[3], Haingomalala L. Tidahy[1], Cédric Gennequin[1,*] and Edmond Abi-Aad[1]

[1] *Unité de Chimie Environnementale et Interactions sur le Vivant (UCEIV, E.A. 4492), MREI, Université du Littoral Côte d'Ople (ULCO), 59140, Dunkerque, France*

[2] *Materials Science and Engineering Postgraduate Program, Federal University of Paraiba (UFPB), 58051-900, João Pessoa, Brazil*

[3] *Centre of Mechanical Technology and Automation (TEMA), Department of Mechanical Engineering, University of Aveiro, 3810-193, Aveiro, Portugal*

Abstract: $Ni_{1-x}Co_xO$-CGO, $Ni_{1-x}Cu_xO$-CGO, CuOCGO and NiO-CGO composites powders were obtained by a one-step synthesis method and their catalytic activities in dry reforming of methane (CH_4+CO_2) for hydrogen production were evaluated. X-ray diffraction (XRD) analysis of composite powders confirmed NiO, Co_3O_4, CuO and $Ce_{0.9}Gd_{0.1}O_{1.95}$ cubic phases. XRD results also showed the formation of nanocrystalline materials. Cobalt-based materials showed higher surface area values (S_{BET}) than copper-based materials. According to temperature programmed reduction (TPR) analyses, $Ni_{0.6}Co_{0.4}O$-CGO ($NiCo_{0.4}$) and $Ni_{0.4}Co_{0.6}O$-CGO ($NiCo_{0.6}$) catalysts have higher reduction capacity and stronger metal/support interaction than $Ni_{0.6}Cu_{0.4}O$-CGO ($NiCu_{0.4}$) and $Ni_{0.4}Cu_{0.6}O$-CGO ($NiCu_{0.6}$) materials. Rietveld refinement analyses for $NiCo_{0.4}$, $NiCo_{0.6}$ and $NiCu_{0.4}$ reduced catalysts, confirmed the presence of Ni-Co and Ni-Cu alloys. These factors are important for enhanced catalytic activity avoiding carbon deposition. $NiCo_{0.4}$ and $NiCo_{0.6}$ catalysts had higher conversions of CH_4 and CO_2 than Ni-CGO and Cu-based catalysts. NiCo-based catalyst showed a better resistance to carbon deposition. $NiCo_{0.4}$ had high H_2/CO ratio and the best reaction selectivity below 600 °C. The electrocatalytic activity of $NiCo_{0.4}$/CGO/$NiCo_{0.4}$ screen-printed symmetrical cells from hydrogen and synthetic biogas (CH_4+CO_2) electro-oxidation reaction was studied by impedance spectroscopy in the temperature range between 650 and 850 °C. The polarization resistance was influenced by the atmosphere conditions. Total polarization resistances (R_p) of 0.96 and 36.10 Ω cm^2 were obtained at 750 °C for

* **Corresponding authors Moisés Cesario, Cédric Gennequin:** Unité de Chimie Environnementale et Interactions sur le Vivant (UCEIV, E.A. 4492), MREI, Université du Littoral Côte d'Opale (ULCO), Dunkerque, France, Tel: +33(0)781559594, +55(84)96397875; E-mails: moises.cesario@univ-littoral.fr, moisesrcesario@gmail.com, cedric.gennequin@univ-littoral.fr

Daniel Macedo: Materials Science and Engineering Postgraduate Program, Federal University of Paraiba (UFPB), 58051-900 João Pessoa, Brazil, Tel: +33328658261; E-mail: damaced@gmail.com

H$_2$ and biogas atmospheres, respectively. The activation energy (E_a) was lower when H$_2$ was used (0.92 eV). The hydrogen oxidation reaction occurs more easily than the dry reforming of methane.

Keywords: Alloy, Carbon Deposition, Dry Reforming of Methane, One Step Synthesis, Electro-Oxidation Reaction, Hydrogen, Impedance Spectroscopy Surface Area, Metal/Support Interaction, Reduction Capacity.

INTRODUCTION

The ecological problems generated by the consumption and production of fossil fuels reinforce the need for new sustainable means of energy production. Hydrogen production becomes an interesting alternative for the industry given the availability of raw materials and techniques. However, the production of hydrogen still lacks processes and materials optimized to become a well-established technology and widely diffused in the society's daily life. This chapter aims at studying new catalytic materials for H$_2$ production processes and electro-oxidation reactions. Hydrogen is a fuel of great interest worldwide that can be used in electrochemical energy conversion devices such as fuel cells. Among the various types of fuel cells, solid oxide fuel cells (SOFCs) stand out as promising because they allow generating high electric power and are environmental friendly.

A typical SOFC is composed mainly of two porous electrodes (anode and cathode) separated by a dense electrolyte membrane. The cathode (a mixed ionic electronic conductor) is usually a perovskite material on which an oxidant gas is reduced to oxygen ions; the electrolyte is an ionic conductor which transports oxygen ions from the cathode to the anode; the anode material may be a cermet (ceramic – metal composite) where hydrogen is oxidized to combine with O^{2-} forming water. The released electrons flow through an external circuit to the cathode side to complete the electrical circuit. These electrochemical (reduction/oxidation) reactions in the overall cell occur mainly in the triple phase boundary (TPB) area between gas, electrolyte and electrode interfaces [1, 2].

Research efforts on SOFCs seek to improve the performance of material's components and long-term microstructural stability. High electrical conductivity, chemical stability, coefficient of thermal expansion matching that of the electrolyte, high catalytic activity to avoid potential losses and adequate porosity (20-40%) are desired properties for the electrode materials [3 - 6]. The chemical composition, microstructure and manufacturing technique are the main factors influencing the electrode performance on both fuel and oxidant atmospheres. This reinforces the importance of developing novel chemical approaches to obtain nanocomposites with improved properties.

Nickel/gadolinium-doped ceria (Ni-CGO) composites are among the best anode materials for electro-oxidation reactions in SOFCs operating at 600-750 °C. This material allows the direct oxidation of gases such as hydrocarbons [5, 6]. One drawback in using hydrocarbon-based fuels is the susceptibility to carbon deposition which decreases electrocatalytic performance or definitively deactivates the anode [7, 8].

The dry reforming of methane or biogas ($CH_4 + CO_2$) reforming into the anodic compartment has been a strategic way for hydrogen production using nickel-based materials [9 - 12].

The Dry Reforming of Methane (DRM) is a reaction between two greenhouse gases (CH_4 and CO_2) for the production of syngas. The syngas derived from the DRM reaction (Equation 1) can be used to develop synthetic fuels by Fischer-Tropsch process or used to obtain hydrogen by further purification processes. The DRM reaction is highly endothermic and thermodynamically favored by high temperature and low pressures, which requires stable and active catalysts to induce a high conversion rate.

$$CH_4 + CO_2 \leftrightarrow 2CO + 2H_2 \quad \Delta H_{298K} = 247 \; KJmol^{-1} \tag{1}$$

Catalysts based on noble metals such as Pt, Rh, Ru, and Ir have high performance for DRM and resistance to carbon deposition [13, 14]. However, the high cost and limited availability of these metals are factors that hinder their applications in DRM.

Nickel has been widely used as the catalyst in DRM due to its low cost, high catalytic activity and selectivity. However, Ni catalysts have a higher susceptibility to deactivation by carbon deposition and sintering [15, 16]. As an alternative to Ni-based materials, cobalt-based catalysts can improve catalytic stability and coke resistance by adjusting the Co content, which probably results from their high affinity for oxygen species [17 - 19]. Ni-Co bimetallic catalysts may be attractive electrodes for fuel electro-oxidation because they combine high catalytic activity, stability and improved resistance to carbon deposition.

In recent years, researchers have studied metal supported catalysts to inhibit their deactivation. The choice of a suitable support can also influence the dispersion of the catalyst and its resistance to carbon deposition. A support with good oxygen storage capacity and surface basicity allows better metal-support interaction [20, 21]. CeO_2 is an interesting candidate as a catalyst support due to its high mobility of oxygen ions.

In this context, this chapter focused on the study of bimetallic materials based on nickel, copper and cobalt synthetized by a one-step polymeric precursor-based method, and further investigation of their catalytic behavior over dry reforming of methane as well as their electrochemical performance as anodes of SOFCs.

Preparation of Ni-Based Materials

One Step Synthesis Method

NiO-CGO, $Ni_{1-x}Co_xO$-CGO and $Ni_{1-x}Cu_xO$-CGO composites (x = 0, 0.2, 0.6, 0.8) with 50 wt.% CGO were obtained by a one-step polymeric precursor-based method. This synthesis approach involves a single thermal treatment to obtain an *in situ* nanocomposite powder derived from the mixture of polymer precursor resins of NiO, CuO, Co_3O_4, and CGO phases [22].

Polymer resins were synthesized simultaneously using separate beakers. A typical experiment starts with the dissolution of citric acid $[C_6H_8O_7.H_2O]$ into distilled water at 50 °C. For the synthesis of the CGO phase (beaker I), a stoichiometric amount of cerium nitrate $[Ce(NO_3)_3.6H_2O]$ was dissolved into the citric acid solution. A cation/citric acid molar ratio of 1/3.5 was used. After 1 h under stirring (@) at 60-70 °C, gadolinium nitrate $[Gd(NO_3)_3.6H_2O]$ was added and the resulting solution remained at this temperature for 2 h to form the metal chelate. Then, ethylene glycol $[C_2H_6O_2]$ was added to the solution in a mass ratio of 60:40 (citric acid/ethylene glycol) in order to enable the esterification reaction. The reactional temperature was increased to 75-80 ° C under constant stirring. For the removal of excess water and formation of the CGO precursor resin, the solution remained for a further 2 h in the range of 75-80 ° C with constant stirring.

In beaker II, the precursor resins of NiO, CuO and Co_3O_4 phases were prepared using nickel nitrate $[Ni(NO_3)_2. 6H_2O]$, copper nitrate $[Cu(NO_3)_2. 3H_2O]$ and cobalt nitrate $[Co(NO_3)_2. 6H_2O]$ as starting materials. The metal/citric acid and citric acid/ethylene glycol ratios were similar to those used to prepare the CGO precursor resin. Table **1** shows the nomenclature adopted for the prepared samples.

Table 1. Nomenclature adopted for the prepared samples.

Samples	Nomenclature
$NiO.Ce_{0.9}Gd_{0.1}O_{1.95}$	NiCGO
$Ni_{0.8}Co_{0.2}O.Ce_{0.9}Gd_{0.1}O_{1.95}$	$NiCo_{0.2}$
$Ni_{0.6}Co_{0.4}O.Ce_{0.9}Gd_{0.1}O_{1.95}$	$NiCo_{0.4}$
$Ni_{0.4}Co_{0.6}O.Ce_{0.9}Gd_{0.1}O_{1.95}$	$NiCo_{0.6}$

(Table 1) cont.....

Samples	Nomenclature
$Ni_{0.8}Cu_{0.2}O.Ce_{0.9}Gd_{0.1}O_{1.95}$	$NiCu_{0.2}$
$Ni_{0.6}Cu_{0.4}O.Ce_{0.9}Gd_{0.1}O_{1.95}$	$NiCu_{0.4}$
$Ni_{0.4}Cu_{0.6}O.Ce_{0.9}Gd_{0.1}O_{1.95}$	$NiCu_{0.6}$
$CuO.Ce_{0.9}Gd_{0.1}O_{1.95}$	CuCGO

The precursor resins in beakers I and II were mixed together at room temperature for 5 min aiming to obtain a homogeneous solution containing cerium, gadolinium, nickel, copper, and cobalt (NiO, CuO and Co_3O_4) cations.

The resulting resins were heat-treated at 350 °C under air for 1 h using a heating rate of 1 °C min^{-1} to produce precursor powders that were further calcined at 700 °C for 1 h using a heating rate of 3 °C min^{-1}. Fig. (**1**) shows the methodology for obtaining NiO-CGO or NiCu-CGO or NiCo-CGO composite powders.

Thermal, Structural and Textural Study

Precursor powders previously calcined at 350 ° C were ground with agate mortar and thermally analyzed by differential scanning calorimetry (DSC) and thermogravimetry (TG) using a Netzsch STA apparatus. TG-DTA analyses were performed in air from room temperature to 1200 °C with heating rate of 5 °C min^{-1}.

Powders calcined at 700°C were characterized by X-ray diffraction (XRD) using a Shimadzu XDR-7000 diffractometer. Diffraction patterns were obtained in the angular range of 10–90° using step-scanning mode (0.02°/step) with a counting time of 2 s/step.

The values of specific surface area were acquired by using the BET method (Brunauer, Emmett and Teller) *via* N_2 adsorption-desorption at -196 °C in a Quanta Sorb Junior analyzer. The catalysts were degassed for 30 min at 150 °C before being analyzed.

Temperature-programmed reduction (TPR) was carried out on an Altamira AMI 200 apparatus to evaluate the reducibility of the catalytic materials and the temperatures at which it takes place. A mass of 10-20 mg was placed in a quartz U-tube (4 mm internal diameter) and pretreated under Ar (30 mL min^{-1}) from room temperature to 150 °C for 30 min. The samples were cooled and then heated to 900 °C (using a heating rate of 5 °C min^{-1}) in a mixture of 95% argon and 5% hydrogen under a total gas flow of 30 mL min^{-1}. A thermal conductivity detector (TCD) allowed the quantitative determination of hydrogen consumption.

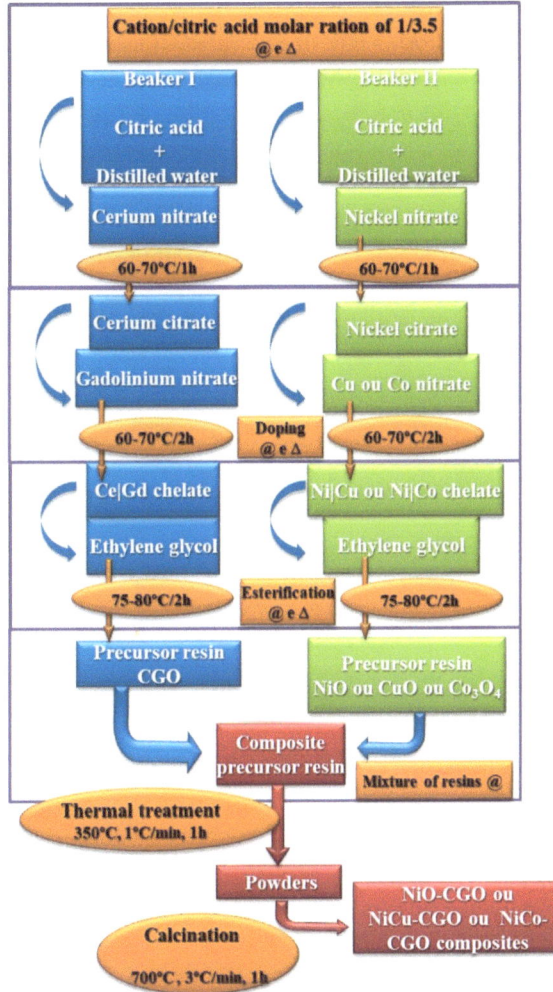

Fig. (1). Methodology used for obtaining NiO-CGO, NiCo-CGO and NiCu-CGO composite powders.

Reduced catalysts were characterized by X-ray diffraction (XRD) using a Bruker D8-Advance diffractometer, with Cu Kα radiation (λ=1.5406 Å), at room temperature. Diffraction powder patterns were obtained in the angular range of 10–90° using step-scanning mode (0.02°/step) with a counting time of 2 s/step. The lattice parameters were determined by Rietveld refinement analysis of the XRD data, which was performed using Materials Analysis Using Diffraction (MAUD) software (version 2.71).

Fig. (**2a**) shows TG/DSC profiles of the pre-calcined NiCGO precursor powder at 350 °C. TG/DSC profiles present a discrete endothermic behavior in the temperature range 100-200 °C, with weight loss of approximately 6 wt.% attributed to water evaporation. The most intense exothermic phenomenon is observed around 350 °C with a weight loss of 13 wt.% which can be attributed to the oxidation of organic matter (citric acid and ethylene glycol) and crystallization of CGO and NiO phases [23, 24]. The weight loss stabilizes around 550 °C, with total loss of around 21 wt.%. TG/DSC analysis allowed to choice the calcination temperature for crystallization of the desired phases (NiO and CGO). For the CuCGO composite, the weight loss behavior consists of two stages (Fig. **2b**): the first one near 180 °C can be associated with dehydration, with weight loss of 6 wt.%; and the second one (at 350-600 °C) with 30 wt.% of weight loss can be related to the degradation of organic matter from chemicals used in the synthesis (citric acid and ethylene glycol) [25, 26]. A total weight loss of 36 wt.% is stabilized around 600 °C.

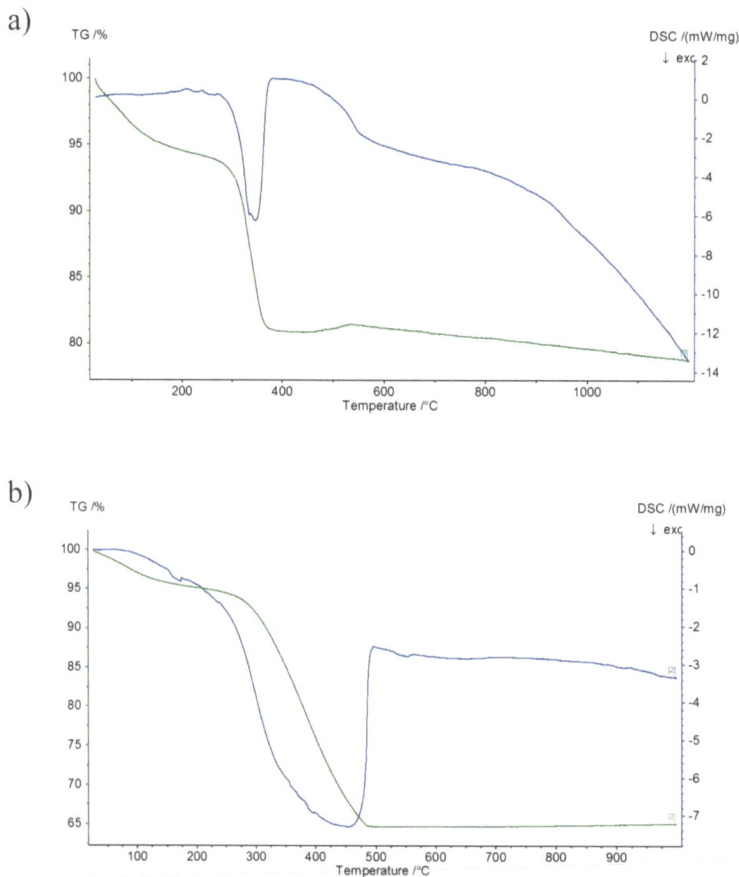

Fig. (2). TG/DSC profiles of the precursor powders: **(a)** NiCGO and **(b)** CuCGO.

X-ray diffraction patterns of calcined $Ni_{1-x}Cu_xO$-CGO and $Ni_{1-x}Co_xO$-CGO composite powders are shown in Fig. (**3**). CGO, NiO and CuO phases were identified using JCPDS file numbers 75-0161 (CGO, fluorite-type structure, a = 5.4180), 78-0429 (NiO, NaCl-type, a = 4.1771 Å), and 78-0428 (CuO, tenorite-type structure, a = 4.2450 Å), respectively. For cobalt-based samples, Co_3O_4 was identified using JCPDS file n° 43-1003 (spinel-type, a = 8.0840 Å).

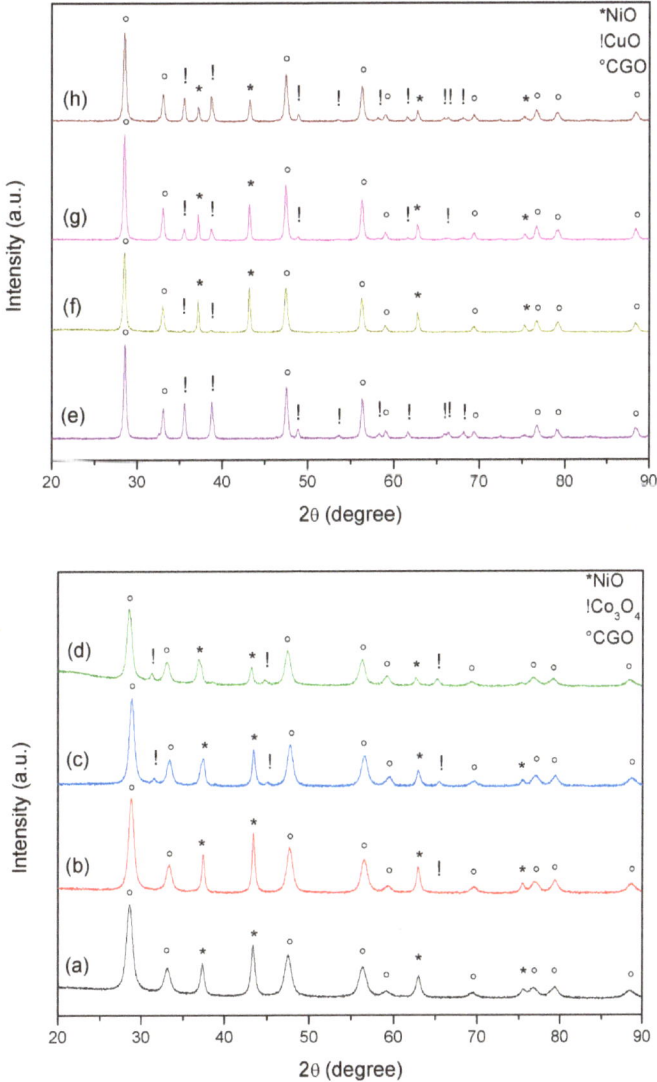

Fig. (3). X-ray diffraction patterns of composite powders: (**a**) NiCGO, (**b**) $NiCo_{0.2}$, (**c**) $NiCo_{0.4}$, (**d**) $NiCo_{0.6}$, (**e**) CuCGO, (**f**) $NiCu_{0.2}$, (**g**) $NiCu_{0.4}$ and (**h**) $NiCu_{0.6}$.

Table **2** presents the specific surface area values (S_{BET}) of calcined powders. Cobalt-based catalysts have higher S_{BET} values than copper-based ones, which can suggest a better dispersion of the metal phase.

Table 2. Specific surface areas of calcined powders before reforming reaction.

Catalysts	BET Surface Area (m² g⁻¹)
NiCGO	20.8
$NiCo_{0.2}$	18.0
$NiCo_{0.4}$	16.9
$NiCo_{0.6}$	18.5
CuCGO	7.2
$NiCu_{0.2}$	11.6
$NiCu_{0.4}$	10.9
$NiCu_{0.6}$	8.4

Fig. (**4**) shows the TPR profile of the NiCGO catalyst. The NiCGO composite showed a peak that can be associated with the NiO reduction to metallic nickel (Equation 2) or overlapping of two peaks, indicating a possible interaction of NiO with the CGO support [27].

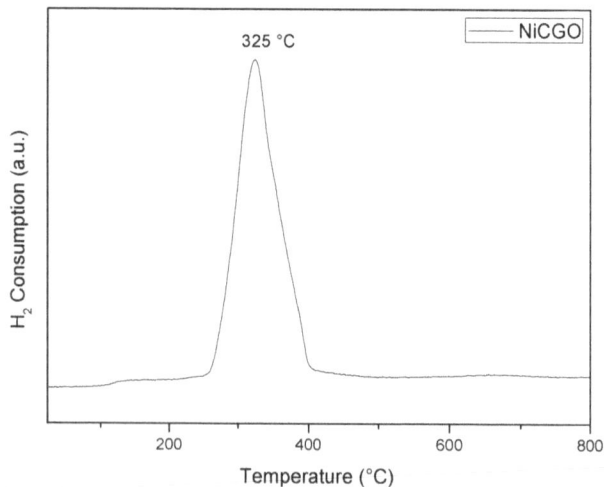

$$NiO + H_2 \rightarrow Ni^0 + H_2O \tag{2}$$

Fig. (4). Temperature-programmed reduction (TPR) profile of NiCGO.

Fig. (**5**) shows the H_2 consumption profile obtained for the CuCGO composite. A single peak is centered at 210 °C which can be associated with the CuO reduction to Cu° (Equation 3) [28, 29]. $NiCu_{0.2}$ bimetallic composite profile shows three well-defined peaks centered at: 180 °C, 290 °C and 462 °C (Fig. **6**). The first reduction peak can be associated with the reduction of CuO to Cu°, while the second peak would be related to the reduction of NiO to Ni at lower temperature in comparison to the reduction profile of the NiCGO sample, due to the interaction between CuO and NiO particles [28]. The third peak of H_2 consumption can be related to the strong interaction of NiO with CGO [28].

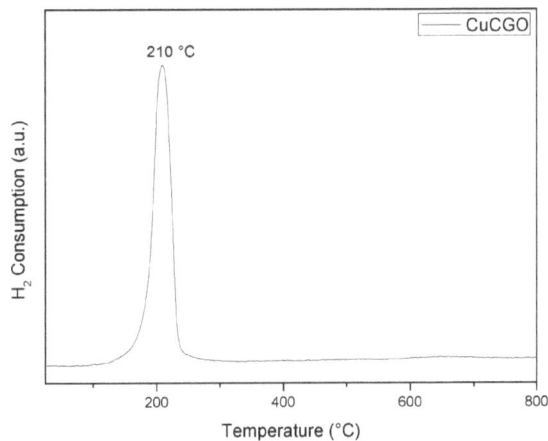

$$CuO + H_2 \rightarrow Cu^0 + H_2O \qquad (3)$$

Fig. (5). Temperature-programmed reduction (TPR) profile of CuCGO.

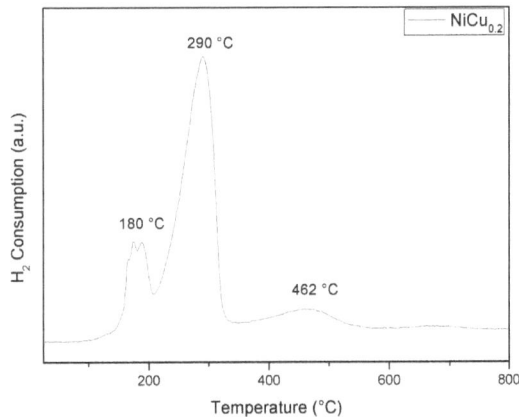

Fig. (6). Temperature-programmed reduction (TPR) profile of $NiCu_{0.2}$.

Figs. (**7** and **8**) show reduction profiles of $NiCu_{0.4}$ and $NiCu_{0.6}$ catalysts, respectively. These profiles exhibit two events: The first one is the reduction peak of CuO, more intense, that has shifted to higher temperatures due to increase of the availability of the reducible species; the second one is characterized by the shift of the less intense NiO reduction peak to lower temperatures due to the influence of the high copper reducibility on the nickel reduction kinetics [28].

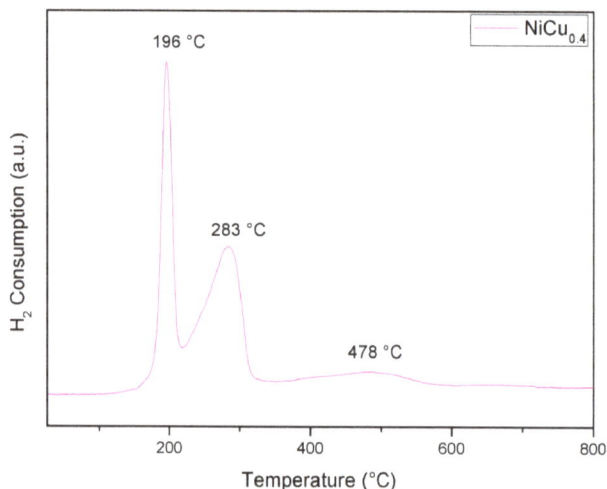

Fig. (7). Temperature-programmed reduction (TPR) profile of $NiCu_{0.4}$.

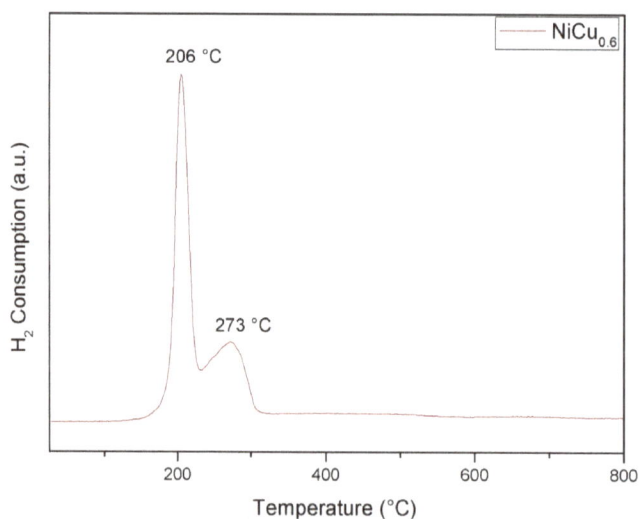

Fig. (8). Temperature-programmed reduction (TPR) profile of $NiCu_{0.6}$.

TPR profiles of the bimetallic catalysts $NiCo_{0.2}$, $NiCo_{0.4}$ and $NiCo_{0.6}$ are shown in Figs. (**9-11**), respectively. The reduction profile of $NiCo_{0.2}$ presented a discrete peak at 198 °C associated with partial reduction of Co_3O_4 (Equation 4) and another asymmetric peak with one shoulder in the temperature range 300-350 ° C due to the reduction of Co^{+2} to Co^0 (Equation 5). The maximum consumption signal of H_2 at 363 °C, related to the reduction of NiO to metallic nickel has also been observed [30, 31].

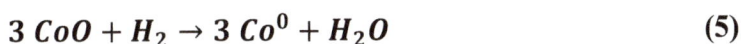

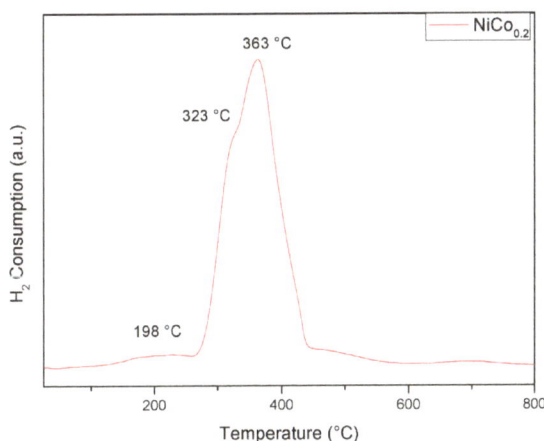

$$Co_3O_4 + H_2 \rightarrow 3\ CoO + H_2O \qquad\qquad (4)$$

$$3\ CoO + H_2 \rightarrow 3\ Co^0 + H_2O \qquad\qquad (5)$$

Fig. (9). Temperature-programmed reduction (TPR) profile of $NiCo_{0.2}$.

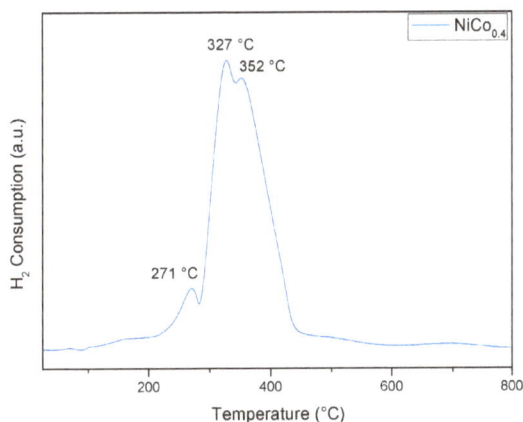

Fig. (10). Temperature-programmed reduction (TPR) profile of $NiCo_{0.4}$.

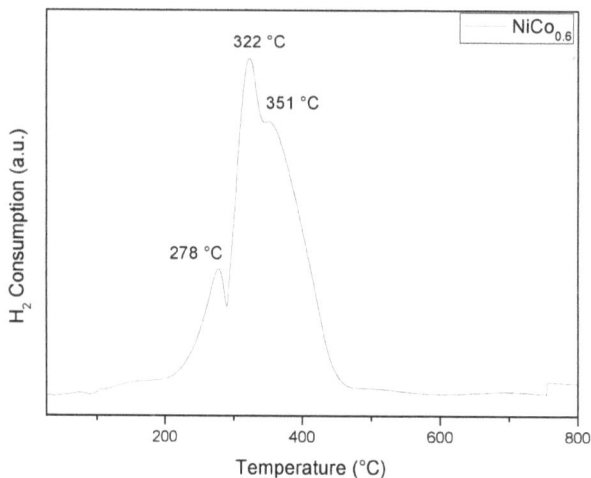

Fig. (11). Temperature-programmed reduction (TPR) profile of $NiCo_{0.6}$.

For the other reduction profiles of cobalt-based catalysts, an increase in the peak area associated with the Co_3O_4 reduction and its shift to higher temperatures as cobalt content increases were observed. The second peak associated to reduction of CoO to Co^0 remained practically the same temperature [31]. In addition, the shift of the NiO reduction peak to higher temperatures (350 - 370 °C) has been observed.

Comparing cobalt and copper-based catalysts, Ni-Co catalysts exhibited stronger catalyst/support interaction than Ni-Cu catalysts. The reason for this different metal/support interaction is related to the shift of reduction peaks to higher temperatures. In addition, cobalt-based catalysts showed higher phase's reduction capacity than copper-based catalysts, which may lead to higher sintering resistance and better catalytic activity. In fact, the total amount of H_2 consumption of the Ni-Co catalysts was higher than that of Ni-Cu catalysts, as shown in Table **3**.

Table 3. H_2 total consumption of the catalysts.

Catalysts	H_2 Total Consumption ($\mu mol\ g^{-1}$)
NiCGO	8012
$NiCo_{0.2}$	8326
$NiCo_{0.4}$	7943
$NiCo_{0.6}$	9416

(Table 3) cont.....

Catalysts	H_2 Total Consumption ($\mu mol\ g^{-1}$)
CuCGO	7010
$NiCu_{0.2}$	7091
$NiCu_{0.4}$	7325
$NiCu_{0.6}$	8011

Fig. (**12**) shows the X-ray diffraction patterns from the Rietveld refinement of $NiCo_{0.4}$ and $NiCu_{0.4}$ catalysts after a reduction test. X-ray diffraction patterns of $NiCo_{0.4}$ and $NiCu_{0.4}$ catalysts indicated the presence of Ni-Co and Ni-Cu alloys, respectively.

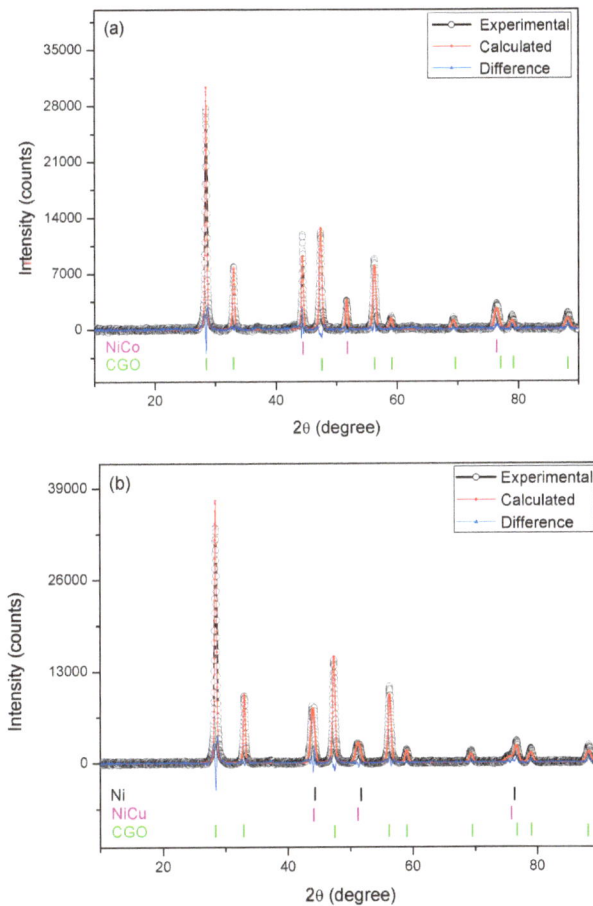

Fig. (12). X-ray diffraction patterns from Rietveld refinement of (**a**) $NiCo_{0.4}$ and (**b**) $NiCu_{0.4}$ catalysts upon reduction.

Crystallographic parameters obtained from Rietveld refinement are presented in Tables **4** and **5** The Rietveld refinement analyses for all catalysts confirm the crystalline structure and lattice parameters are in good agreement with JCPDS data.

Table 4. Crystallographic parameters and refinement indexes for Ni-CGO and NiCo-based catalysts upon reduction, evaluated by Rietveld refinement.

Parameters	NiCGO		NiCo$_{0.2}$				NiCo$_{0.4}$		NiCo$_{0.6}$	
	Ni	CGO	NiO	Ni	Co	CGO	NiCo	CGO	NiCo	CGO
a = b = c (Å)	3.5265	5.4250	4.1771	3.5267	3.5441	5.4203	3.5332	5.4228	3.5379	5.4245
R_{wp} (%)	38.15		42.43				43.72		45.80	
R_{exp} (%)	3.62		3.93				4.10		4.28	
χ^2	10.54		10.80				10.66		10.70	

Table 5. Crystallographic parameters and refinement indexes for NiCu-based catalysts upon reduction, evaluated by Rietveld refinement.

Parameters	CuCGO			NiCu$_{0.2}$			NiCu$_{0.4}$			NiCu$_{0.6}$		
	CuO	Cu	CGO	Ni	Cu	CGO	Ni	NiCu	CGO	Ni	Cu	CGO
a = b = c (Å)	4.245	3.6173	5.4212	3.5411	3.6078	5.4267	3.5276	3.5627	5.4212	3.5524	3.6113	5.4263
R_{wp} (%)	40.30			41.48			39.48			39.54		
R_{exp} (%)	3.56			3.69			3.68			3.66		
χ^2	11.32			11.24			10.73			10.80		

Catalytic Performance

The experimental tests of carbon dioxide reforming of methane were carried out in the temperature range 400-800 °C in atmospheric pressure. The powders were pelletized with 350 and 800 μm sizes to avoid mass and heat transfer limitations and eventual reactor clogging due to carbon deposition [32].

A mass of 100 mg of catalyst was placed in the U-shaped fixed bed quartz reactor. It was reduced *in situ* with H$_2$ (30mL min^{-1}) for 1 h at 800 °C and then purged with N$_2$.

The reactor was fed with a gas mixture consisting of CH$_4$ and CO$_2$ (molar ratio 1:1) diluted in Ar. The total flow was 100 mL min^{-1}. The reaction products were analyzed by Varian 3800 gas chromatography with TCD detection.

The conversions of CH_4 (Equation 6) and CO_2 (Equation 7), selectivity of H_2 (Equation 8) and CO (Equation 9), H_2/CO (Equation 10), H_2 yield (Equation 11) and CO (Equation 12) and carbon balance (Equation 13) were calculated as follows:

$$(CH_4\ Conversion) = \frac{(CH_4,in - CH_4,out)}{CH_4,in} \times 100 \tag{6}$$

$$(CO_2\ Conversion) = \frac{(CO_2,in - CO_2,out)}{CO_2,in} \times 100 \tag{7}$$

$$(H_2\ Selectivity) = \frac{(H_2,out)}{2 \times (CH_4,in - CH_4,out)} \times 100 \tag{8}$$

$$(CO\ Selectivity) = \frac{(CO,out)}{(CH_4,in + CO_2,in) - (CH_4,out + CO_2,out)} \times 100 \tag{9}$$

$$\frac{H_2}{CO} = \frac{S_{H_2}}{S_{CO}} \tag{10}$$

$$(H_2\ Yield) = \frac{(H_2,out)}{2 \times (CH_4,in)} \times 100 \tag{11}$$

$$(CO\ Yield) = \frac{(H_2,out)}{2 \times (CH_4,in)} \times 100 \tag{12}$$

$$(Carbon\ balance) = \frac{(CO_2,out + CH_4,out + CO,out)}{(CO_2,in + CH_4,in)} \times 100 \tag{13}$$

After the catalytic tests the catalysts were characterized by XRD and thermogravimetry/differential scanning calorimetry (TG-DSC). TG-DSC analyses were performed in air (75 ml min⁻¹) from room temperature to 1000 °C with heating rate of 5 °C min⁻¹ using a Netzsch STA 409 apparatus.

The average crystallite size of metallic nickel particles was calculated from the broadening of the main diffraction rays using the Scherrer equation (Equation 14) for the catalysts which underwent a reduction or the catalytic test.

$$L = \frac{0.9\lambda}{\beta \cos\theta} \tag{14}$$

Where L is the crystallite size, λ is the X-ray wavelength, β is the line broadening and θ is the Bragg angle.

Fig. (**13**) shows CH_4 and CO_2 conversions, H_2 and CO selectivity, H_2 and CO yield, carbon balance, and C_2H_4 concentration during the DRM reaction.

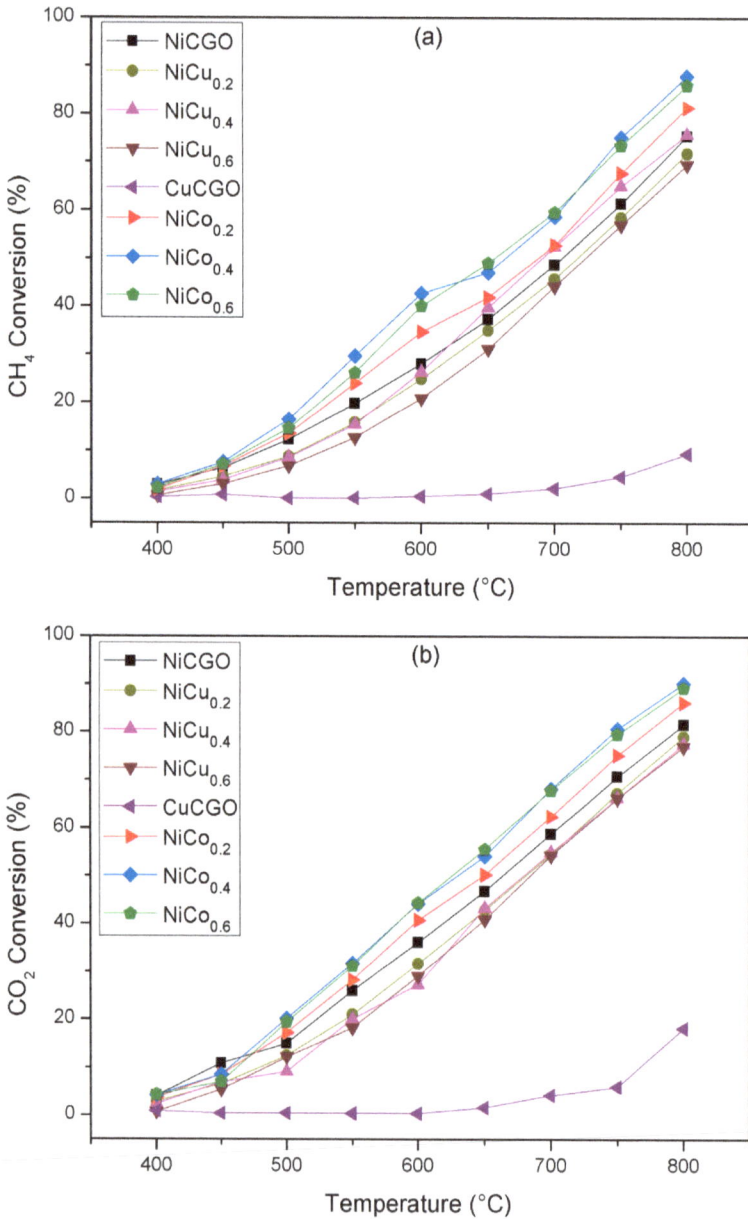

Fig. 13 cont.....

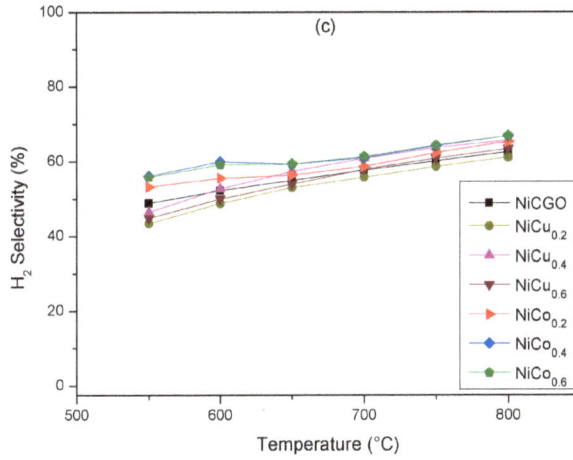

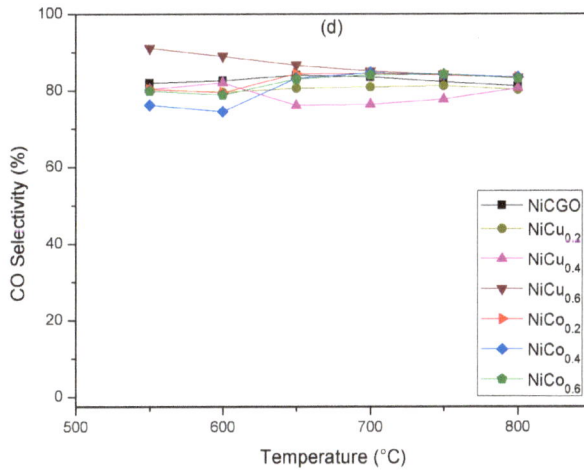

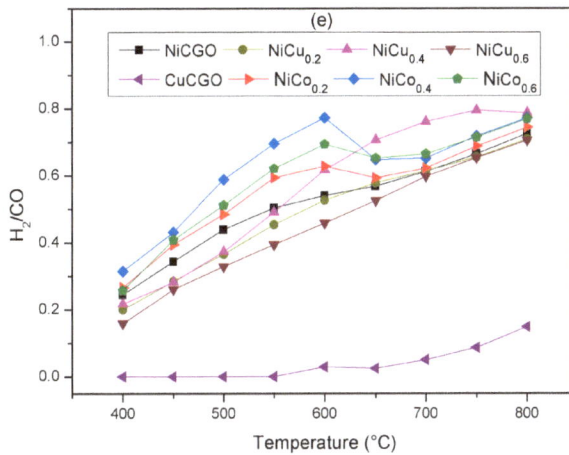

Fig. 13 cont.....

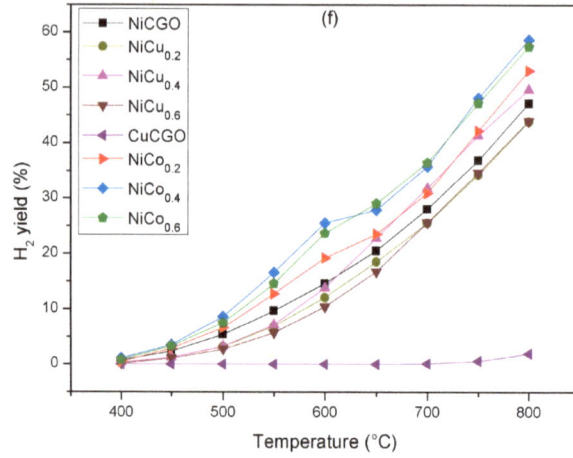

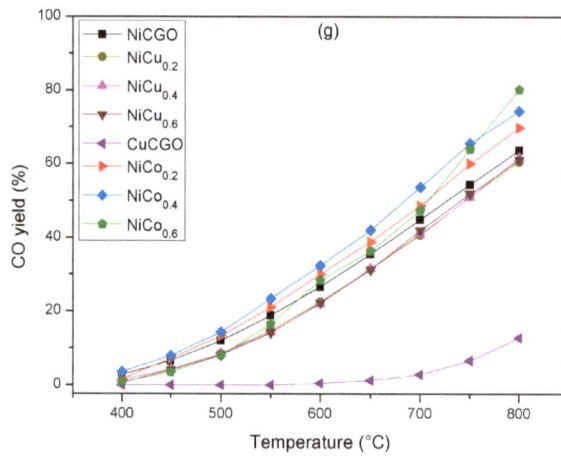

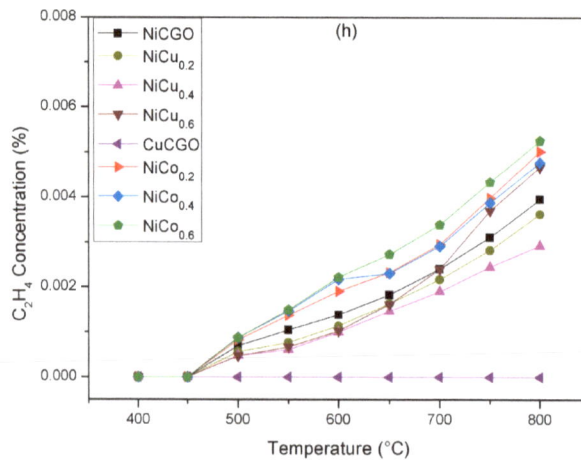

Fig. 13 cont.....

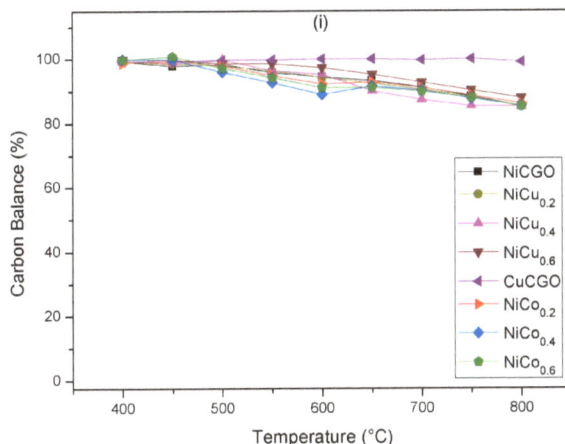

Fig. (13). (**a**) CH_4 conversion, (**b**) CO_2 conversion, (**c**) H_2 selectivity, (**d**) CO selectivity, (**e**) H_2/CO ratio, (**f**) H_2 yield, (**g**) CO yield, (**h**) C_2H_4 concentration, and (**i**) carbon balance obtained during the DRM reaction over the different catalysts.

The conversions of CH_4 and CO_2 increased with increasing reaction temperature, as expected, due to the endothermic character of the dry reforming of methane (Figs. **13a** and **b**). Cobalt-based catalysts showed higher conversions of CH_4 and CO_2 than copper-based and NiCGO catalysts. The $NiCo_{0.4}$ showed CH_4 conversion similar to that of $NiCo_{0.6}$. The $NiCu_{0.4}$ catalyst showed conversion of CH_4 close to the NiCGO catalyst, between 650 and 800 °C. $NiCu_{0.2}$ and $NiCu_{0.6}$ exhibited the lower CH_4 conversions at 650-800 °C. CuCGO catalyst had a very poor catalytic performance.

Regarding the selectivity of H_2 (Fig. **13c**), NiCo catalysts and $NiCu_{0.4}$ are the most promising materials. However, NiCo catalysts exhibited CO selectivity relatively close to the NiCGO catalyst (Fig. **13d**).

Higher conversion of CO_2 in comparison to conversion of CH_4 (Figs. **13a** and **b**), higher CO selectivity (and yield) than of selectivity and yield of H_2, and consequently H_2/CO ratio below 1 (Figs. 13c-g), suggest the simultaneous occurrence of reactions that lead to a given product as function of the consummation of another.

These results indicate the contribution of Reverse Water Gas Shift (RWGS) and methanation reactions (Equation 15 and 16) which increase CO_2 conversion and decrease H_2 selectivity, as expected. However, the occurrence of Boudouard (Equation 17), reverse of carbon gasification (Equation 18) and methane decomposition (Equation 19 and 20) reactions are also possible.

Reverse Water Gas Shift reaction (favorable > 600 °C):

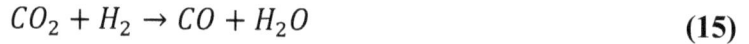

$$CO_2 + H_2 \rightarrow CO + H_2O \tag{15}$$

Methanation (favorable < 750 °C):

$$CO_2 + 4\,H_2 \rightarrow CH_4 + 2\,H_2O \tag{16}$$

Boudouard reaction (favorable < 750 °C):

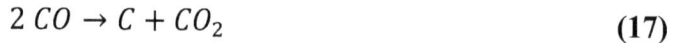

$$2\,CO \rightarrow C + CO_2 \tag{17}$$

Reverse of carbon gasification (favorable < 750 °C):

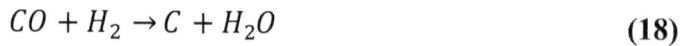

$$CO + H_2 \rightarrow C + H_2O \tag{18}$$

Methane decomposition (favorable > 500 °C):

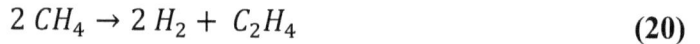

$$CH_4 \rightarrow C + 2\,H_2 \tag{19}$$

$$2\,CH_4 \rightarrow 2\,H_2 + C_2H_4 \tag{20}$$

NiCo-based catalysts showed higher H_2/CO ratios than NiCu and NiCGO catalysts at temperatures below 600 °C (Fig. **13e**). However, $NiCo_{0.4}$ catalyst has higher H_2/CO ratio than $NiCo_{0.6}$, which may be related to its higher activity in the methane decomposition reaction (Equation 19 and 20), leading to an increase in the H_2 production, as well as carbon formation. The occurrence of the Boudouard reaction (Equation 17) cannot be excluded. The strong decrease in H_2/CO ratio after 600 °C can be related to the RWGS reaction (Equation 15) which is favored at high temperature. The increase in CO selectivity confirms this hypothesis (Fig. **13d**). H_2 can also be consumed through the methanation reaction (Equation 16) which is confirmed by the decrease in CH_4 conversion.

$NiCo_{0.4}$ and $NiCo_{0.6}$ catalysts showed similar H_2/CO ratios in the temperature range 650-800 °C. While $NiCu_{0.4}$ has higher H_2/CO ratio at 650-750 °C. However, $NiCo_{0.4}$ and $NiCo_{0.6}$ catalysts favor more the CH_4 decomposition reaction for H_2 and C_2H_4 production (Equation 20) than the $NiCu_{0.4}$ catalysts due to its higher H_2 yield and C_2H_4 concentration (Figs. **13f** and **13h**).

The CH$_4$ decomposition reaction for carbon formation (Equation 10) is possible, but occurs with low intensity in the presence of the NiCo$_{0.6}$ catalysts (confirmed by TG/DSC). In addition, the reverse carbon gasification and Boudouard reactions can also occur to high extent in the presence of the NiCu$_{0.4}$-based catalyst, due to its low CO yield (Fig. **13g**).

Fig. (**13i**) shows the carbon balance over the catalysts. The carbon balance is close to 100% at 400 and 450 °C because catalysts showed low conversions. The carbon balance below 100% means that some input carbon atoms are not present in the outgoing gas, but are deposited on the catalysts [17]. In general, the carbon balance values of CuCGO and NiCu$_{0.6}$ catalysts are relatively higher than those of cobalt-based and NiCu$_{0.4}$ catalysts due their low activity. The amount of carbon produced during the methane dry reforming reaction was determined by the TG/DSC profiles shown in Fig. (**14**).

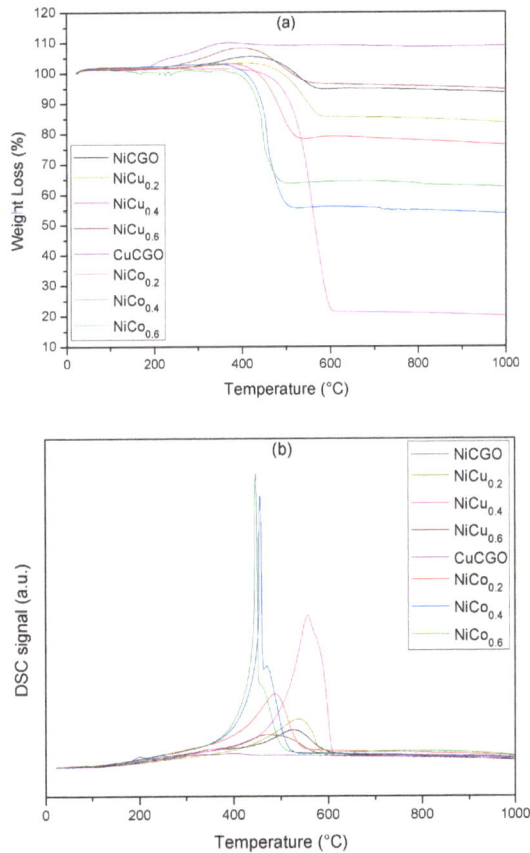

Fig. (14). TG **(a)** and DSC **(b)** profiles of the catalysts upon DRM reaction.

The $NiCo_{0.4}$ and $NiCo_{0.6}$ catalysts appear to be the most promising for dry reforming of methane. These catalysts have significantly higher surface areas than $NiCu_{0.4}$ and $NiCu_{0.6}$, which allows more dispersion of the metal phase on the support. In addition, $NiCo_{0.4}$ and $NiCo_{0.6}$ catalysts have high reduction capacity (Table **3**). $NiCu_{0.4}$ has lower crystallite size stability during the reaction (Table **6**). The formation of the Ni-Co alloy (confirmed by XRD) can promote high reducibility and strong metal-support interactions. All these factors are important to avoid the active phase sintering and carbon deposition, thus improving the catalytic properties [33 - 36].

Table 6. Crystallite sizes of Ni, Ni-Co and Ni-Cu upon reduction and DRM reaction.

Catalysts	Crystallite Size after Reduction (nm)	Crystallite Size after DRM Reaction (nm)
NiCGO	47.3	61.1
$NiCo_{0.2}$	51.3	55.6
$NiCo_{0.4}$	45.2	50.3
$NiCo_{0.6}$	42.7	42.3
$NiCu_{0.2}$	47.0	42.3
$NiCu_{0.4}$	49.4	32.4
$NiCu_{0.6}$	27.0	26.7

TG-DSC profiles of the catalysts upon DRM reaction are shown in (Fig. **14**).

NiCGO, CuCGO and $NiCu_{0.6}$ catalysts gave slight weight gain (4-9%) between 200 and 400 ° C (Fig. **13a**) due to the uptake of oxygen by metallic nickel present in the catalysts after the catalytic tests (Equation **21**). All catalysts had weight loss between 400 and 600 °C and exothermic phenomena due to carbon oxidation (Equation 22) [37].

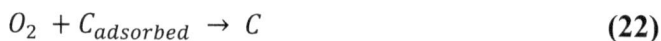

$$Ni^0 + \tfrac{1}{2}O_2 \rightarrow NiO \qquad (21)$$

$$O_2 + C_{adsorbed} \rightarrow C \qquad (22)$$

The following order is described for weight losses (%): $NiCu_{0.4}$ (77.8%) > $NiCo_{0.4}$ (47.5%) > $NiCo_{0.6}$ (37.0%) > $NiCo_{0.2}$ (24.9%) > $NiCu_{0.6}$ (11.8%) > NiCGO (10.8%) > CuCGO (0.9%).

$NiCu_{0.4}$ has higher carbon deposition than NiCo catalysts (confirmed by XRD). On the other hand, NiCGO, $NiCu_{0.6}$ and CuCGO composite materials showed lower carbon deposition due to their low catalytic activities. The stronger interaction between Ni and Co metals allowed better carbon resistance on the catalyst surface [38]. Therefore, NiCo catalysts are promising to enhance the CO_2 adsorption on CGO support, which reduces carbon formation through Boudouard reverse reaction (Equation 23).

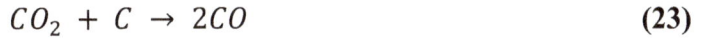

$$CO_2 + C \rightarrow 2CO \tag{23}$$

X-ray diffraction patterns of catalysts after catalytic tests are shown in Fig. (**15**).

All catalysts exhibited diffraction peaks of the CGO phase. The NiCGO catalyst showed additional peaks of the Ni cubic phase (JCPDS file n° 01-1260, a = 3.5240 Å). According to the respective crystallographic JCPDS files, crystalline phases of Co and Ni presented main diffraction peaks at $2\theta = 44.2°$ and $44.6°$, respectively. $NiCo_{0.4}$ and $NiCo_{0.6}$ catalysts had a single main peak at $2\theta = 44.4°$ that was attributed to the Ni-Co alloy. The presence of this alloy was evidenced after the catalysts be reduced and remained stable after the catalytic test. A less intense peak at $2\theta = 51.6°$ is also characteristic of the Ni-Co alloy. These results are in agreement with those previously reported in literature [35, 39, 40]. The profile of $NiCo_{0.2}$ presented the metallic Co (JCPDS file n° 15-0806, a = 3.5447 Å) and metallic Ni phases. In addition, the NiCGO and cobalt-based catalysts after reaction contain rhombohedral (29123-ICSD, a = 3.635 Å, alpha = 39.82°) and cubic carbon (66464-ICSD, a = 3.5667 Å) phases.

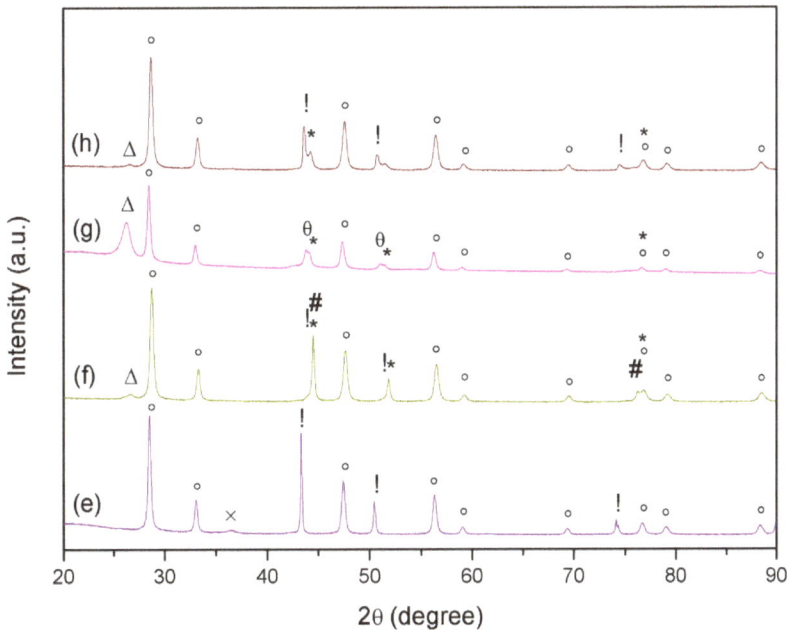

Fig. 15 cont.....

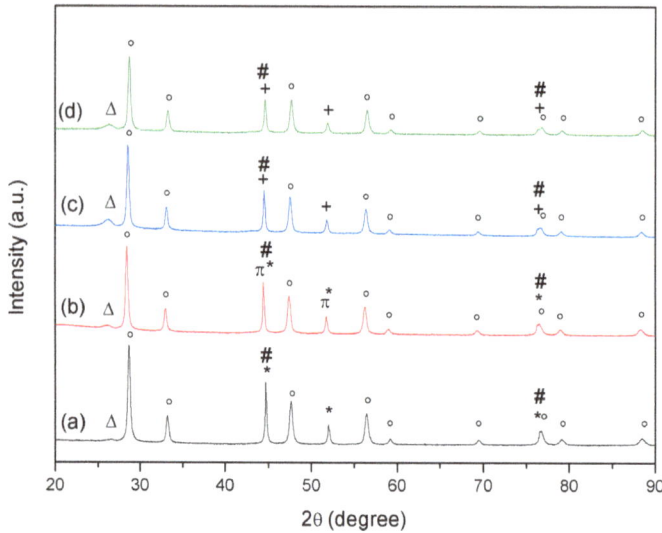

Fig. (15). XRD patterns of catalysts upon DRM: **(a)** NiCGO, **(b)** NiCo$_{0.2}$, **(c)** NiCo$_{0.4}$, **(d)** NiCo$_{0.6}$, **(e)** CuCGO, **(f)** NiCu$_{0.2}$, **(g)** NiCu$_{0.4}$ and **(h)** NiCu$_{0.6}$. Phases: * Ni, + NiCo, π Co, ! Cu, θ NiCu, × CuO, ° CGO, # C cubic, Δ C rhombohedral.

The X-ray diffraction profile of the NiCu$_{0.4}$ catalyst indicated the formation of the Ni-Cu alloy (JCPDS file n° 65-7246, a = 3.5615 Å) [28] and Ni metal phase. While NiCu$_{0.2}$ and NiCu$_{0.6}$ catalysts contain metallic Cu (JCPDS file n° 89-2838, a = 3.6151 Å) and Ni phases. The CuCGO catalyst presented Cu and CuO phases (JCPDS file n° 78-0428, a = 4.2450 Å). Concerning the presence of carbon, NiCu$_{0.4}$ catalyst exhibited an intense rhombohedral carbon peak, in good agreement with TG-DSC data.

Fig. (**16**) shows X-ray diffraction patterns from Rietveld refinement analyses of NiCo$_{0.4}$ and NiCu$_{0.4}$ catalysts as obtained of catalytic tests. The patterns confirm the presence of the Ni-Co alloy. Crystallographic parameters evaluated by Rietveld refinement are given in Table **7**. Rietveld refinements confirm crystal structure and lattice parameters. These results are in good agreement with those in the respective JCPDS cards.

Electrochemical Performance

The screen-printing technique has been used for the deposition of the NiCo$_{0.4}$ electrodes, in its oxidized form, on both faces of Gd-doped ceria (CGO) substrates to give symmetrical cells.

Table 7. Crystallographic parameters and refinement indexes for NiCGO, NiCo and NiCu-based catalysts upon DRM reaction, evaluated by Rietveld refinement.

Parameters	NiCGO				NiCo$_{0.4}$				NiCu$_{0.4}$			
	Ni	CGO	C*	C**	NiCo	CGO	C*	C**	Ni	NiCu	CGO	C**
a = b = c (Å)	3.5277	5.4256	3.5667	3.635	3.5343	5.4239	3.5742	3.7780	3.5276	3.5653	5.4200	3.6748
alpha (°)				39.82				43.42				38.78
R_{wp} (%)	37.97				39.84				44.30			
R_{exp} (%)	3.68				3.99				3.87			
χ^2	10.32				9.98				11.45			

*C cubic. **C rhombohedral.

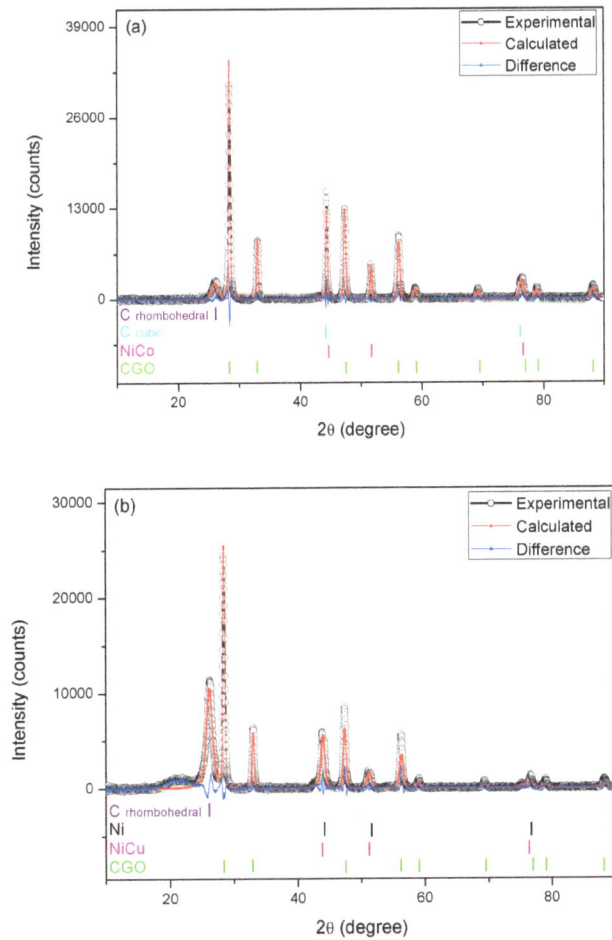

Fig. (16). X-ray diffraction patterns from Rietveld refinement of (a) NiCo$_{0.4}$ and (b) NiCu$_{0.4}$ catalysts upon DRM reaction.

Screen-printing is a mechanical process to obtain thick films (> 1μm), on flat substrates, with controlled characteristics. In this process, an ink consisting of the particulate material and organic additives, is forced to flow through the openings of a screen using a squeegee. This technique is much used in the manufacture of electrode materials for SOFCs due to its low cost and ability to produce films with homogeneous thickness that can vary from 1 to 100 μm, depending on the opening size of the used screen, rheological properties of the ink and other factors.

In a first step, CGO powders were uniaxially pressed into pellets using 300 MPa and then sintered in air atmosphere. Conventional sintering was performed at 1550 °C with dwell time of 4 h and using a heating rate of 2 °C min^{-1} in order to obtain dense CGO substrates.

Then, the inks were prepared by mixing $NiCo_{0.4}$ powders with polyethylene glycol (Alfa Aesar, Germany) in a ratio of 1:1 (g/mL). $NiCo_{0.4}$ composite anode-based slurries were screen printed onto both sides of CGO substrates and sintered at 1350 °C for 4 h in air, resulting in a $NiCo_{0.4}$/CGO/$NiCo_{0.4}$ symmetrical cell configuration used for the electrochemical performance measurements. The reduction of the metallic oxide was carried out in an atmosphere of hydrogen ($10\% \ H_2 + 90\% \ N_2$), before the electrochemical characterization.

The electrical characterization of electrode materials in an SOFC is generally carried out using the Electrochemical Impedance Spectroscopy (EIS) technique. Impedance spectroscopy offers the possibility of correlation between electrical properties and microstructure of materials.

Impedance spectroscopy (Z) consists in studying the response of an electrochemical system to a sinusoidal voltage of variable frequency. Therefore, the electrical impedance is the ratio between the applied AC voltage and the resulting current intensity. There are several representations to describe the impedance, in this chapter we adopted the Nyquist plots.

The Nyquist diagram can be decomposed into different semicircles according to the used frequency range. The high frequency domain is associated with the intrinsic properties of the material (bulk). At medium frequencies, the presence of conduction-blocking elements is distinguished. These various blockages to the displacements of the charge carriers can be induced by microstructural defects (grain boundaries, cracks, pores) or presence of secondary phase. The low frequency domain corresponds to the electrochemical phenomena which occur at the electrodes. The separation of these contributions is possible due to different relaxation constants resulting from individual polarizations, which change with temperature.

Electrochemical impedance spectroscopy was performed using an Autolab (Model PGStat30) equipment under open circuit conditions in temperature range between 650 and 850 °C in hydrogen (10% H_2 + 90% N_2) and synthetic biogas (25% CH_4 + 75% CO_2). Two Pt meshes were attached to the symmetrical cells as current collectors before electrochemical measurements. A signal of 50 mV amplitude and frequency range between 0.1 Hz and 1 MHz was applied for EIS measurements.

The impedance contributions ascribed to electrode were obtained after deconvolution of impedance spectra by fitting to conventional equivalent circuits using Zview (version 2.8).

Fig. (**17**) depicts the impedance spectra of the $Ni_{0.6}Co_{0.4}$-CGO cermet anode obtained in the temperature range 650-850 °C. The spectra were modelled using the equivalent circuit $L_1R_{ohm}(R_2CPE_2)(R_3CPE_3)$. In this circuit, L_1 corresponds to the inductance created by the equipment connection cables and platinum wires. R_{ohm} is the electrolyte ohmic resistance in series with two distinct electrode contributions consisting of resistances (R_2 and R_3) in parallel with constant phase elements (CPE_2 and CPE_3). The total anode polarisation resistance (R_p), characterizing the anode electrochemical performance, was obtained from the sum of the overall electrode resistance ($R_2 + R_3$) multiplied by the electrode surface area (0.2 cm^2) and divided by 2, to take into account the symmetrical cell configuration.

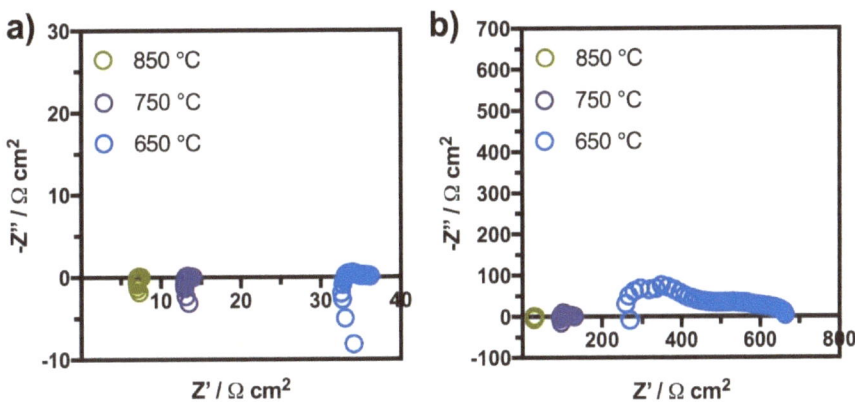

Fig. (17). Nyquist plots obtained in a) H_2 and in b) biogas. Measurements were performed in the temperature range 650-850 °C.

From the first inspection, one can easily observe that the polarization resistances in biogas are higher than those obtained in H_2 atmosphere. Total polarization resistances (R_p) of 0.96 and 36.10 Ω cm^2 were obtained at 750 °C for, respectively, H_2 and biogas atmospheres. In terms of activation energy (E_a), values of 0.92 and 2.17 eV were obtained for, respectively, H_2 and biogas atmospheres (Table **8**), in agreement with the literature [6].

Table 8. . Calculated parameters from the electrochemical measurements, obtained in H_2 and in biogas.

$T / °C$	ASR / ohm cm^2	E_a */ eV
	H_2	
650	3.18	
750	0.96	0.92
850	0.46	
	Biogas	
650	394.59	
750	36.10	2.17
850	3.06	

This tendency is expected as the hydrogen oxidation reaction occurs more easily than the reforming of methane [41]. Moreover, it is expected that the oxygen partial pressure (pO_2) in biogas will be higher than that in H_2, thus reducing the concentration of electronic species in Equation 24 and 25 and consequently increasing the resistivity of the anode processes in such conditions [42, 43]. This effect originates from the variation in the oxygen non-stoichiometry in the CGO phase, due to the increased concentration of oxygen vacancies (v_o) in reducing atmospheres, as well as the formation of n-type conductivity ($C\grave{e}_{ce}$) [4]:

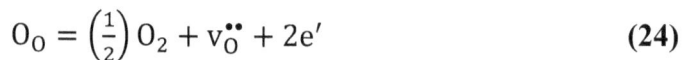

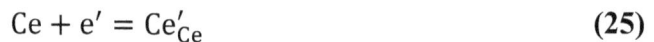

$$O_O = \left(\tfrac{1}{2}\right) O_2 + v_O^{\bullet\bullet} + 2e' \tag{24}$$

$$Ce + e' = Ce'_{Ce} \tag{25}$$

CONCLUDING REMARKS

Co and Cu-based Ni composite materials were successfully prepared by a one-step polymer-based synthesis method. NiCo$_{0.4}$, NiCo$_{0.6}$ and NiCu$_{0.4}$ reduced catalysts contain Ni-Co and Ni-Cu alloys. NiCo-based materials exhibited higher surface area (S_{BET}) than NiCu-based materials. Based on TPR analyses, NiCo$_{0.4}$ and NiCo$_{0.6}$ catalysts exhibit high reducibility and strong metal-support interaction. NiCGO, CuCGO, NiCo and NiCu-based catalysts were tested in CO_2 reforming of methane. Cobalt-additions played a key role in the enhancement of the catalytic activity and suppression of carbon deposition. NiCo-based catalysts showed higher CH_4 and CO_2 conversions than NiCu, CuCGO and NiCGO catalysts. $NiCo_{0.4}$ and $NiCo_{0.6}$ demonstrated high carbon resistance than $NiCu_{0.4}$. $NiCo_{0.4}$ catalyst had the highest H_2/CO ratio and the best reaction selectivity below 600 °C. The electrochemical activity of a $NiCo_{0.4}/CGO/NiCo_{0.4}$ symmetrical cell towards electro-oxidation reactions at 650-850 °C showed to be influenced by the fuel type during test with H_2 and biogas. As anode material for SOFC, $NiCo_{0.4}$-CGO presented polarization resistance (R_p) of 0.96 Ω cm^2 at 750 °C and activation energy of 0.92 eV at 650-750 °C in H_2 atmosphere. R_p of 36.10 Ω cm^2 at 750 was obtained with biogas as fuel. It is important to mention that $R_p < 1$ cm^2 is the target for efficient fuel cell operation. The electrochemical characterization of the $NiCo_{0.4}$-CGO electrode showed that hydrogen electro-oxidation reaction occurs more easily (lower R_p and activation energy) than the dry reforming of methane.

CONSENT FOR PUBLICATION

Not applicable.

CONFLICT OF INTEREST

Declare None.

ACKNOWLEDGEMENT

Francisco J. A. Loureiro acknowledges financial support from the FCT grant PD/BDE/114353/2016, project grant PTDC CTM-ENE/6319/2014, UID/CTM/50011/2013, CENTRO-01-0145-FEDER-022083, QREN, FEDER and COMPETE Portugal and the Europea.

REFERENCES

[1] Carrete, L.; Friedrich, K.A.; Stimming, U. Fuel cells – fundamentals and applications. *Fuel Cells (Weinh.),* **2001**, *1*, 5-39.
[http://dx.doi.org/10.1002/1615-6854(200105)1:1<5::AID-FUCE5>3.0.CO;2-G]

[2] Brett, D.J.L.; Atkinson, A.; Brandon, N.P.; Skinner, S.J. Intermediate temperature solid oxide fuel cells. *Chem. Soc. Rev.,* **2008**, *37*, 1568-1578.
[http://dx.doi.org/10.1039/b612060c]

[3] Macedo, D.A.; Cesario, M.R.; Souza, G.L.; Cela, B.; Paskocimas, C.A.; Martinelli, A.E.; Melo, D.M.A.; Nascimento, R.M. Infrared spectroscopy characterization of SOFC functional ceramics.*Infrared Spectroscopy - Materials Science, Engineering and Technology*; Theophile, T., Ed.; IntechOpen: Rijeka, **2012**, Vol. 1, pp. 383-404.
[http://dx.doi.org/10.5772/34884]

[4] Martinelli, A.E.; Macedo, D.A.; Cesário, M.R.; Cela, B.; Nicodemo, J.P.; Paskocimas, C.A.; Melo, D.M.A.; Nascimento, R.M. Synthesis of functional ceramic materials for application in 2 kW stationary SOFC stacks. *Mater. Sci. Forum,* **2013**, *730-732*, 147-152.
[http://dx.doi.org/10.4028/www.scientific.net/MSF.730-732.147]

[5] Moura, C.G.; Grilo, J.P.F.; Nascimento, R.M.; Macedo, D.A. A Brief Review on Anode Materials and Reactions Mechanism in Solid Oxide Fuel Cells*Frontiers in Ceramic Science – Functional materials*

for solid oxide fuel cells: processing, microstructure and performance; Cesário, Moisés R; de Macedo, Daniel A, Eds.; Sharjah UAE26-41.Bentham Science Publishers, **2017**, 1, pp.

[6] Sousa, A.R.O.; Araujo, A.J.M.; Souza, G.S.; Grilo, J.P.F.; Loureiro, F.J.A.; Fagg, D.P.; Macedo, D.A. Electrochemical assesment of one-step Cu-CGO cermets under hydrogen and biogas fuels. *Mater. Lett.,* **2017**, *191*, 141-144.
[http://dx.doi.org/10.1016/j.matlet.2016.12.087]

[7] Lanzini, A.; Leone, P.; Guerra, C.; Smeacetto, F.; Brandon, N.P.; Santarelli, M. Durability of anode supported Solid Oxides Fuel Cells (SOFC) under direct dry-reforming of methane. *Chem. Eng. J.,* **2013**, *220*, 254-263.
[http://dx.doi.org/10.1016/j.cej.2013.01.003]

[8] Yin, W.; Chuang, S.S.C. CH$_4$ internal dry reforming over a Ni/YSZ/ScSZ anode catalyst in a SOFC: A transient kinetic study. *Catal. Commun.,* **2017**, *102*, 62-66.
[http://dx.doi.org/10.1016/j.catcom.2017.08.027]

[9] Johnson, G.B.; Hjalmarsson, P.; Norrman, K.; Ozkan, U.S.; Hagen, A. Biogas catalytic reforming studies on nickel-based solid oxide fuel cell anodes. *Fuel Cells (Weinh.),* **2016**, *16*, 219-234.
[http://dx.doi.org/10.1002/fuce.201500179]

[10] Fonseca, R. O.da.; Silva, A.A.A.da.; Signorelli, M.R.M.; Rabelo-Neto, R.C.; Noronha, F.B.; Simões, R.C.C.; Mattos, L.V. Nickel/doped ceria solid oxide fuel cell anodes for dry reforming of methane. *J. Braz. Chem. Soc.,* **2014**, *25*, 2356-2363.

[11] Wang, Y.; Yao, L.; Wang, S.; Mao, D.; Hu, C. Low-temperature catalytic CO$_2$ dry reforming of methane on Ni-based catalysts: A review. *Fuel Process. Technol.,* **2018**, *169*, 199-206.
[http://dx.doi.org/10.1016/j.fuproc.2017.10.007]

[12] Abdullah, B.; Ghani, N.A.A.; Vo, D-V.N. Recent advances in dry reforming of methane over Ni-based catalysts. *J. Clean. Prod.,* **2017**, *162*, 170-185.
[http://dx.doi.org/10.1016/j.jclepro.2017.05.176]

[13] Pakhare, D.; Spivey, J. A review of dry (CO$_2$) reforming of methane over noble metal catalysts. *Chem. Soc. Rev.,* **2014**, *43*, 7813-7837.
[http://dx.doi.org/10.1039/C3CS60395D]

[14] Tsyganok, A.I.; Inaba, M.; Tsunoda, T.; Hamakawa, S.; Suzuki, K.; Hayakawa, T. Dry reforming of methane over supported noble metals: a novel approach to preparing catalysts. *Catal. Commun.,* **2003**, *4*, 493-498.
[http://dx.doi.org/10.1016/S1566-7367(03)00130-4]

[15] Shiraz, M.H.A.; Rezaei, M.; Meshkani, F. Microemulsion synthesis method for preparation of mesoporous nanocrystalline γ-Al$_2$O$_3$ powders as catalyst carrier for nickel catalyst in dry reforming reaction. *Int. J. Hydrogen Energy,* **2016**, *41*, 6353-6361.
[http://dx.doi.org/10.1016/j.ijhydene.2016.03.017]

[16] Ozdemir, H.; Oksuzomer, M.A.F.; Gurkaynak, M.A. Preparation and characterization of Ni based catalysts for the catalytic partial oxidation of methane: effect of support basicity on H$_2$/CO ratio and carbon deposition. *Int. J. Hydrogen Energy,* **2010**, *35*, 12147-12160.
[http://dx.doi.org/10.1016/j.ijhydene.2010.08.091]

[17] Tanios, C.; Bsaibes, S.; Gennequin, C.; Labaki, M.; Cazier, F.; Billet, S.; Tidahy, H.L.; Nsouli, B.; Aboukaıs, A.; Abi-Aad, E. Syngas production by the CO$_2$ reforming of CH$_4$ over Ni-Co-Mg-Al catalysts obtained from hydrotalcite precursors. *Int. J. Hydrogen Energy,* **2017**, *42*, 12818-12828.
[http://dx.doi.org/10.1016/j.ijhydene.2017.01.120]

[18] Fan, M-S.; Abdullah, A.Z.; Bhatia, S. Hydrogen production from carbon dioxide reforming of methane over Ni-Co/MgO-ZrO$_2$ catalyst: Process optimization. *Int. J. Hydrogen Energy,* **2011**, *36*, 4875-4886.
[http://dx.doi.org/10.1016/j.ijhydene.2011.01.064]

[19] Zhang, X.; Yang, C.; Zhang, Y.; Xu, Y.; Shang, S.; Yin, Y. Ni-Co catalyst derived from layered

double hydroxides for dry reforming of methane. *Int. J. Hydrogen Energy,* **2015**, *40*, 16115-16126.
[http://dx.doi.org/10.1016/j.ijhydene.2015.09.150]

[20] Luisetto, I.; Tuti, S.; Bartolomeo, E.D. Co and Ni supported on CeO_2 as selective bimetallic catalyst for dry reforming of methane. *Int. J. Hydrogen Energy,* **2012**, *37*, 15992-15999.
[http://dx.doi.org/10.1016/j.ijhydene.2012.08.006]

[21] Liu, Z.; Lustemberg, P.; Gutiérrez, R.A.; Carey, J.J.; Palomino, R.M.; Vorokhta, M.; Grinter, D.C.; Ramírez, P.J.; Matolín, V.; Nolan, M.; Ganduglia-Pirovano, M.V.; Senanayake, S.D.; Rodriguez, J.A. *In situ* Investigation of Methane Dry Reforming on Metal/Ceria(111) Surfaces: metal–support interactions and C–H bond activation at low temperature. *Angew. Chem. Int. Ed. Engl.,* **2017**, *42*, 13041-13046.
[http://dx.doi.org/10.1002/anie.201707538]

[22] Cela, B.; Macedo, D.A.; Souza, G.L.; Martinelli, A.E.; Nascimento, R.M.; Paskocimas, C.A. NiO-CGO *in situ* nanocomposite attainment: one step synthesis. *J. Power Sources,* **2011**, *196*, 2539-2544.
[http://dx.doi.org/10.1016/j.jpowsour.2010.11.026]

[23] Araujo, A.J.M.; Sousa, A.R.O.; Grilo, J.P.F.; Campos, L.F.A.; Loureiro, F.J.A.; Fagg, D.P.; Dutra, R.P.S. Macedo, D.A. Preparation of one-step NiO/Ni-CGO composites using factorial design. *Ceram. Int.,* **2016**, *42*, 18166-18172.
[http://dx.doi.org/10.1016/j.ceramint.2016.08.131]

[24] Muccillo, E.N.S.; Souza, E.C.C.; Muccillo, R. Synthesis of reactive neodymia doped zirconia powders by the sol–gel technique. *J. Alloys Compd.,* **2002**, *344*, 175-178.
[http://dx.doi.org/10.1016/S0925-8388(02)00335-3]

[25] Lima, C.G.M.; Santos, T.H.; Grilo, J.P.F.; Dutra, R.P.S.; Nascimento, R.M.; Rajesh, S.; Fonseca, F.C.; Macedo, D.A. Synthesis and properties of CuO-doped $Ce_{0.9}Gd_{0.1}O_{2-\delta}$ electrolytes for SOFCs. *Ceram. Int.,* **2015**, *41*, 4161-4168.
[http://dx.doi.org/10.1016/j.ceramint.2014.12.093]

[26] Dong, Y.; Hampshire, S. Zhou, J-er.; Meng, G. Synthesis and sintering of Gd-doped CeO_2 electrolytes with and without 1 at.% CuO dopping for solid oxide fuel cell applications. *Int. J. Hydrogen Energy,* **2011**, *36*, 5054-5066.
[http://dx.doi.org/10.1016/j.ijhydene.2011.01.030]

[27] Marrero-Jerez, J.; Chinarro, E.; Moreno, B.; Colomer, M.T.; Jurado, J.R.; Núñez, P. TPR studies on NiO-CGO composites prepared by combustion synthesis. *Ceram. Int.,* **2014**, *40*, 3469-3475.
[http://dx.doi.org/10.1016/j.ceramint.2013.09.083]

[28] Bonura, G.; Cannilla, C.; Frusteri, F. Ceria-gadolinia supported NiCu catalyst: A suitable system for dry reforming of biogas to feed a solid oxide fuel cell (SOFC). *Appl. Catal. B,* **2012**, •••, 121-122, 135-147.

[29] Marrero-Jerez, J.; Murugan, A.; Metcalfe, I.S.; Núñez, P. TPR–TPD–TPO studies on CGO/NiO and CGO/CuO ceramics obtained from freeze-dried precursors. *Ceram. Int.,* **2014**, *40*, 15175-15182.
[http://dx.doi.org/10.1016/j.ceramint.2014.06.132]

[30] Gómez-Cuaspud, J.A.; Schmal, M. Effect of metal oxides concentration over supported cordierite monoliths on the partial oxidation of ethanol. *Appl. Catal. B,* **2014**, *148-149*, 1-10.
[http://dx.doi.org/10.1016/j.apcatb.2013.10.036]

[31] Wang, X.; Wen, W.; Mi, J.; Li, X.; Wang, R. The ordered mesoporous transition metal oxides for selective catalytic reduction of NO_x at low temperature. *Appl. Catal. B,* **2015**, *176*, 454-463.
[http://dx.doi.org/10.1016/j.apcatb.2015.04.038]

[32] Estephane, J.; Aouad, S.; Hany, S.; Khoury, B.E.; Gennequin, C.; Zakhem, H.E.; Nakat, J.E.; Aboukaïs, A.; Abi-Aad, E. CO_2 reforming of methane over Ni–Co/ZSM5 catalysts. Aging and carbon deposition study. *Int. J. Hydrogen Energy,* **2015**, *40*, 9201-9208.
[http://dx.doi.org/10.1016/j.ijhydene.2015.05.147]

[33]   Usman, M.; Daud, W.M.A.W.; Abbas, H.F. Dry reforming of methane: Influence of process parameters-A review. *Renew. Sustain. Energy Rev.,* **2015**, *45*, 710-744.
[http://dx.doi.org/10.1016/j.rser.2015.02.026]

[34]   Takanabe, K.; Nagaoka, K.; Aika, K-I. Improved resistance against coke deposition of titania supported cobalt and nickel bimetallic catalysts for carbon dioxide reforming of methane. *Catal. Lett.,* **2005**, *102*, 153-157.
[http://dx.doi.org/10.1007/s10562-005-5848-4]

[35]   Arbag, H.; Yasyerli, S.; Yasyerli, N.; Dogu, G.; Dogu, T. Enhancement of catalytic performance of Ni based mesoporous alumina by Co incorporation in conversion of biogas to synthesis gas. *Appl. Catal. B,* **2016**, *198*, 254-265.
[http://dx.doi.org/10.1016/j.apcatb.2016.05.064]

[36]   Sidik, S.M.; Triwahyono, S.; Jalil, A.A.; Majid, Z.A.; Salamun, N.; Talib, N.B.; Abdullah, T.A.T. CO_2 reforming of CH4 over NiCo/MSN for syngas production: role of Co as a binder and optimization using RSM. *Chem. Eng. J.,* **2016**, *295*, 1-10.
[http://dx.doi.org/10.1016/j.cej.2016.03.041]

[37]   Perez-Lopez, O.W.; Senger, A.; Marcilio, N.R.; Lansarin, M.A. Effect of composition and thermal pretreatment on properties of Ni–Mg–Al catalysts for CO_2 reforming of methane. *Appl. Catal. A.,* **2006**, *303*, 234-244.
[http://dx.doi.org/10.1016/j.apcata.2006.02.024]

[38]   Ay, H.; Üner, D. Dry reforming of methane over CeO_2 supported Ni, Co and Ni-Co catalysts. *Appl. Catal. B,* **2015**, *179*, 128-138.
[http://dx.doi.org/10.1016/j.apcatb.2015.05.013]

[39]   Ahmed, J.; Sharma, S.; Ramanujachary, K.V.; Lofland, S.E.; Ganguli, A.K. Microemulsion-mediated synthesis of cobalt (pure fcc and hexagonal phases) and cobalt–nickel alloy nanoparticles. *J. Colloid Interface Sci.,* **2009**, *336*, 814-819.
[http://dx.doi.org/10.1016/j.jcis.2009.04.062]

[40]   Takanabe, K.; Nagaoka, K.; Nariai, K.; Aika, K-I. Titania-supported cobalt and nickel bimetallic catalysts for carbon dioxide reforming of methane. *J. Catal.,* **2005**, *232*, 268-275.
[http://dx.doi.org/10.1016/j.jcat.2005.03.011]

[41]   Hanna, J.; Lee, W.Y.; Shi, Y.; Ghoniem, A.F. Fundamentals of electro- and thermochemistry in the anode of solid-oxide fuel cells with hydrocarbon and syngas fuels. *Pror. Energy Combust. Sci.,* **2014**, *40*, 74-111.
[http://dx.doi.org/10.1016/j.pecs.2013.10.001]

[42]   Frade, J.R.; Kharton, V.V.; Yaremchenko, A.; Naumovich, E. Methane to syngas conversion: Part I. Equilibrium conditions and stability requirements of membrane materials. *J. Power Sources,* **2004**, *130*, 77-84.
[http://dx.doi.org/10.1016/j.jpowsour.2003.11.067]

[43]   Navarro, L.; Marques, F.; Frade, J. n-Type Conductivity in Gadolinia-Doped Ceria. *J. Electrochem. Soc.,* **1997**, *144*, 267-273.
[http://dx.doi.org/10.1149/1.1837395]

CHAPTER 5

NiO-Ce$_{0.9}$Gd$_{0.1}$O$_{1.95}$ Composites and their Cermets as Anode Materials for SOFC

Allan J.M. de Araújo[1], Gabriel M. Santos[2], João P. de F. Grilo[3], Rubens M. do Nascimento[1], Carlos A. Paskocimas[1], Moisés R. Cesário[4] and Daniel A. de Macedo[*, 5]

[1] *Department of Materials Engineering, Federal University of Rio Grande do Norte, Natal, 59078-970, Brazil*

[2] *Department of Petroleum Engineering, University of Campinas, Campinas, 13083-970, Brazil*

[3] *CICECO, Department of Materials & Ceramic Engineering, University of Aveiro, 3810-193Aveiro, Portugal*

[4] *Unité de Chimie Environnementale et Interactions sur le Vivant (UCEIV, E.A. 4492), MREI, Université du Littoral Côte d'Opale (ULCO), 59140, Dunkerque, France*

[5] *Department of Materials Engineering, Federal University of Paraíba, João Pessoa, 58051-900, Brazil*

Abstract: The anode of a Solid Oxide Fuel Cell (SOFC) is the component responsible for releasing electrons. It must present some prerequisites such as porosity (30-40 wt.%) and mixed conductivity (ionic and electronic). To achieve that, it often consists of a ceramic-metal composite (cermet) material. Ni-YSZ cermet is the material most commonly used as anode SOFC, however it presents low electrochemical efficiency below 750 °C. With this in mind, new materials have been studied, and Ni-GDC cermet has shown promising results. This chapter focused mainly on a brief review of NiO-Ce$_{0.9}$Gd$_{0.1}$O$_{1.95}$ (NiO-GDC) composites and their Ni-GDC cermets as SOFC anode materials. The chapter reports the history, concept, operation principle, reaction mechanisms and components of SOFCs, addressing recent works in the field. The synthesis of NiO-GDC composites by a one-step synthesis method (polymer-based method) is compared with the conventional mechanical mixing method. Impedance spectroscopy is shown to be a crucial technique to investigate the electrical properties of SOFC functional materials. The chapter ends with a literature review on the electrical properties of NiO-GDC composites and their cermets.

Keywords: Impedance Spectroscopy, Microstructure, NiO-Ce$_{0.9}$Gd$_{0.1}$O$_{1.95}$, NiO-GDC, Ni-GDC, One-Step Synthesis, SOFC, SOFC Anode.

---

[*] **Corresponding author Daniel A.de Macedo:** Materials Science and Engineering Postgraduate Program, Federal University of Paraíba (UFPB), 58051-900 João Pessoa, Brazil; Tel: +55(84)3215-3826; E-mail: damaced@gmail.com

INTRODUCTION

Solid oxide fuel cell (SOFC) is an electrochemical device with high electrical efficiency and significant environmental benefits in clean and efficient (~80% with fuel regeneration) electric power generation [1]. SOFC generates electricity by the oxidation reaction between fuel and oxygen ions *via* diffusion of oxide ions (or protons) through an ion-conducting solid-electrolyte layer [2].

Fuel gas is fed to the anode, passes through an oxidation reaction, and releases electrons to an external circuit. SOFC anodes must be porous and show mixed ionic-electronic conductivity for the electrochemical oxidation reaction occurs more easily. The electronic conductivity is needed to transport electrons released in the fuel oxidation reaction. Porous Ni-YSZ cermets (YSZ: yttria stabilized zirconia) are among the most common SOFC anode materials [3]. However, Ni-YSZ anodes are not suitable for SOFC operating temperatures below 750 °C. Therefore, other materials have to be developed for this application, and Ni/Gd-doped ceria has emerged as a promising candidate [4 - 6].

Besides an appropriate material selection, the adopted powder synthesis method plays a crucial role in the microstructure and, in turn, electrical properties of SOFC functional materials. One of the methods that have been gaining prestige is the one-step synthesis method (Pechini method). With this in mind, the current book chapter addresses a one-step synthesis method that allows obtaining NiO-GDC nanocomposites with remarkable particulate properties [7 - 9]. Impedance spectroscopy (IS) is shown to be a powerful technique for measuring electrical properties and correlate them with samples microstructure [10, 11].

Fuel Cells

Fuel cells are electrochemical devices capable of converting energy from a chemical reaction directly into electricity, generating electricity and water steam as a by-product [12]. The electrical energy conversion in a fuel cell occurs through two electrochemical reactions. The oxidation of a fuel (typically hydrogen or hydrogen-rich fuels) takes place at the anode. This reaction generates an electron flow through an external circuit toward the cathode, where the oxidant (typically oxygen gas) is reduced to oxygen ions. However, unlike batteries, oxidant and fuel gases must be replenished continuously to allow uninterrupted operation [13, 14].

The fundamental principles of the fuel cell were discovered in 1839 by Sir William Robert Grove. While investigating the electrolysis of water, Grove noticed that after the current was switched off, a small amount of current flowed through the circuit in the opposite direction. It occurred due to the reaction between the products of electrolysis, hydrogen and oxygen, catalyzed by platinum electrodes. Grove recognized the possibility of stacking several cells in a series to form a gaseous voltaic battery. In addition, he made a crucial observation that there should be a "notable surface of action" among the gas, the electrode and the electrolyte phases in a cell. The maximization of the contact area among gaseous reagent, electrolyte and electrode (the electrocatalytic conductor) is called three-phase boundary (TPB). This was proposed by Grove and remains ahead of the research in fuel cells and in its development [15].

The term "fuel cell" emerged only 50 years after Grove's "gas battery" in 1889 by Mond and Langer. They used it to describe their device that had a porous platinum electrode structure and used a diaphragm made by a non-conductive porous substance to hold the electrolyte [15].

Ludwig Mond (1839-1909) spent most of his career developing industrial chemical technology. Mond and his assistant Carl Langer (d. 1935) described their experiments with a hydrogen-oxygen fuel cell of 6 amps per square foot (the surface area of the measuring electrode) at 0.73 V. Friedrich Wilhelm Ostwald (1853-1932), the pioneer of the physical chemistry field, largely collaborated with the theoretical understanding of how fuel cells operate. Emil Baur (1873-1944), from Switzerland, conducted extensive research on different types of fuel cells during the first half of the 20th century. Baur's work included high-temperature devices (using molten silver as an electrolyte) and a unit that used a solid electrolyte of clay and metallic oxides. Francis Thomas Bacon (1904-1992) began his research in alkaline electrolyte fuel cells at the end of 1930. In 1939, he built his first cell. Since 1945, 3 research groups (USA, Germany and the former USSR) have conducted studies on some major types of generators, improving their technologies for industrial development. These works yielded the real concepts at Siemens and Pratt & Wittney. With the Apollo space program in 1960, NASA spent tens of millions of dollars on a successful program that used hydrogen fuel cells to feed the electrical systems on board at the Apollo journey to the moon. In the mid-80's government agencies in the United States, Canada and Japan increased their funding for fuel cell research and development. Today, fuel cells are common in spaceflights (Space Shuttle, Skylab and Gemini spacecrafts), means of transport, used as portable energy, in residences and for power generation [16].

Fuel Cells: Types and Applications

There are several types of fuel cells and several are being developed. The characteristics of each type vary mainly in terms of electrolyte and electrode materials, operating temperature, generated power density, tolerance to thermal cycles and impurities. These characteristics determine the most appropriate application of the device. The main criterion of classification of fuel cells is which electrolyte is employed, making its identification widely recognized around the world [15, 17]. Based on the electrolyte material, there are five main types of fuel cells (Fig. **1**).

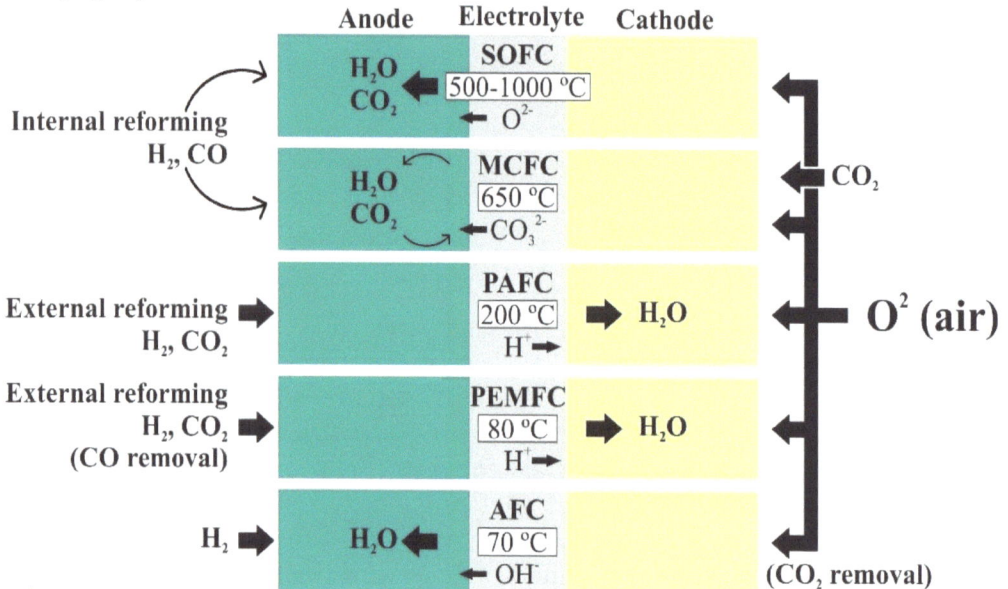

Fig. (1). Fuel cell types. SOFC: solid oxide fuel cell; MCFC: molten carbonate fuel cell; PAFC: phosphoric acid fuel cell; PEMFC: Proton exchange membrane fuel cells; and AFC: alkaline fuel cell. Reaction products are formed in the anode for SOFC, MCFC and AFC types, and in the cathode for PAFC and PEMFC types.

These cells operate mainly powered by pure hydrogen (PEMFC, AFC and PAFC), diluted methanol (DMFC) and hydrogen gas mixed with CO-rich fuel, obtained from a renovation or partial oxidation of hydrocarbons [15, 17].

Alkaline fuel cell (AFC) has been used as the main source of energy in Space Shuttles. Although it has been operated efficiently and reliably in space missions for more than 40 years, it has not been used for other purposes, mainly because of the high cost. In addition, they work at a low operating temperature and require a relatively complicated process to remove water from the electrolyte. However, this type of cell was the forerunner of the most modern cells [18].

The potential applications of the fuel cells in society are growing due to the different benefits that these devices can provide, in addition to environmental considerations (non-emission of NO$_X$, SO$_X$ or hydrocarbons and null or low CO$_2$ emission) especially in urban areas where localized pollution is a priority issue. The possibility of combining reliability, low maintenance cost, silent operation and modularity makes fuel cells suitable for decentralized power generation, free distribution networks, power supply continuously and, on a smaller scale, mobile applications [19, 20].

Stationary fuel cells designed for residential and industrial applications can use different kinds of fuels to obtain hydrogen *in situ*, through an internal reaction. In the case of low temperature fuel cells (T < 600 °C), an external reformer must be used to convert natural gas, for example, into hydrogen and CO_2. In a high temperature, fuel cell reactions such as dry reforming and steam reforming of methane occur internally at temperatures above 700 °C [21].

The high fuel flexibility of fuel cells, especially those operating at high temperatures, is attractive to the global interest of decreasing carbon dioxide emissions. The use of hydrogen as a gas feeder on the basis of renewable energies leads to special benefits with respect to resource conservation and climate protection, however this represents a medium-to long-term perspective [15, 22].

The most important attributes of fuel cells for stationary power generation are the high efficiency and the possibility of distributed energy. To achieve commercial success for all systems in order to compete with conventional technologies costs should be reduced. The cost depends mainly on the size (power) of the fuel cell system and on its applications. Combined heat and power (CHP) can be used in residences (<10 kWh) or even for energy generation plants (>1000 kWh). Fuel cell systems can be applied in automobiles or for portable applications [16, 23 - 25].

Among all types of cells, PEMFCs (Proton Exchange Membrane Fuel Cells) and SOFCs (Solid Oxide Fuel Cells) are the most attractive technologies because the high current densities they can achieve [16]. Solid oxide fuel cells are the most demanding from the point of view of materials. The following aspects make SOFC still competitive compared to other types of fuel cells:

- SOFCs are the most efficient fuel cell electricity generators to be developed worldwide.
- they have fuel flexible, being possible to use carbon-based fuels
- it is the most suitable technology to be applied in the distributed generation market (*i.e.*, stationary power), because its high conversion efficiency provides the greatest outcome when fuel costs are higher, due to long fuel delivery systems to customer premises
- it has a modular and solid-state construction and do not present any moving parts, thereby are quiet enough to be installed indoors
- the high operating temperature of SOFCs produces quality heat as byproduct, it can be used for cogeneration, or it can be used in combined cycle applications

- absence of noble metals which can be problematic in the case of availability and price issue in high volume manufacture
- its friendly management and absence of risks to the operator. Liquid electrolytes in the other hand are corrosive and difficult to handle
- its extremely low emissions eliminate the danger of carbon monoxide in the exhaust gases, as all CO produced is converted to CO_2 at the high operating temperature
- its potential long-life expectancy, approximately 40,000 – 80,000 h

Solid Oxide Fuel Cell (SOFC)

Solid Oxide Fuel Cells (SOFCs) were used only 60 years after its discovery by Sir William Grove. The first electrolyte material used to produce these cells was created by Sir Walther Hermann in 1899, and the operation of the first SOFC-type cell, at 1000 °C, was made by Baur and Preis in 1937. Since that time, ceramic fuel cell technology has progressed considerably. Fuel cells based on stabilized zirconia electrolyte are able to operate for thousands of hours at excellent performance. Recently, the development and research of ceramic fuel cells has received a lot of attention, reflecting the growing interest in this technology [26].

SOFC is a complete solid-state device that uses a ceramic material (solid oxide) oxygen ion conductor as electrolyte. Thus, it has a simpler concept when compared with all other fuel cell systems mentioned above and only two phases (solid-gas) are necessary to operate it, in contrast to the other types of cells, which require phase zones (solid-liquid-gas) to be operated. PEMFCs models are made with solid electrolyte, but the electrolyte needs to be hydrated during the operation, characterizing the need for a three-phase zone, making SOFC simpler to operate. Usually, the use of three-phase zone operations is related with many problems, such as efficiency decline caused by corrosion or decrease of the active area of catalysts, due to the presence of a liquid phase in its constitution. In the other hand those problems are minimized in the SOFCs models, due to the absence of a liquid phase [24, 27].

SOFCs are entirely constituted by solid components and distinct designs can be found in literature, such as tubular, planar or monolithic [2, 28]. Their designs differ in the extension of dissipative losses within the cell, in the form of the seal between fuel and oxidant, and the electrical connections made cell by cell through stacking. The kind of manufacture and assembly varies depending on projects [26].

Generally, drawbacks related to the electrolytes can be found on AFC and MCFC fuel cells. On the other hand, SOFC models have solid electrolytes, which most of times are chemically and physically compatible to the other cell components. In these fuel cells, the use of noble metals as catalysts, at high temperatures, is not necessary, decreasing the costs of manufacturing and both hydrogen and carbon monoxide can act as fuels. Recently, ethanol has attracted attention as a fuel for SOFCs, as they are liquid fuels with high energy densities and are readily available from industrial processes (or renewable biomass) [29]. The operation of SOFCs using ethanol (for example) is similar to the operation of MCFC models in which a negatively charged ion is transferred from the cathode through the electrolyte to the anode. Within this process, electrons are released and transported by an external circuit performing the electrical work of the cell. Thus, water is formed in the anode as a byproduct [30].

A SOFC consists essentially of two porous electrodes (cathode and anode) separated by a solid, dense electrolyte [31]. The oxidant (usually air) passes through the pores of the cathode where it is reduced to O^{-2}. The formed ions cross the electrolyte towards the anode. The fuel (hydrogen, for example) passes through the anode and reacts with oxygen ions (oxidation process) from the electrolyte, releasing electrons and forming water. The electrons generated from these reactions are transported from the anode, through an external circuit, to the cathode, restarting the cycle [32], as shown in Fig. (**2**).

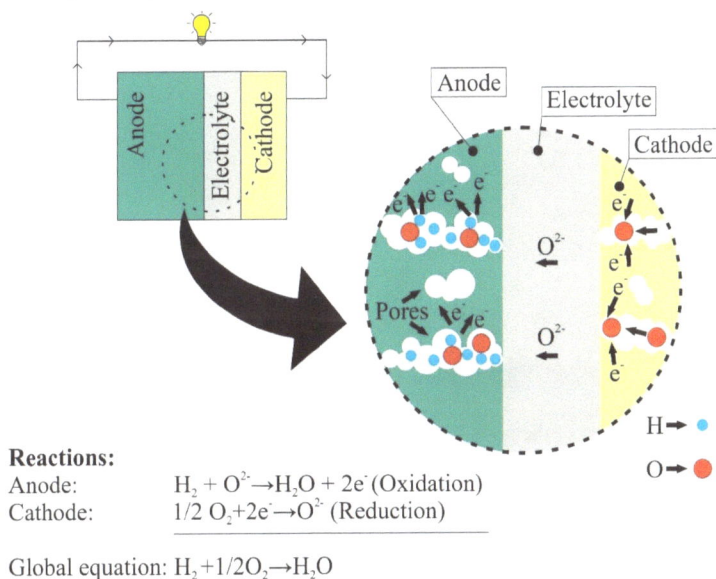

Reactions:
Anode: $H_2 + O^{2-} \rightarrow H_2O + 2e^-$ (Oxidation)
Cathode: $1/2\ O_2 + 2e^- \rightarrow O^{2-}$ (Reduction)

Global equation: $H_2 + 1/2 O_2 \rightarrow H_2O$

Fig. (2). Schematic diagram of the operation and reactions in SOFC using electrolytes oxygen ion conductors (adapted figure - (Florio *et al.*, 2007) [31]).

In SOFCs the electrochemical reactions of fuel oxidation and oxidant reduction occur in the interface gas (fuel or oxidant) electronic conductor/ionic conductor, the so-called triple phase boundary (TPB).

A schematic illustration of the region between the electrolyte and the anode where the TPB exists is shown in Fig. (**3**). If there is a problem in any of the three phases, the reaction may not occur. If the electrolyte ions cannot reach the reaction site, if the gas fuel molecules fail to reach the site, or if the electrons cannot be removed from the site, the cell performance is diminished. Although the structure and composition clearly affect the size of the TPB, various theoretical and experimental methods have been used to estimate that it occupies no more than, approximately, 10 μm from the electrolyte to the electrode. The concept of TPB has important implications with the optimization of both anodes and cathodes. The last generation of fuel cell electrodes has a complex micro/nanostructure involving an electronic and ionic interconnected conduction phase, enough porosity to allow gas penetration and active catalytic surfaces [33 - 37].

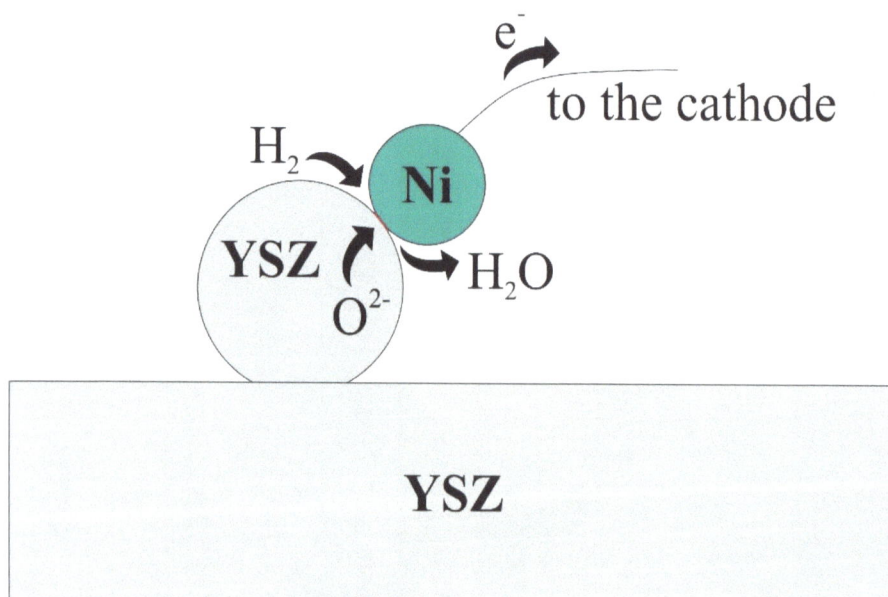

Fig. (3). Three-phase boundary of the Ni-YSZ anode.

SOFCs are efficient energy conversion devices and can achieve electrical efficiency of 45-50%. Due to its high operating temperatures and high CHP, this model of fuel cells can be effectively employed with total efficiency higher than 80%. Allied to its high efficiency, SOFCs can produce electricity by electrochemical reaction of a fuel gas (hydrogen or even methane) with oxygen in the air generating low emissions of pollutants. Due to it, this concept is very promising to be used as stationary plants as well as auxiliary power units for mobile applications [38, 39].

However, the practical use (domestic and industrial use) of SOFC still faces resistance, because of the high cost of materials (electrodes and electrolyte), the high cell complexity and manufacturing process, degradation issues arising from high operating temperatures, safety-related problems, difficulties to handle gases used in the process and maintenance routine. In addition, high operating temperatures (800 – 1000 °C) of SOFCs can lead to various material problems including electrodes coarsening, catalyst poisoning, interfacial diffusion between electrolyte materials and electrodes, thermal instability and mechanical (or thermal) tensions due to different thermal expansion coefficients (TECs) of the cell components. Such problems have limited the development and use of SOFCs largely. However, many experiences are being made to overcome such limitations, mainly focused on material selection of the device to improve the feasibility of the technology. In addition, the reduction of operating temperatures may allow selecting interconnectors materials among more varied options (including metals/alloys). Attempts to minimize resistive losses through electrolyte consist to replace yttrium-stabilized zirconia (YSZ) material by an alternative electrolytic material (such as gadolinium-doped ceria, samarium-doped ceria, among others) with higher electrical conductivity, and/or producing a fine solid oxide electrolyte sandwich. The electrolyte is considered as the heart of the system and its properties play a key role in the performance of the cell. Currently, YSZ is the most widely used electrolyte, because it has an adequate conductivity of oxide ions (0.13 S/cm at 1000 °C), and also shows a desirable phase stability in both oxidation-reduction environments [1, 40, 41].

The main challenge to improve the performance of SOFCs is the choice of the material used, which can be based on the main properties of fuel cells materials (high electron/ionic conductivity of electrodes, catalysts, sealants and interconnectors, and high ionic conductivity of electrolytes). In addition, low polarities are expected with high electrochemical activity (fast kinetics) at lower temperatures. The performance of SOFCs is also based on the processing of the fuel cell materials due to the evolution of the associated phase, grain size and phase distribution. Thus, the optimization of various synthesis processes for such materials used as electrodes, electrolytes and interconnector are the main concerns in making SOFC more feasible to real life applications. High operating temperature exerts many limitations on the selection of materials for electrodes, electrolytes, interconnect, and sealing materials. This has limited the final application of the SOFC technology [42, 43].

The cost is clearly the most important barrier to widespread SOFC implementation, there are also some technical barriers to its implementation, being the most important those related to electrodes, both anodes and cathodes [29].

SOFC Components

The SOFC is composed primarily of two electrodes (anode and cathode) separated by an electrolyte. Electrodes are the components where chemical reactions occur, releasing and conducting electrons. They must be porous, mixed electronic and ionic conductor and chemically/thermally stable. The electrolyte is the pathway that paves the flux of ions from the cathode to the anode. The electrolyte material needs to be dense, good ionic conductor and shows high thermal/chemical stability [12, 26]. The sealant is the component that avoids the fuel gas to mix with oxygen.

Each material used not only has to function ideally by meeting its own requirements, but it also has to function properly with the other components of the cell [44]. The high operating temperature imposes significant restrictions on electrodes, electrolytes and interconnection materials for SOFCs. The materials should not be reactive with neighboring components at high operating temperatures and must have compatible thermal expansion coefficients. Interconnectors and electrolytes should be impermeable to gas, to provide high electrical conductivity minimizing ohmic losses (electron - interconnectors and ionic - electrolytes), and to be stable in both atmospheres of reduction and oxidation. The ideal microstructure of a SOFC is mechanically, chemically and thermally stable during its operation [44, 45].

Electrolyte

The hole of the electrolyte is to drive O^{2-} ions towards the anode. Its main functions are: to separate the reagents between anode and cathode preventing them to react spontaneously and because of that its structure must be very dense; blocking all electronic current to flow internally, forcing it to flow in an external circuit, raising the efficiency of the cell; and promote conduction of ionic charge carriers [13, 24, 26, 46 - 48].

To perform these functions, the materials used as electrolyte must present some characteristics, such as: being a perfect electronic insulator; being good ionic conductor (anions O^{2-} conductor) at typical SOFC operating temperatures (> 500 °C); being impermeable to gases, avoiding mixing the gaseous constituents of the anode and cathode compartments; chemical stability under a wide range of partial oxygen pressures, structural stability at the temperature of use, high density, coefficient of thermal expansion compatible with the other components of the cell and being mechanically resistant [49].

The pure form of zirconia (ZrO_2) presents an insignificant ionic conductivity, however doping with appropriate amounts of oxides (Y_2O_3, Yb_2Os, Sc_2O_3, CaO,

MgO, and others) can stabilize its, increasing the concentration of oxygen vacancies and, as a result, increasing the ionic conductivity. For instance, yttria (Y_2O_3, 8 mol%) doped zirconia exhibits cubic crystalline phase at room temperature. Oxygen vacancies are created to maintain charge compensation, according to Eq. 1, using the notation of Kröger-Vink:

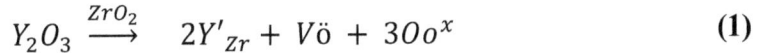

$$Y_2O_3 \xrightarrow{ZrO_2} 2Y'_{Zr} + V\ddot{o} + 3Oo^x \qquad (1)$$

where: Y'_{Zr} represents the yttrium ion that replaces the zirconium ion in the crystalline lattice of zirconia; $V\ddot{o}$ represents the avoidance of oxygen created to compensate the difference in the valences between the cations; and Oo^x represents the oxygen anion, of zero effective load, occupying a site of the lattice.

The oxygen vacancies concentration gives rise to high mobility of the oxygen ion. The conduction of oxygen ions occurs in the stabilized ZrO_2 *via* oxygen vacancies.

Other materials are being studied for electrolyte SOFC applications, such as gadolinium-doped ceria (GDC); and strontium and magnesium-doped lanthanum manganite (LSGM). However, the YSZ is the most used as electrolyte due to its high ionic conductivity and chemical and physical stability in the redox atmosphere [13, 24, 26, 46 - 48].

Cathode

The cathode is the interface between the oxidant (air or oxygen) and the electrolyte. Materials used as cathode should present mixed conductivity (electronic and ionic) and high catalytic activity for the oxidant reduction reaction; structural and chemical stability; electrolyte compatible thermal expansion; minimum reactivity between electrolyte and interconnector materials, with which the electrode is in contact and, should be sufficiently porous to facilitate the oxidant transport to the electrode/electrolyte interface. The conventional materials used are lanthanum manganite ($LaMnO_3$), which presents structure of the perovskite type. The doping of $LaMnO_3$ with Sr ($La_{1-x}Sr_xMnO_3$ - LSM) is preferable to be used in SOFC because of its high electron conductivity in oxidizing atmospheres [50]. Doping with Sr increases the electronic conductivity of $LaMnO_3$ due to the increase of Mn^{3+}/Mn^{4+} ratio by the substitution

of the ions La^{3+} by Sr^{2+}, which creates electronic carriers, ensuring the electrical neutrality of the structure, as shown in Eq. 2 [1, 13, 23, 26, 51].

$$LaMnO_3 \xrightarrow{SrO} La_{1-x}^{3+}Sr_x^{2+}Mn_{1-x}^{3+}Mn_x^{4+}O_3 \qquad (2)$$

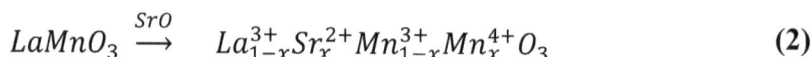

In the same way as electrical conductivity, the thermal expansion coefficient of LSM also grows with the increase of the quantity of Sr^{2+} ions replaced. Due to this, the composition of the La$_{1-x}$Sr$_x$MnO$_3$ of interest to SOFC are $0.4 \leq x \leq 0.2$. Such composition avoids different coefficients of thermal expansion in the electrolyte. Depending on the level of doping the family of compounds La$_{1-x}$Sr$_x$MnO$_3$ can present three different crystalline structures: rhombohedral ($0 \leq x \leq 0.5$), tetragonal ($x = 0.5$), and cubic ($x = 0.7$) [1, 13, 23, 26, 51].

Anode

The fuel cell anode is the interface between the fuel and the electrolyte, where the electrochemical fuel oxidation occurs. To minimize polarization losses in oxidation reactions, anode materials should present mixed conductivity (ionic and electronic), sufficient catalytic activity for the fuel oxidation and sufficient porosity (30-40 wt.%) to facilitate the transport of the fuel to the anode/electrolyte interface and the output of the reaction products. Due to operation requirements in reducing atmospheres and high electronic conductivity, metals can be used as anodic materials in SOFC [26, 52, 53].

SOFCs of high operating temperatures (~ 900 °C) and coupled to the conventional YSZ electrolyte (yttrium-stabilized zirconia) should use metals such as nickel (Ni), cobalt (Co), or noble metals like platinum (Pt), palladium (Pd) and ruthenium (Ru). Considering volatility, chemical stability, catalytic activity and cost, nickel is the best candidate to be used as metallic anode. However, the difference between thermal expansion coefficients of metals and ceramic electrolyte avoids its use as a porous layer, because metals tend to take off from the electrolyte during the thermal cycle, causing damage to the cell operation. To solve this problem an ionic conductor electrolyte is added to the anode, giving rise a composite anode material [26, 52, 53].

Ni-YSZ cermets are the most commonly anode materials for SOFC. This ceramic/metal composite can be obtained by *in situ* reduction of the composite NiO-YSZ. Ni acts as a catalyst for the oxidation reaction and electron conductor (1.38×10^6 Scm^{-1} at 1000 °C). A well-known problem of Ni at temperatures near 1000 °C is its agglomeration and loss of conductivity. Another disadvantage is the mismatch of TEC in relation with other SOFC components. The Ni-YSZ cermet has TEC of 13.3×10^{-6} K^{-1}, while the TEC of YSZ is 10.5×10^{-6} K^{-1}. Thus, the introduction of the YSZ phase into the anode composition decreases the difference between TEC of anode and electrolyte materials. Besides that, it decreases the formation of nickel pellets due to sintering [26, 52, 53].

Materials for SOFC Anodes

Cermets are used as SOFC anodes because they are more stable than a single phase metal anode. They maintain the desirable porous structure (microstructural stability) of the fuel electrode and provide better electrical properties than a pure, Ni anode. The ceramic phase works supporting Ni particles, inhibiting the metallic phase coarsening during the cell operation. It also avoids loss of conductivity, because all nickel particles are covered by the ceramics structure. Additional functions of the ceramic phase are: contribute to the ionic transport and makes the thermal expansion coefficient of the anode to closer to those of other SOFC components, avoiding delamination [26, 54].

The electrical conductivity of the Ni-YSZ composite is strongly dependent on the Ni content. A good electrical property depends on the Ni phase percolation, in this case nickel grains must act as a continuous metallic interface allowing the migration of electrons through the composite material. This requires an appropriate microstructure that depends on phase distribution and quantity. In cermets, electrical properties can be reached by two conduction mechanisms: electronic and ionic pathways through nickel and YSZ phases, respectively. The percolation threshold for the electrical conductivity occurs when the Ni amount is around 30 vol.%. Below this concentration, cermet conductivity is similar to that of the YSZ, indicating that ionic conduction dominates the total electrical response. Above this concentration, electrical conductivity is dominated by an electron conduction mechanism. This is supported by the fact that conductivity of Ni-YSZ cermets with more than 30 vol.% nickel decreases with increasing the temperature, a behavior similar to that observed for metals. In addition, the activation energy for conduction in these kind of composites is similar to that of pure nickel (5.38 KJ/mol). The cermet conductivity also depends on the composite microstructure [55].

Incorporating of YSZ in the microstructure of the anode increases the number of active sites responsible for oxidation reactions. These catalytic sites are the three-phase boundary regions. Initially, this region would be limited to the electrode/electrolyte interface, but with the introduction of the YSZ it extends throughout the electrode [56].

The three-phase boundary (TPB) area is essential for electromechanical reactions, because its length is intrinsically connected to the performance of the cell, as bigger it is better will be the performance. Normally, the length of the TPB is increased by adding a thin layer of small particles of Ni and YSZ with porosity between 10-30% at the electrolyte/anode support interface. This thin layer causes rapid kinetics of load transfer. Moreover, the density of this thin layer is bigger than that of the anode support, making the TEC of the anode support closer to that of the YSZ. It improves the adhesion between the electrolyte and the anode. However, a smaller porosity causes a decrease of diffusion of gases (fuel intake and reaction products output). During manufacture process, the thickness of this layer should be determined by considering these two factors: increased TPB and easy gas diffusion [57].

Although Ni-YSZ cermet anodes are the most used in SOFC, they are inefficient in the oxidation of other fuels such as natural gas and ethanol. These fuels contain traces of sulfur, which causes degradation of the cermet. YSZ has the limitation of having low conductivity at operating temperatures below 800 °C [58]. In the search for materials that have high performance at intermediate temperatures, reduced costs and problems allied to properties compatible with materials used to manufacture auxiliary components, as well as enhancing materials as a whole, the research focuses mainly on replacing zirconia oxide by rare-earth doped cerium oxide [59].

Cerium is the most abundant chemical element among the ones known as rare earths. Cerium has two valence states, +III and +IV. Cerium (IV) oxide (CeO_2), also known as ceria, is the most stable cerium oxide in ambient atmosphere. Within its typical fluorite-type cubic structure, the cerium ion is coordinated with eight oxygen anions at the vertices of a cube. Each anion, in turn, coordinates four Ce^{4+} cations at the vertices of a tetrahedron (Fig. **4**). The cerium oxide lattice parameter is 0.541134 nm at ambient temperature. Sub-stoichiometric compositions, however, have larger lattice parameters due to the difference between the ionic radius of Ce^{4+} (0.097 nm) and Ce^{3+} (0.1143 nm) [60 - 63].

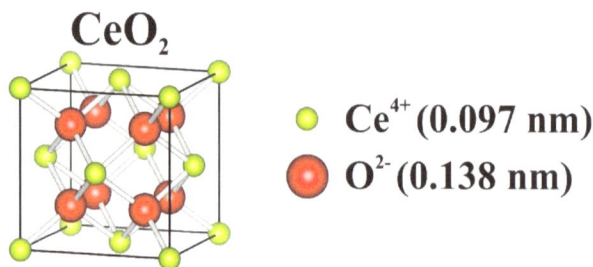

CeO_2

○ Ce^{4+} (0.097 nm)

● O^{2-} (0.138 nm)

Fig. (4). Fluorite-type structure of the cerium oxide.

Ceria can display intrinsic and extrinsic defects. The formation of intrinsic defects may appear during exposure to reducing atmospheres or due to heat agitation. Extrinsic defects are generally formed by doping with aliovalent elements. Intrinsic (Schottky or Frenkel) defects arising from thermal agitation occur at very small concentrations, not causing significant changes in the stoichiometry of the cerium oxide. However, intrinsic defects are more frequent when the CeO_2 is exposed to reducing atmospheres. Under these conditions, ceria tends to loss oxygen atoms to the environment, forming oxygen vacancies and inducing the Ce oxidation state to decrease ($Ce^{4+} \rightarrow Ce^{3+}$), but keeping the electrical neutrality of the lattice [61, 62, 64 - 67]. The equilibrium equation describing the reduction of ceria can be written as:

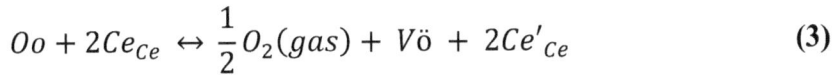

$$Oo + 2Ce_{Ce} \leftrightarrow \frac{1}{2}O_2(gas) + V\ddot{o} + 2Ce'_{Ce} \qquad (3)$$

where, according to Kröger-Vink notation, Oo and Ce_{Ce} represent the oxygen and cerium ions in their respective positions, $V\ddot{o}$ refers to the oxygen vacancy created on the oxygen site and Ce'_{Ce} represents the Ce^{3+} cation in the position of the Ce^{4+}.

The ionic conductivity of pure ceria is low, but it increases and achieves a maximum value when doping is done with cations having ionic radius near to that of Ce [64, 68 - 70].

The doping of ceria with divalent or trivalent metal oxides, or the presence of impurities in its crystalline lattice, can lead to the formation of extrinsic defects. When ions of valence 2+ or 3+ are inserted into the cerium oxide lattice, the substitution of Ce^{4+} cations by the dopant cations induces the formation of oxygen vacancies to balance the electric charges [61].

In the ceria structure, metal ions are surrounded by eight O^{2-} anions, which form the vertices of the cube. This structure type is relatively open, which makes it susceptible to the insertion of a wide variety of doping elements. Doping materials with an ion of analogous valence (M^{x-1}) of the replaced ion and similar ionic radius ($\pm$ 15% of the replaced ion) generate a greater number of oxygen vacancies compared to the same material without dopant. Gd is among the main metal ions usually used to dope the ceria structure [62, 64 - 67], it occurs as follows:

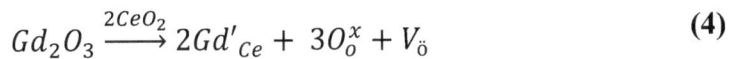

$$Gd_2O_3 \xrightarrow{2CeO_2} 2Gd'_{Ce} + 3O_o^x + V_{\ddot{o}} \qquad (4)$$

where, according to Kröger-Vink notation, Gd′Ce represents the replacement of the Ce^{4+} ion by a Gd^{3+} ion, O_o^x represents the oxygen ion in its normal position in the crystalline lattice, and $V\ddot{o}$ is the avoidance of doubly ionized oxygen [62, 64 - 67].

Methods for Obtaining NiO-GDC Composites

Anode materials are usually obtained by mechanical mixture of commercial and/or synthesized powders [71 - 75], being subsequently sintered in air and further reduced in hydrogen to form the cermet. This material must contain an electronically conductive metallic phase and a ceramic phase with ionic or mixed conductivity. The chemical composition during the cermet manufacturing is simple and easy to control, but it may cause concerns regarding limited uniform distribution of the elements, possibly resulting in inhomogeneous microstructures. In Ni-GDC cermets, nickel particles must be linked to the ceramic matrix particles and exposed to the fuel gas. To obtain an anode with efficient electrical (mixed ionic and electronic) performance, it is necessary to obtain NiO-GDC nanocomposites with uniform distribution of NiO and GDC phases [76 - 78].

Several synthesis methods are used to significantly decrease the particle size of NiO-GDC composites, which is expected to enhance the anode electrochemical performance. NiO and GDC powders may be separated prepared by a glycine–nitrate solution method and mix together by ball milling. Alternative synthetic methods are spray pyrolysis, co-precipitation, hydrothermal synthesis, combustion, sol-gel process, polymeric organic complex, and cellulose precursor method [79 - 87].

Most of these techniques involve complicated chemical routes and are time consuming. An interesting, one-step polymer precursor-based method was developed by Cela *et al.* [7] to make NiO-GDC as an *in situ* nanocomposite.

Polymeric Precursor Method (Pechini Method)

The polymeric precursor method was proposed by Pechini in 1967 [88]. It presents several advantages in the synthesis of nanoparticles, being highlighted as an alternative and promising process to obtain nanoscale materials with the following benefits: chemical homogeneity, high purity, stoichiometric control, and good sinterability [89 - 94].

Pechini method is based on the formation of a polymer precursor, by means of condensation reactions using a hydroxycarboxylic acid (such as citric, lactic and glycolic acids) combined to an alcohol (such as ethylene glycol (EG)), as shown in Fig. (**5**). The method created by Pechini involves two basic reactions: (1) chelation between complex cations and citric acid (CA), leading to a metallic citrate; and (2) chelate polyesterification with ethylene glycol in a slightly acidified solution, leading to the formation of a polyester resin [95, 96].

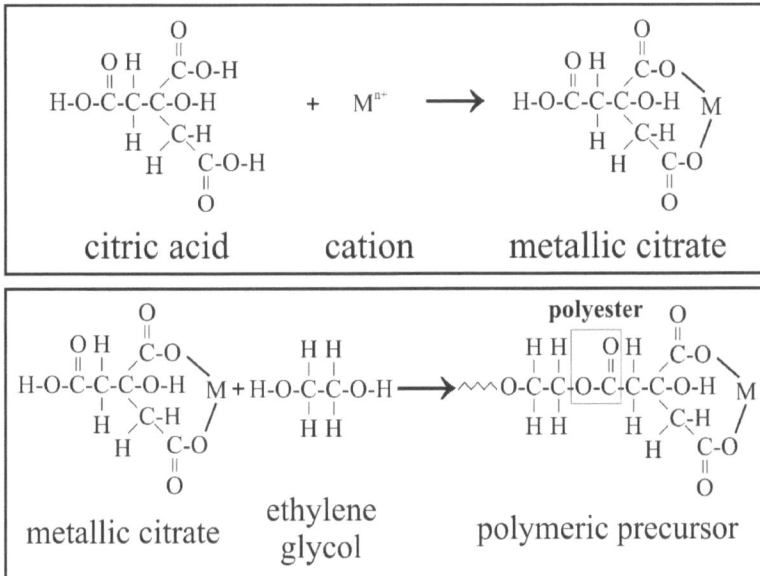

Fig. (5). Reactions involved in the Pechini method.

The acid is a chelating agent responsible to bind cations. These cations, normally salts, are mixed with hydrocarboxylic acids in an aqueous solution. Several cation salts can be used, such as chlorides, hydroxides, carbonates, nitrates or any other salts that ensure purity, solubility in solution and easy anion elimination. Metallic ions are assigned by carboxylic groups and remain homogeneously distributed in the polymer structure [96].

The polymerization reaction occurs through a chelation between metallic citrate and an alcohol, in presence of heat, obtaining a complete solubilization of all components, characterized by a translucent solution. An intermediate porous and rough resin is required to obtain non-agglomerated powders, in a single phase and with fine grains. The control of the process parameters is necessary. These parameters include mass ratio of organic substances and metallic salts, water content in the mixture and spraying before the calcination [96, 97].

The obtained polymer resin has the ability to confine cations in its polymer structure, reducing the cation mobility and, in turn, collaborating to a good chemical homogeneity [96].

The resin is subjected to thermal treatments at 300-1000 °C to break polymeric chains, induce organic matter pyrolysis, eliminate organic waste, and promote oxidation of desired crystalline phases [96].

NiO-Ce$_{0.9}$Gd$_{0.1}$O$_{1.95}$ nanocomposite powders prepared by the one-step synthesis method offer better microstructural features, such as improved contact area between the electronic conductive phase (Ni) and the ionic conductive phase (GDC). To produce this nanocomposite, precursor resins of the individual phases (NiO and GDC) are separated obtained by the polymeric precursor method and then mix together to become a single homogeneous resin [7].

This process comprises eight fundamental stages. The 1st stage refers to the dissolution of cerium cations into distilled or deionized water, followed by a complexion step with an organic acid. The 2nd stage refers to the dissolution of cations used to dope the ceria structure. The 3rd stage refers to the dissolution of nickel nitrate (in a second beaker) followed by complexion with an organic acid. The 4th stage is related to the poly alcohol addition aiming to polymerize both solutions (NiO and GDC precursors). The 5th stage involves the elimination of water excess from solutions. The 6th stage is related to the formation of the gel mixture composed by Ce-Gd-Ni cations. The 7th stage comprises the carbon oxidation to obtain an amorphous, precursor material. In the 8th stage occurs the crystallization of the desired phases [98].

Sintering

Sintering is a thermally activated process that connects crystalline particles, in a compacted form to give a coherent solid structure, *via* mass transport events occurring at atomic scale. It is followed by many microstructural modifications taking place in ceramic, metal or composite materials. Depending on the sintering conditions (temperature, atmosphere, pressure and so on) the following properties may change: density, surface area, mechanical strength, and thermal conductivity [99]. Sintering occurs when particles in close contact with each other and the system temperature is between 50-90% of the material melting temperature. The feasibility of a given material to the sintering process is dependent on powder characteristics such as particle size and phase composition. Particles of different sizes tend packaging better and form denser ceramics after sintering. Therefore, processing conditions also exert a great influence on the sintering process [100].

The sintering process may be divided into three stages. In the early stage, changes are quickly obtained, being characterized by the begin of neck formation generated by the connection between particles. This rearrangement consists of slight movements or adjacent particle rotation, causing an increase in the number of contact points. At the begin of the process, there is no grain growth, causing small changes in the material density (when it occurs). Porosity can be reduced substantially and its characteristics are dependent on the next stages of the process. At the intermediate stage there is a considerable grain growth that causes porosity reduction. The final stage is characterized by preeminent grain growth and formation of spherical/isolated pores [85, 101 - 104].

Impedance Spectroscopy

Impedance spectroscopy (IS) is an analytical technique used to measure the electrical response of solid or liquid materials (ionic, semiconductor and even dielectric) and electronic devices. It explores the frequency dependent response of materials and electrical devices. This technique is used to separate electrical contributions of different microstructural components in the frequency domain. The use of this technique is relatively simple since the obtained results can be related to dielectric properties, polarization effects, crystalline defects, microstructure, and electrical conductivity [11, 105 - 111].

Many types of stimulus can be used to perform impedance spectroscopy measurements, however the most common, and considered the standard process, is to use a sinusoidal alternating current (a.c.). It allows to measure the complex (real and imaginary contributions) impedance as a function of the frequency. The graph showing the relationship between real and imaginary contribution of impedance and frequency is the base to form the impedance spectrum of any device or sample [106, 110, 112].

Electrical impedance can be considered as a measure of opposition to an alternating current signal. Furthermore, impedance is a more general concept when compared to electrical resistance, because it allows to decompose many terms concerning dissipation and polarization phenomena. The magnitude of the impedance is mapped into an orthogonal axis system where the sum of elements along the axes reproduces the measured magnitude, in the form of $Z = a + jb$, being $j = \sqrt{-1} = \exp(j\pi/2)$. An impedance $Z(\omega) = Z' + jZ''$ is the greatness representation of this property in a cartesian coordinate plan, with the real part represented by $Re(z) = Z' = |Z|\cos\theta$ and the imaginary part given by $Im(Z) = Z'' = |Z|\mathrm{sen}\theta$, being $\theta = \mathrm{tg}^{-1}(Z''/Z')$ the respective phase angle. In general Z is conditioned to the frequency, so that in a scan of $Z(\omega)$ *versus* ω it is acceptable to obtain a complete data set of electrical properties (interfaces and volumes) [106, 108, 110].

Fig. (6) shows a simplified scheme of an impedance analyzer. The method consists essentially in applying a known disturbance (a potential or an electrical current) on the electrodes observing the response (the resulting current and/or voltage) [113].

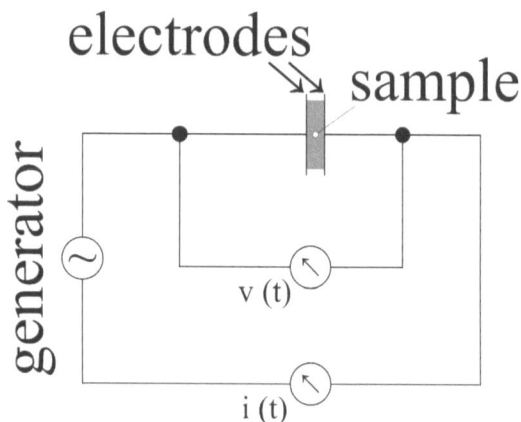

Fig. (6). Experimental arrangement of an impedance analyzer.

The general objective of the impedance spectroscopy is determining the materials electrical properties and correlate them with controlled operating variables such as pressure, temperature and applied voltage (in case of polarization conditions) [113, 114].

The response of any system to an applied disturbance may differ in phase angle and amplitude, when compared to the signal applied over the system. Measurements of the phase difference, regarding current and voltage amplitude signals and the real and imaginary contributions of impedance allow obtaining the impedance spectrum [114].

From an experimental point of view, the study of a given system usually has a set of measured values (experimental data), which represents the system response to a certain type of disturbance applied during the test. In this situation, the graphical representation of experimental data is an important artifice because it provides a visual representation of the system behavior, allowing the acquisition of parameters related to the studied phenomenon. In particular, the graphical representation is a very employed way to compare a specific model of the system and the experimental data [114].

As mentioned above, the impedance of a circuit usually has real and imaginary parts, and the graphical representation of these quantities is called the impedance spectrum. The most common representations for impedance, in the context of electrical circuits, are the Nyquist (or Argand) and Bode plots [110, 115].

In the Nyquist plot (complex plane), the imaginary part Z'' of the impedance is represented as a function of the real part Z', for several values of the frequency. In the Bode plot, the impedance module, |Z|, and the phase angle ϕ are represented as a function of the angular frequency [110].

The Nyquist plot is defined by a series of dots with each one representing the magnitude and direction of the impedance vector for a particular frequency. The graph is a complex plan of Cartesian coordinates, where the real part is represented in the abscissa (resistive part) and the imaginary part (capacitive and inductive parts) are shown in the axis of ordinates. Impedance data are usually plotted in a large frequency range to represent all electrochemical mechanisms of a given device or sample. Fig. (**7**) shows a typical Nyquist plot accompanied by its equivalent circuit.

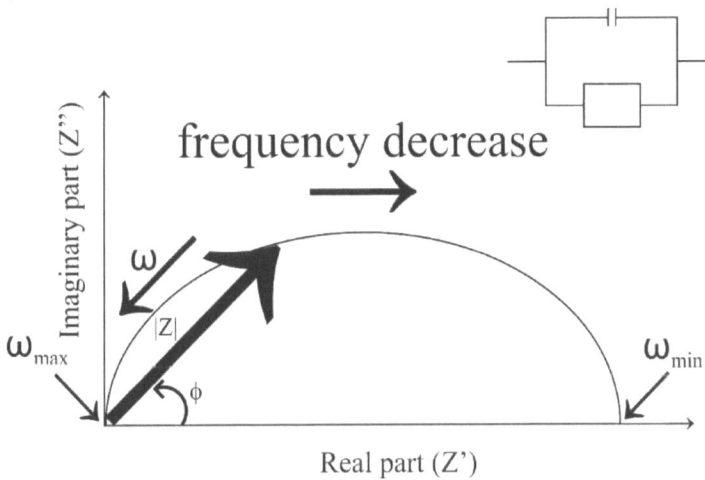

Fig. (7). Representation of a typical Nyquist plot.

The Bode plot consists of an orthogonal axes plane, where the ordinate axis represents two magnitudes: the logarithm of the impedance ($\log|Z|$) in ohms (Ω) and phase angle (ϕ) in degrees; and on the abscissa axis, the angular frequency logarithm ($\log\omega$) with ω in radians per second (rad/s). The logarithmic frequency ($\log f$) can be represented in the abscissa axis, with f in Hertz (Hz). In the $\log\omega$ *versus* impedance module |Z| configuration, both R_s and R_p can be determined, as shown in Fig. (**8**). By means of $\log\omega$ *versus* phase angle, it is possible to determine the dual-layer electric capacitance Cdl, knowing that $R_p = 2|Z|.\tan\phi_{max}$ and $\omega^\phi_{max} = 1/C_{dl}R_p(1+R_p/R_s)^{1/2}$, where:

R_s: solution resistance;

R_p: polarization resistance;

ϕ_{max}: is the system maximum impedance phase angle;

ω^{ϕ}_{max}: is the angular frequency corresponding to the maximum ϕ; and

$|Z|$: is the impedance module corresponding to the maximum ϕ.

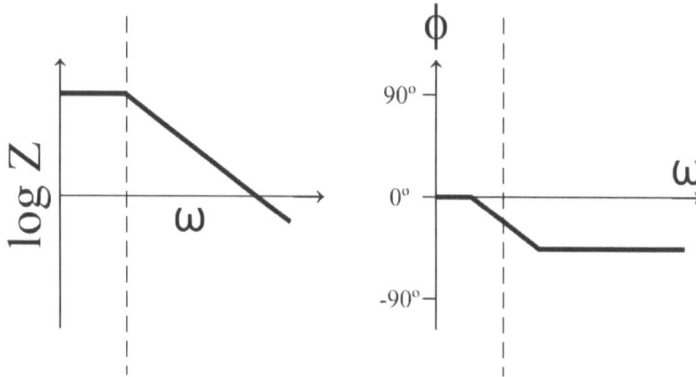

Fig. (8). Representation of a typical Bode plot.

The proper interpretation of a given impedance spectrum involves three basic steps. Initially, some relevant electrochemical knowledge about the sample under investigation must help to define an appropriate equivalent circuit model. Then, preliminary parameter values of all circuit components need to be determined. Finally, parameters related to the electrochemical reactions are calculated using the equivalent model and preliminary parameters (Fig. **9**) [110].

The possibility of associating to a particular set of results more than an equivalent circuit, is one of the outcomes of this generalist approach. However, the procedures involved to build an equivalent circuit consistent with a physical model. The most important effects impacting the choice of an appropriate equivalent circuit are the experimenter physical intuition and exhaustive tests under different conditions, allowing sustaining the mentioned physical model. Stablishing a similarity relationship between resistive and capacitive phenomena and understanding the behavior of resistance (R) and capacitance (C) allows interpreting impedance spectra through equivalent RC circuits [115, 116].

The resistance values are obtained from the interceptions with the real axis. The semicircles are often centered at a point below the real axis, receiving the denomination of flat semicircles. The decentralization angle, α, of these semicircles presents two real roots, moving away from the ideal behavior. In these situations, the capacitor (C) of an RC element can be replaced by a constant phase element (CPE) (Fig. **9**) [117].

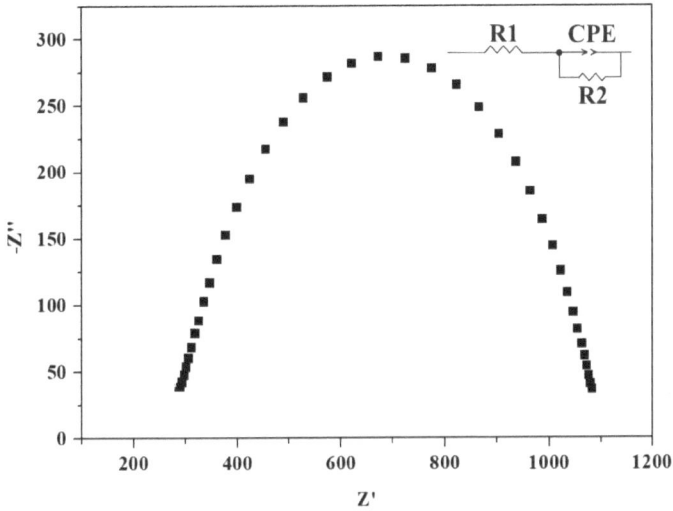

Fig. (9). Schematic diagram of a typical Nyquist plot and its respective equivalent circuit showing imaginary (Z") and real (Z') impedance parts.

CPE is a generalized capacitor that considers the "no ideality" of experimentally observed semicircles. The parameter "n" measures the imperfection of the arc. For n = 1, CPE is identical to a capacitance, corresponding to a perfect semicircle in the impedance spectrum. When n < 1, flats arcs are observed. Assuming a constant phase element as an input data it is possible to achieve an exact fit for non-ideality arcs [110]. Using the adjustment parameters CPE and n, capacitance (C) can be calculated by equation: $C = (R^{1-n}EFC)^{1-n}$.

Making use of obtained data from impedance diagrams is possible to calculate the electrical resistivity, ρ, or conductivity, σ: $R = \rho(L/S) = (1/\sigma)(L/S)$, in which L is the sample thickness and S is the cross-sectional area [110].

The advantage of this a.c. (alternated current) method is the ability to measure sample resistance variation, depending on temperature and frequency, while other techniques only vary the temperature. In addition, this technique is the only one that correlates microstructure and electrical properties [110].

The dependence between electrical conductivity and temperature is calculated by Arrhenius equation (Eq. 5):

$$\sigma = \frac{A}{T} \exp\left(\frac{-Ea}{kT}\right) \tag{5}$$

where A is the pre-exponential factor, Ea is the activation energy for the conduction process; K is the Boltzmann constant; and T is the measured temperature [110]. The ln (σT) x 1000/T graph shows a straight-line that allows to calculate activation energy of the conduction process.

The electrical resistance of the anode is composed mainly of internal resistance, contact resistance, concentration and activation polarization. The internal one measures the resistance of electrons to be transported within the anode. The contact resistance is related to the adherence at the anode/electrolyte interface. The concentration polarization measures the difficulty of gases to flow through electrode pores. This phenomenon is related to the electrode microstructure, specifically its porosity, pores size and tortuosity. Activation polarization is associated to the charge transfer process that depends on the TPB extension [3].

Impedance Spectroscopy of NiO-GDC Composites

The impedance spectroscopy technique has very limited bibliographic references about the interpretation of NiO and CeO$_2$ composite curves and how each phase contributes to the electrical conductivity. The work published by Grilo *et al.* [118] presented impedance spectra obtained in air atmosphere, between 300 and 650 °C, for NiO-GDC composites with NiO content varying from 0 to 100 wt.%. Impedance spectra preserve the typical form of spectra for crystalline solid electrolytes. However, authors approached only the total electrical resistance of the composites, making no distinction between grain/grain boundary resistances of each phase. They observed that increasing the level of NiO composites the total resistivity increases and decreases depending on the level of NiO phase percolation. The conductivity of composites with 10 wt.% NiO is lower than that of pure GDC, while the conductivity of the composite with 20 wt.% NiO is comparable to that of GDC. It is probably explained by the smaller specific gravity of those composites when compared to the GDC density. Additionally, total conductivity in this composition range (10-20 wt.% NiO) is found to be predominantly ionic. With 30 wt.% NiO the composites conductivity at low temperatures (up to 450 °C) reached higher values. The increase in conductivity was mainly due to the contacts (percolation) between the grains of the electronic conductive phase (NiO), even fewer in the composites with less than 30 wt.% NiO.

Araujo *et al.* [119] also carried out the electrical characterization of the NiO-GDC composite containing 50 wt.% NiO, sintered at 1400 °C and free of pore former (open porosity of 22%) by impedance spectroscopy measurements in air at 100-650 °C. The study addressed only the total electrical resistance of the composite, making no distinction between grain resistance and grain boundary resistance.

Based on that, the total resistance values for each temperature were obtained from the interception with the Z' axis (real impedance) in the low frequency region (0.01 Hz).

The relatively high porosity of the composite, showing few NiO-NiO and GDC-GDC contacts, affected the electrical properties at high (dominated by the GDC phase) and low (dominated by the NiO phase) temperatures. The Arrhenius plots presented straight lines for both samples, suggesting the existence of a single conduction mechanism (a single activation energy value, Ea) in the temperature range between 150 and 650 °C. The activation energy of the process in the studied composite was very close to that of GDC, indicating that this phase dominated the total conductivity in the NiO-GDC composite.

Wandekar *et al.* [120] showed impedance spectra in atmospheric air for NiO-GDC composites containing 66.7; 75; 81.8; and 87.5 mol% NiO. Regardless to the NiO level, the impedance spectra revealed two overlapped semicircles at 135 °C. The grain boundary resistance was smaller than the grain one for all composites studied. This characteristic is related to larger grain size and lack of obstruction in the grain boundary. Is was obtained by a good dispersion of NiO particles between GDC grains, thus avoiding its agglomeration. Activation energies were found to be 0.222; 0.189; 0.165 and 0.178 eV for samples with 66.7, 75, 81.8 and 87.5 mol% NiO, respectively. The plot of $\ln\sigma$ *versus* NiO concentration (at three different temperatures) showed a constant increase in the grain conductivity while the content of NiO is 66.7 or 81.8 mol%, but felt slightly for the sample with 87.5 mol% NiO. All composites with higher NiO content are expected to show predominantly electronic conduction. The authors commented that more work is needed to elucidate the conduction mechanisms in these composites.

Macedo *et al.* [8] showed impedance spectra at 250 °C for NiO-GDC samples sintered at 1350-1450°C. As previously reported by Wandekar *et al.* [120] the presence of two overlapping semicircles spanning the whole frequency range was also noted. These impedance spectra preserved the typical form of polycrystalline solid electrolytes consisting of high and low frequency arcs approximately related to grain (R_g) and grain boundary (R_{gb}) resistances, respectively. Therefore, each semicircle was simulated by a combination of a resistor and a constant phase element in parallel (R-CPE). It was observed that composites derived from powders obtained by one-step synthesis show R_{gb} much lower than R_g, especially for samples sintered at the two lowest temperatures. The gradual decrease in total resistance could be explained by the decrease in porosity (from 14 to 11% for sintered samples at 1350 and 1400 °C, respectively), achieving a stepped shape for samples sintered at 1450 °C (porosity 2%) corresponding to the elimination of percolated pores (usually reached by about 93% of the theoretical density). It was discussed that the significant grain agglomeration of conventional samples (derived from commercial post mixes from GDC and NiO) resulted in composites with much higher R$_{gb}$ than R$_g$, which was actually confirmed by the impedance spectra. Generally, composites obtained from commercial powders clearly display higher R$_g$ and R$_{gb}$ when compared to one-step synthesized materials. Another relevant difference in the spectra of conventional composites was the opposite evolution of semicircles amplitude, which decreased to the sintered samples at 1400 °C (in relation to sintered at 1350 °C) because of the reduction in density, as expected. For samples sintered at 1450°C the semicircle amplitude unexpectedly increased substantially. In addition, the shift towards low frequency of the relaxation frequency of the so-called grain boundary semicircle clearly indicated a major change in the transport mechanism. Authors mentioned that behavior was probably related to microstructural features of these composites, nominally a very coarse and heterogeneous distribution of clusters of both phases.

Chavan *et al.* [121] reported that electrical conductivity in NiO$_x$-GDC$_{(1-x)}$ composites, with x = 0.1; 0.2; 0.3; 0.4; 0.5; and 0.6 (mol%), occurs through two mechanisms: ionic (through GDC phase) and electronic (through nickel oxide phase). The d.c. electrical conductivity was measured using a two-probe method in air at a temperature range of 200-700 °C. Conductivity of all nanocomposites at 600 °C presented a S type behavior with an increase in conductivity above 40 mol% NiO. When the NiO concentration is below 40 mol%, the conductivity is predominantly ionic. Above 40 mol% NiO the conductivity is predominantly electronic (metal type). The conductivity almost saturates above 40 mol% NiO, indicating this concentration is the percolation threshold. It was observed that the activation energy for 10 mol% NiO was 0.8 eV, decreasing to 0.55 eV in composites with 40 mol% NiO. The activation energy has steadily decreased when NiO levels were increased. For low NiO contents, this phase was clearly distributed throughout the composite, affecting its connectivity with GDC phase and consequently reducing ionic conductivity. Because of that, composites with low NiO contents showed electronic conductivity prevailing over the system. Despite of this characteristic of low NiO content composites, they had activation energies smaller than 1 eV. For further additions of NiO, NiO and GDC phases are distributed homogeneously, thus forming a three-dimensional connectivity. NiO/GDC contacts would increase leading conduction path along the interface between NiO and GDC. Authors commented that more information is needed on microstructure and impedance spectroscopy.

Pyda [122] conducted a study of electrical properties in Ni-YSZ cermets. A continuous increase in the absolute value of impedance was observed with increasing the measuring temperature, which is typical of metals. The phase angle slightly changed with temperature. Notable changes of | Z | and phase angle were observed for the cermet containing 30 wt.% Ni. Some different Bode plots (shape and values) were observed in the case of a sample with 30% Ni in argon atmosphere. The Nyquist plots on the imaginary impedance were shown for samples containing 40 and 30 wt.% Ni. The electrical resistivity increased with increasing the temperature.

Pinheiro, Martinelli and Fonseca [123] carried out measurements of electrical properties in NiO/ZrO_2 composites: 8 mol% of Y_2O_3 (YSZ)/CeO_2 (60/20/20 wt.%) sintered at 1200 °C for 3 h. Impedance spectroscopy was carried out in the frequency range from 1 Hz to 30 MHz and varying the temperature from 100 to 700 °C. Composites presented a relatively low electrical resistance due to the high level of NiO (~ 60 vol.%), which resulted in a parasitic inductive effect more at high frequencies. Impedance spectroscopy analyses were limited to the total electrical resistance at the entire temperature range studied. The impedance plots showed a decrease in electrical resistivity with increasing temperature. The calculated activation energies were ~ 0, 65 eV for T < 450 °C and ~ 0,39 eV for T > 450 °C. These values were lower than that expected for YSZ (~ 1 eV), indicating that NiO dominated the electrical transport in the composite.

Recently, Grilo *et al.* [124] assessed the electrical properties of composites obtained by mechanical mixing of commercial powders and by a one-step synthesis method. Although the work focused on the one-step composite with 50 wt.% NiO, the distinct electrical characteristics provided by impedance spectroscopy indicates the contribution of the NiO phase for the total electrical conductivity of this kind of composite. The qualitative study highlights the separation of the distinct conducting pathways by using a circuit model where the electronic resistance is in parallel with the ionic conductor contribution (association in series of a parallel association of constant phase element - CPE - and resistance). The predictable electrical performance was correlated to the composite microstructure, directly influencing the magnitude of intermediate frequency (IF) arcs. The role of NiO could be shifted from a well percolated phase (composite obtained by one-step synthesis) providing well electronic conducting pathway (with lower magnitude IF arcs), to a dispersed ion-blocking phase in composites obtained by mechanical mixing (with higher magnitude of IF arcs).

CONCLUDING REMARKS

This chapter presents a brief review of the solid oxide fuel cell technology and the current stage of development of nickel-based cermet anodes, with emphasis on those containing Ni and GDC or YSZ as ceramic components. We report the main synthesis methods available to obtain anode precursor powders. Moreover, a review on the effects of processing on the microstructure and electrical properties of these materials is described in detail. Emphasis is given to powder preparation by chemical synthesis and electrical characterization by impedance spectroscopy.

CONSENT FOR PUBLICATION

Not applicable.

CONFLICT OF INTEREST

The authors declare no conflict of interest, financial or otherwise.

ACKNOWLEDGEMENT

Allan J. M. de Araújo and Gabriel M. Santos thank Coordenação de Aperfeiçoamento de Pessoal de Nível Superior - Brazil (CAPES) - Finance Code 001.

REFERENCES

[1]   Mahato, N.; Banerjee, A.; Gupta, A.; Omar, S.; Balani, K. Progress in Material Selection for Solid Oxide Fuel Cell Technology: A Review. *Prog. Mater. Sci.,* **2015**, *72*, 141-337.
        [http://dx.doi.org/10.1016/j.pmatsci.2015.01.001]

[2]   Minh, N.Q. Solid Oxide Fuel Cell Technology - Features and Applications. *Solid State Ion.,* **2004**, *174*, 1-4, 271-277.
        [http://dx.doi.org/10.1016/j.ssi.2004.07.042]

[3]   Zhu, W.Z.; Deevi, S.C. A Review on the Status of Anode Materials for Solid Oxide Fuel Cells. *Mater. Sci. Eng. A,* **2003**, *362*, 1-2, 228-239.
        [http://dx.doi.org/10.1016/S0921-5093(03)00620-8]

[4]   Timmermann, H.; Fouquet, D.; Weber, A.; Ivers-Tiffée, E.; Hennings, U.; Reimert, R. Internal Reforming of Methane at Ni/YSZ and Ni/CGO SOFC Cermet Anodes. *Fuel Cells (Weinh.),* **2006**, *6*, 3-4, 307-313.
        [http://dx.doi.org/10.1002/fuce.200600002]

[5]   Othman, M. H. D.; Wu, Z.; Droushiotis, N.; Kelsall, G.; Li, K. Morphological Studies of Macrostructure of Ni-CGO Anode Hollow Fibres for Intermediate Temperature Solid Oxide Fuel Cells *J. Memb. Sci,* **2010**, *360*(1–2), 410-417.
        [http://dx.doi.org/10.1016/j.memsci.2010.05.040]

[6]   Macedo, D. A.; Figueiredo, F. M. L.; Paskocimas, C. A.; Martinelli, A. E.; Nascimento, R. M.; Marques, F. M. B. Ni-CGO Cermet Anodes from Nanocomposite Powders: Microstructural and Electrochemical Assessment *Ceram. Int,* **2014**, *40*(8 PART B), 13105-13113.
        [http://dx.doi.org/10.1016/j.ceramint.2014.05.010]

[7]   Cela, B.; De MacEdo, D.A.; De Souza, G.L.; Martinelli, A.E.; Do Nascimento, R.M.; Paskocimas, C.A. NiO-CGO *in situ* Nanocomposite Attainment: One Step Synthesis. *J. Power Sources,* **2011**, *196*(5), 2539-2544.
        [http://dx.doi.org/10.1016/j.jpowsour.2010.11.026]

[8]   Macedo, D.A.; Souza, G.L.; Cela, B.; Paskocimas, C.A.; Martinelli, A.E.; Figueiredo, F.M.L.; Marques, F.M.B.; Nascimento, R.M. A Versatile Route for the Preparation of Ni-CGO Cermets from Nanocomposite Powders. *Ceram. Int.,* **2013**, *39*(4), 4321-4328.
        [http://dx.doi.org/10.1016/j.ceramint.2012.11.014]

[9]   Choo, C. K.; Horri, B. A.; Salamatinia, B. NiO-CGO *in situ* Nanocomposite Attainment: One Step Synthesis *J. Power Sources,* **2011**, *196*(5), 2539-2544. [http://dx.doi.org/10.1016/j.jpowsour.2010.11.026]

[10]   Nielsen, J.; Hjelm, J. Impedance of SOFC Electrodes: A Review and a Comprehensive Case Study on the Impedance of LSM. *YSZ Cathodes. Electrochim. Acta,* **2014**, *115*, 31-45. [http://dx.doi.org/10.1016/j.electacta.2013.10.053]

[11]   Huang, Q.A.; Hui, R.; Wang, B.; Zhang, J. A Review of AC Impedance Modeling and Validation in SOFC Diagnosis. *Electrochim. Acta,* **2007**, *52*(28), 8144-8164. [http://dx.doi.org/10.1016/j.electacta.2007.05.071]

[12]   Fuel Cell Handbook. EG&G technical services, Inc., Albuquerque, NM, DOE/NETL-2004/1206, **2004**.

[13]   de Florio, D.Z.; Fonseca, F.C.; Muccillo, E.N.S. R. M. Materiais Cerâmicos Para Células a Combustível. *Ceramica,* **2004**, *50*(594), 275-290. [http://dx.doi.org/10.1590/S0366-69132004000400002]

[14]   Singhal, S. Advances in Solid Oxide Fuel Cell Technology. *Solid State Ion.,* **2000**, *135*(1), 305-313. [http://dx.doi.org/10.1016/S0167-2738(00)00452-5]

[15]   Ormerod, R.M. Solid oxide fuel cells. *Chem. Soc. Rev.,* **2003**, *32*(1), 17-28. [http://dx.doi.org/10.1039/b105764m] [PMID: 12596542]

[16]   Stambouli, A.B.; Traversa, E. Solid Oxide Fuel Cells (SOFCs): A Review of an Environmentally Clean and Efficient Source of Energy. *Renew. Sustain. Energy Rev.,* **2002**, *6*(5), 433-455. [http://dx.doi.org/10.1016/S1364-0321(02)00014-X]

[17]   Steele, B.C.; Heinzel, A. Materials for fuel-cell technologies. *Nature,* **2001**, *414*(6861), 345-352. [http://dx.doi.org/10.1038/35104620] [PMID: 11713541]

[18]   Sharaf, O.Z.; Orhan, M.F. An Overview of Fuel Cell Technology: Fundamentals and Applications. *Renew. Sustain. Energy Rev.,* **2014**, *32*, 810-853. [http://dx.doi.org/10.1016/j.rser.2014.01.012]

[19]   Loges, B.; Boddien, A.; Junge, H.; Beller, M. Controlled Generation of Hydrogen from Formic Acid Amine Adducts at Room Temperature and Application in H_2 / O_2. *Fuel Cells (Weinh.),* **2008**, *4*, 3962-3965. [http://dx.doi.org/10.1002/anie.200705972]

[20]   Wang, Y.; Chen, K.S.; Mishler, J.; Chan, S.; Cordobes, X. A Review of Polymer Electrolyte Membrane Fuel Cells : Technology, Applications, and Needs on Fundamental Research. *Appl. Energy,* **2011**, *88*(4), 981-1007. [http://dx.doi.org/10.1016/j.apenergy.2010.09.030]

[21]   Neef, H.J. International Overview of Hydrogen and Fuel Cell Research. *Energy,* **2009**, *34*(3), 327-333. [http://dx.doi.org/10.1016/j.energy.2008.08.014]

[22]   Kundu, A.; Jang, J. H.; Gil, J. H.; Jung, C. R.; Lee, H. R.; Kim, S.; Ku, B.; Oh, Y. S. Micro-Fuel Cells — Current Development and Applications **2007**, *170*, 67-78. [http://dx.doi.org/10.1016/j.jpowsour.2007.03.066]

[23]   Carrette, B. L.; Friedrich, K. A.; Stimming, U. Stimming, U. Fuel Cells ± Fundamentals and Applications **2001**, *1*, 5-39. [http://dx.doi.org/10.1002/1615-6854(200105)1:1<5::AID-FUCE5>3.0.CO;2-G]

[24]   Carrette, L.; Friedrich, K.A.; Stimming, U. Fundamentals and Applications. In: *Fuel Cells*; , **2001**; 1, pp. 5-39. [http://dx.doi.org/10.1002/1615-6854(200105)1:1<5::AID-FUCE5>3.0.CO;2-G]

[25]   Steinberger-Wilckens, R.; Blum, L. Overview of the Development of Solid Oxide Fuel Cells at Forschungszentrum Juelich. *Int. J. Appl. Ceram. Technol.,* **2006**, *476*, 470-476.

[http://dx.doi.org/10.1111/j.1744-7402.2006.02102.x]

[26]   Minh, N.Q. Ceramic Fuel Cells. *J. Am. Ceram. Soc.*, **1993**, *76*(3), 563-588.
        [http://dx.doi.org/10.1111/j.1151-2916.1993.tb03645.x]

[27]   Kendall, K.; Kendall, M. *High-Temperature Solid Oxide Fuel Cells for the 21st Century:
        Fundamentals*; Design and Applications, Second Edi, **2015**.

[28]   Song, C. Fuel Processing for Low-Temperature and High-Temperature Fuel Cells: Challenges, and
        Opportunities for Sustainable Development in the 21st Century. *Catal. Today,* **2002**, *77*, 1-2, 17-49.
        [http://dx.doi.org/10.1016/S0920-5861(02)00231-6]

[29]   Atkinson, A.; Barnett, S.; Gorte, R.J.; Irvine, J.T.; McEvoy, A.J.; Mogensen, M.; Singhal, S.C.; Vohs,
        J. Advanced anodes for high-temperature fuel cells. *Nat. Mater.,* **2004**, *3*(1), 17-27.
        [http://dx.doi.org/10.1038/nmat1040] [PMID: 14704781]

[30]   Yamamoto, O. Solid Oxide Fuel Cells: Fundamental Aspects and Prospects. *Electrochim. Acta,* **2000**,
        45, 15-16, 2423-2435.
        [http://dx.doi.org/10.1016/S0013-4686(00)00330-3]

[31]   De Florio, D.Z.; Varela, J.A.; Fonseca, F.C.; Muccillo, E.N.S.; Muccillo, R. Direcionamentos Da
        Tecnologia Industrial de Células a Combustível de Óxidos Sólidos. *Quim. Nova,* **2007**, *30*(5), 1339-
        1346.
        [http://dx.doi.org/10.1590/S0100-40422007000500047]

[32]   Blomen, L.J.M.J.; Mugerwa, M.N. *Fuel Cell Syst.,* **2013**.

[33]   Brown, M.; Primdahl, S.; Mogensen, M. Structure / Performance Relations for Ni / Yttria-Stabilized
        Zirconia Anodes for Solid Oxide Fuel Cells. *J. Electrochem. Soc.,* **2000**, *147*(2), 475-485.
        [http://dx.doi.org/10.1149/1.1393220]

[34]   Gorte, R.J.; Vohs, J.M. Novel SOFC Anodes for the Direct Electrochemical Oxidation of
        Hydrocarbons. *J. Catal.,* **2003**, *216*, 1-2, 477-486.
        [http://dx.doi.org/10.1016/S0021-9517(02)00121-5]

[35]   Horita, T.; Yamaji, K.; Sakai, N.; Xiong, Y.; Kato, T.; Yokokawa, H.; Kawada, T. Imaging of Oxygen
        Transport at SOFC Cathode/Electrolyte Interfaces by a Novel Technique. *J. Power Sources,* **2002**,
        106, 1-2, 224-230.
        [http://dx.doi.org/10.1016/S0378-7753(01)01017-5]

[36]   Sun, C.; Stimming, U. Recent Anode Advances in Solid Oxide Fuel Cells. *J. Power Sources,* **2007**,
        171(2), 247-260.
        [http://dx.doi.org/10.1016/j.jpowsour.2007.06.086]

[37]   Wang, X.; Nakagawa, N.; Kato, K. Anodic Polarization Related to the Ionic Conductivity of Zirconia
        at Ni-Zirconia/Zirconia Electrodes. *J. Electrochem. Soc.,* **2001**, *148*(6), A565-A569.
        [http://dx.doi.org/10.1149/1.1369364]

[38]   Gross, S.M.; Koppitz, T.; Remmel, J.; Bouche, J.B.; Reisgen, U. Joining Properties of a Composite
        Glass-Ceramic Sealant. *Fuel Cells Bull.,* **2006**, *2006*, 9-, 12-15.
        [http://dx.doi.org/10.1016/S1464-2859(06)71320-7]

[39]   Yokokawa, H.; Tu, H.; Iwanschitz, B.; Mai, A. Fundamental Mechanisms Limiting Solid Oxide Fuel
        Cell Durability. *J. Power Sources,* **2008**, *182*(2), 400-412.
        [http://dx.doi.org/10.1016/j.jpowsour.2008.02.016]

[40]   Lamp, P.; Tachtler, J.; Finkenwirth, O.; Mukerjee, S.; Shaffer, S. Development of an Auxiliary Power
        Unit with Solid Oxide Fuel Cells for Automotive Applications. *Fuel Cells (Weinh.),* **2003**, *3*(3), 146-
        152.
        [http://dx.doi.org/10.1002/fuce.200332107]

[41]   Singhal, S.C. Solid Oxide Fuel Cells for Stationary, Mobile, and Military Applications. *Solid State
        Ion.,* **2002**, *152–153*, 405-410.

[http://dx.doi.org/10.1016/S0167-2738(02)00349-1]

[42] Brett, D.J.L.; Atkinson, A.; Brandon, N.P.; Skinner, S.J.; Brandon, N.P.; Brandon, N.P.; Skinner, S.J.; Skinner, S.J. Intermediate temperature solid oxide fuel cells. *Chem. Soc. Rev.,* **2008**, *37*(8), 1568-1578.
[http://dx.doi.org/10.1039/b612060c] [PMID: 18648682]

[43] Shao, Z.; Zhou, W.; Zhu, Z. Advanced Synthesis of Materials for Intermediate-Temperature Solid Oxide Fuel Cells. *Prog. Mater. Sci.,* **2012**, *57*(4), 804-874.
[http://dx.doi.org/10.1016/j.pmatsci.2011.08.002]

[44] Badwal, S.P.S.; Foger, K. Solid Oxide Electrolyte Fuel Cell Review. *Ceram. Int.,* **1996**, *22*(3), 257-265.
[http://dx.doi.org/10.1016/0272-8842(95)00101-8]

[45] Irvine, J.T.S.; Neagu, D.; Verbraeken, M.C.; Chatzichristodoulou, C.; Graves, C.; Mogensen, M.B. Evolution of the Electrochemical Interface in High-Temperature Fuel Cells and Electrolysers. *Nat. Energy,* **2016**, *1*(January), 15014.
[http://dx.doi.org/10.1038/nenergy.2015.14]

[46] Singhal, S.C.; Kendall, K. *High-Temperature Solid Oxide Fuel Cells: Fundamentals - Design and Applications*; Elsevier, **2003**, p. 406.

[47] Tietz, F.; Buchkremer, H.-P.; Stöver, D. Components Manufacturing for Solid Oxide Fuel Cells *Solid State Ionics,* **2002**, *152–153*, 373-381.
[http://dx.doi.org/10.1016/S0167-2738(02)00344-2]

[48] Tucker, M.C. Progress in Metal-Supported Solid Oxide Fuel Cells: A Review. *J. Power Sources,* **2010**, *195*(15), 4570-4582.
[http://dx.doi.org/10.1016/j.jpowsour.2010.02.035]

[49] Zhu, B. Solid Oxide Fuel Cell (SOFC) Technical Challenges and Solutions from Nano-Aspects. *Int. J. Energy Res.,* **2009**, *33*, 1126-1137.
[http://dx.doi.org/10.1002/er.1600]

[50] Garcia, L.M.P.; Macedo, D.A.; Souza, G.L.; Motta, F.V.; Paskocimas, C.A.; Nascimento, R.M. Citrate-Hydrothermal Synthesis, Structure and Electrochemical Performance of La0.6Sr0.4Co0.2Fe0.8O3-x Cathodes for IT-SOFCs. *Ceram. Int.,* **2013**, *39*(7), 8385-8392.
[http://dx.doi.org/10.1016/j.ceramint.2013.04.019]

[51] Kilner, J.A.; Burriel, M. Materials for Intermediate-Temperature Solid-Oxide Fuel Cells. *Annu. Rev. Mater. Res.,* **2014**, *44*, 365-393.
[http://dx.doi.org/10.1146/annurev-matsci-070813-113426]

[52] Clemmer, R.M.C.; Corbin, S.F. The Influence of Pore and Ni Morphology on the Electrical Conductivity of Porous Ni/YSZ Composite Anodes for Use in Solid Oxide Fuel Cell Applications. *Solid State Ion.,* **2009**, *180*, 9-10, 721-730.
[http://dx.doi.org/10.1016/j.ssi.2009.02.030]

[53] Wilson, J.R.; Barnett, S.a. Solid Oxide Fuel Cell Ni–YSZ Anodes: Effect of Composition on Microstructure and Performance. *Electrochem. Solid-State Lett.,* **2008**, *11*, B181.
[http://dx.doi.org/10.1149/1.2960528]

[54] Jiang, S. A. N. P.; Chan, S. H. W. A. A Review of Anode Materials Development in Solid Oxide Fuel Cells **2004**, *9*, 4405-4439.
[http://dx.doi.org/10.1023/B:JMSC.0000034135.52164.6b]

[55] Amado, R.S.; Malta, L.F.B.; Garrido, F.M.S.; Medeiros, M.E. Pilhas a Combustível de Óxido Sólido: Materiais, Componentes e Configurações. *Quim. Nova,* **2007**, *30*(1), 189-197.
[http://dx.doi.org/10.1590/S0100-40422007000100031]

[56] McIntosh, S.; Gorte, R.J. Direct hydrocarbon solid oxide fuel cells. *Chem. Rev.,* **2004**, *104*(10), 4845-4865.
[http://dx.doi.org/10.1021/cr020725g] [PMID: 15669170]

[57]   Park, Y.M.; Lee, H.J.; Bae, H.Y.; Ahn, J.S.; Kim, H. Effect of Anode Thickness on Impedance Response of Anode-Supported Solid Oxide Fuel Cells. *Int. J. Hydrogen Energy,* **2012**, *37*(5), 4394-4400.
[http://dx.doi.org/10.1016/j.ijhydene.2011.11.152]

[58]   Ogumi, Z.; Uchimoto, Y.; Tsuji, Y.; Takehara, Z. ichiro. Preparation of Thin Yttria-Stabilized Zirconia Films by Vapor-Phase Electrolytic Deposition. *Solid State Ion.,* **1992**, *58*, 3-4, 345-350.
[http://dx.doi.org/10.1016/0167-2738(92)90139-G]

[59]   Huang, B.; Zhu, X. jian; Hu, W. qi; Wang, Y. yun; Yu, Q. chun. Characterization of the Ni-ScSZ Anode with a LSCM-CeO2 Catalyst Layer in Thin Film Solid Oxide Fuel Cell Running on Ethanol Fuel. *J. Power Sources,* **2010**, *195*(10), 3053-3059.
[http://dx.doi.org/10.1016/j.jpowsour.2009.11.126]

[60]   Adachi Gy, G.; Imanaka, N. The Binary Rare Earth Oxides. *Chem. Rev.,* **1998**, *98*(4), 1479-1514.
[http://dx.doi.org/10.1021/cr940055h] [PMID: 11848940]

[61]   Aneggi, E.; Boaro, M.; Colussi, S.; de Leitenburg, C.; Trovarelli, A. *Ceria-Based Materials in Catalysis: Historical Perspective and Future Trends,* 1st ed; Elsevier B.V, **2016**. Vol. 50.
[http://dx.doi.org/10.1016/bs.hpcre.2016.05.002]

[62]   Mogensen, M.; Sammes, N.M.; Tompsett, G.A. Physical, Chemical and Electrochemical Properties of Pure and Doped Ceria. *Solid State Ion.,* **2000**, *129*(1), 63-94.
[http://dx.doi.org/10.1016/S0167-2738(99)00318-5]

[63]   Shaikh, S.P.S.; Muchtar, A.; Somalu, M.R. A Review on the Selection of Anode Materials for Solid-Oxide Fuel Cells. *Renew. Sustain. Energy Rev.,* **2015**, *51*, 1-8.
[http://dx.doi.org/10.1016/j.rser.2015.05.069]

[64]   Etsell, T.H.; Flengas, S.N. Electrical Properties of Solid Oxide Electrolytes. *Chem. Rev.,* **1970**, *70*(3), 339-376.
[http://dx.doi.org/10.1021/cr60265a003]

[65]   Kröger, F.A.; Vink, H.J. Relations between the Concentrations of Imperfections in Crystalline Solids. Solid State Phys. *Adv. Res. Appl.,* **1956**, *3*(C), 307-435.
[http://dx.doi.org/10.1016/S0081-1947(08)60135-6]

[66]   Naik, I.K.; Tien, T.Y. Small-Polaron Mobility in Nonstoichiometric Cerium Dioxide. *J. Phys. Chem. Solids,* **1978**, *39*(3), 311-315.
[http://dx.doi.org/10.1016/0022-3697(78)90059-8]

[67]   Santha, N.L.; Sebastian, M.T.; Mohanan, P.; Alford, N.M.; Sarma, K.; Pullar, R.C.; Kamba, S.; Pashkin, a; Samukhina, P.; Petzelt, J. Effect of Doping on the Dielectric Properties of Cerium Oxide in the Microwave and Far-Infrared Frequency Range. *J. Am. Ceram. Soc.,* **2004**, *87*(7), 1233-1237.
[http://dx.doi.org/10.1111/j.1151-2916.2004.tb07717_33.x]

[68]   Hayashi, H.; Inaba, H.; Matsuyama, M.; Lan, N. G.; Dokiya, M.; Tagawa, H. Structural Consideration on the Ionic Conductivity of Perovskite-Type Oxides *Solid State Ionics,* **1999**, *122*(1–4), 1-15.
[http://dx.doi.org/10.1016/S0167-2738(99)00066-1]

[69]   Imanaka, N.; Adachi, G. Rare Earth Contribution in Solid State Electrolytes, Especially in the Chemical Sensor Field *J. Alloys Compd,* **1997**, *250*(1–2 pt 2), 492-500.
[http://dx.doi.org/10.1016/S0925-8388(96)02628-X]

[70]   Kim, D. Lattice Parameters, Ionic Conductivities, and Solubility Limits in Fluorite-Structure M0$_2$ Oxide [M = Hf$_{4+}$, Zr$_{4+}$, Ce$_{4+}$, Th$_{4+}$, U$_{4+}$] Solid Solutions. *J. Am. Chem. Soc.,* **1989**, *72*, 1415-1421.

[71]   Choi, S.R.; Bansal, N.P. Mechanical Behavior of Zirconia/Alumina Composites. *Ceram. Int.,* **2005**, *31*(1), 39-46.
[http://dx.doi.org/10.1016/j.ceramint.2004.03.032]

[72]   Gupta, A.; Chatterjee, R. Dielectric and Magnetoelectric Properties of BaTiO$_3$–Co$_{0.6}$Zn$_{0.4}$Fe$_{1.7}$Mn$_{0.3}$O$_4$

Composite. *J. Eur. Ceram. Soc.,* **2013**, *33*(5), 1017-1022.
[http://dx.doi.org/10.1016/j.jeurceramsoc.2012.11.003]

[73]   Li, J.; Wu, Y.; Pan, Y.; Liu, W.; Zhu, Y.; Guo, J. Crystallization of Al_2O_3/ZrO_2 Solid Solution Powders Prepared by Coprecipitation *Materials Letters,* **2008**, *34*(3), 1539-1542.
[http://dx.doi.org/10.1016/S0167-577X(98)00057-3]

[74]   Peter, S. D.; Garbowski, E.; Guilhaume, N.; Perrichon, V.; Primet, M. Catalytic Properties of La_2CuO_4 in the CO + NO reaction *Catal. Letters,* **2000**, *54*(1998), 79-84.
[http://dx.doi.org/10.1023/A:1019063502409]

[75]   Sharif, A.A.; Mecartney, M.L. Superplasticity in Cubic Yttria Stabilized Zirconia with 10 Wt.%. *Alumina. J. Eur. Ceram. Soc,* **2004**, *24*(7), 2041-2047.
[http://dx.doi.org/10.1016/S0955-2219(03)00354-6]

[76]   Cabrelon, M.D.; Zauberas, R.T.; Boschi, A.O. Influência Da Temperatura e Do Método de Mistura Na Formação Do $ZrSiO_4$ *via* Reação Em Estado Sólido. *Ceramica,* **2007**, *53*, 83-88.
[http://dx.doi.org/10.1590/S0366-69132007000100013]

[77]   Ding, C.; Sato, K.; Mizusaki, J.; Hashida, T. A Comparative Study of NiO-Ce0.9Gd0.1O1.95 Nanocomposite Powders Synthesized by Hydroxide and Oxalate Co-Precipitation Methods. *Ceram. Int.,* **2012**, *38*(1), 85-92.
[http://dx.doi.org/10.1016/j.ceramint.2011.06.041]

[78]   Lange, F. F. Powder Processing Science and Technology for Increased Reability *J. Am. Ceram. Soc,* **1989**, *72*, 3-15.
[http://dx.doi.org/10.1111/j.1151-2916.1989.tb05945.x]

[79]   Bošković, S.B.; Djurović, D.R.; Zec, S.P.; Matović, B.Z.; Zinkevich, M.; Aldinger, F. Doped and Co-Doped CeO2: Preparation and Properties. *Ceram. Int.,* **2008**, *34*(8), 2001-2006.
[http://dx.doi.org/10.1016/j.ceramint.2007.07.036]

[80]   Chinarro, E.; Jurado, J.R.; Colomer, M.T. Synthesis of Ceria-Based Electrolyte Nanometric Powders by Urea-Combustion Technique. *J. Eur. Ceram. Soc.,* **2007**, *27*, 13-15, 3619-3623.
[http://dx.doi.org/10.1016/j.jeurceramsoc.2007.02.007]

[81]   Hui, S. (Rob); Roller, J.; Yick, S.; Zhang, X.; Decès-Petit, C.; Xie, Y.; Maric, R.; Ghosh, D. A Brief Review of the Ionic Conductivity Enhancement for Selected Oxide Electrolytes. *J. Power Sources,* **2007**, *172*(2), 493-502.
[http://dx.doi.org/10.1016/j.jpowsour.2007.07.071]

[82]   Hwang, C.C.; Huang, T.H.; Tsai, J.S.; Lin, C.S.; Peng, C.H. Combustion Synthesis of Nanocrystalline Ceria (CeO_2) Powders by a Dry Route. Mater. Sci. Eng. B Solid-State Mater. *Adv. Technol.,* **2006**, *132*(3), 229-238.
[http://dx.doi.org/10.1016/j.mseb.2006.01.021]

[83]   Li, J.G.; Ikegami, T.; Wang, Y.R.; Mori, T. 10-Mol%-Gd_2O_3-Doped CeO2 Solid Solutions *via* Carbonate Coprecipitation: A Comparative Study. *J. Am. Ceram. Soc.,* **2003**, *86*(6), 915-921.
[http://dx.doi.org/10.1111/j.1151-2916.2003.tb03397.x]

[84]   Gil, V.; Tartaj, J.; Moure, C. Chemical and Thermomechanical Compatibility between Ni-GDC Anode and Electrolytes Based on Ceria. *Ceram. Int.,* **2009**, *35*(2), 839-846.
[http://dx.doi.org/10.1016/j.ceramint.2008.03.004]

[85]   Ring, T. *Fundamentals of Ceramic Powder Processing and Synthesis*; Academic Press: San Diego, **1996**, p. 961.

[86]   Singh, P.; Minh, N.Q. Solid Oxide Fuel Cells: Technology Status. *Int. J. Appl. Ceram. Technol.,* **2004**, *1*, 1-, 5-15.
[http://dx.doi.org/0.1111/j.1744-7402.2004.tb00149.x]

[87]   Souza, E.C.C.; Muccillo, E.N.S. Effect of Solvent on Physical Properties of Samaria-Doped Ceria Prepared by Homogeneous Precipitation. *J. Alloys Compd.,* **2009**, *473*, 1-2, 560-566.

[http://dx.doi.org/10.1016/j.jallcom.2008.06.027]

[88]   Pechini, M. P. Method of Preparing Lead and Alkaline Earth Titanates and Niobates and Coating Method Using the Same to Form a Capacitor **1967**.

[89]   Cho, W-S.; Hamada, E. Synthesis of Ultrafine BaTiO$_3$ Particles from Polymeric Precursor: Their Structure and Surface Property. *J. Alloys Compd.,* **1998**, *266*, 1-2, 118-122. [http://dx.doi.org/10.1016/S0925-8388(97)00446-5]

[90]   Ianculescu, A.; Berger, D.; Viviani, M.; Ciomaga, C.E.; Mitoseriu, L.; Vasile, E.; Drăgan, N.; Crişan, D. Investigation of Ba1-XSrxTiO3 Ceramics Prepared from Powders Synthesized by the Modified Pechini Route. *J. Eur. Ceram. Soc.,* **2007**, *27*, 13-15, 3655-3658. [http://dx.doi.org/10.1016/j.jeurceramsoc.2007.02.017]

[91]   Ries, A.; Simões, A.Z.; Cilense, M.; Zaghete, M.A.; Varela, J.A. Barium Strontium Titanate Powder Obtained by Polymeric Precursor Method. *Mater. Charact.,* **2003**, *50*, 2-3, 217-221. [http://dx.doi.org/10.1016/S1044-5803(03)00095-0]

[92]   Stojanović, B.D.; Mastelaro, V.R.; Paiva Santos, C.O.; Varela, J.A. Structure Study of Donor Doped Barium Titanate Prepared from Citrate Solutions. *Sci. Sinter.,* **2004**, *36*(3), 179-188. [http://dx.doi.org/10.2298/SOS0403179S]

[93]   Venkata Saravanan, K.; James Raju, K.C.; Ghanashyam Krishna, M.; Bhatnagar, A.K. Preparation of Barium Strontium Titanate Using a Modified Polymeric Precursor Method. *J. Mater. Sci.,* **2007**, *42*(4), 1149-1155. [http://dx.doi.org/10.1007/s10853-006-1435-3]

[94]   Vinothini, V.; Singh, P.; Balasubramanian, M. Synthesis of Barium Titanate Nanopowder Using Polymeric Precursor Method. *Ceram. Int.,* **2006**, *32*(2), 99-103. [http://dx.doi.org/10.1016/j.ceramint.2004.12.012]

[95]   Grossin, D.; Noudem, J.G. Synthesis of Fine La0.8Sr0.2MnO3 Powder by Different Ways. *Solid State Sci.,* **2004**, *6*(9), 939-944. [http://dx.doi.org/10.1016/j.solidstatesciences.2004.06.003]

[96]   Tai, L-W.; Anderson, H.U. Mixed-Cation Oxide Powders *via* Resin Intermediates Derived from a Water-Soluble Polymer. *J. Am. Ceram. Soc.,* **1992**, *75*(12), 3490-3494. [http://dx.doi.org/10.1111/j.1151-2916.1992.tb04458.x]

[97]   Lee, H.; Hong, M.; Bae, S.; Lee, H.; Park, E.; Kim, K. A Novel Approach to Preparing Nano-Size Co$_3$O$_4$-Coated Ni Powder by the Pechini Method for MCFC Cathodes *J. Mater. Chem,* **2003**, *13*(10)

[98]   Paskocimas, C. A.; Martinelli, A. E.; Nascimento, R. M.; Macedo, D. A. Cela, B. Processo de Produção de Nanocompósito à Base de Óxido de Níquel e Céria Dopada Com Cátions Trivalentes PI 0903688-1 A8, 2009

[99]   Vlack, L. H.; Van, *Propriedades Dos Materiais Cerâmicos*; Edgard Blucher: São Paulo, SP, **1973**.

[100]   da Silva, A.G.P.; Júnior, C.A. Teoria de Sinterização Por Fase Sólida; Uma Análise Crítica de Sua Aplicação. *Ceramica,* **1998**, *44*, 171-176. [http://dx.doi.org/10.1590/S0366-69131998000500004]

[101]   Boch, P.; Niepce, J-C. *Ceramic Materials: Processes, Properties, and Applications*; ISTE: London, **2010**.

[102]   Kang, S-J.L. Sintering: Densification, Grain growth and Microstructure. In: *Book Review/Science of Sintering*; Elsevier: Amsterdam, **2005**; 37, p. 239.

[103]   Rahaman, M.N. *Ceramic Processing*; CRC Press: Boca Raton, Fl, **2006**, p. 526.

[104]   Richerson, D.W. *Modern Ceramic Engineering: Properties*; Processing, and Use in Design, **2005**. [http://dx.doi.org/10.1201/b18952]

[105]   Bauerle, J.E. Study of Solid Electrolyte Polarization by a Complex Admittance Method. *J. Phys.*

Chem. Solids, **1969**, *30*(12), 2657-2670.
[http://dx.doi.org/10.1016/0022-3697(69)90039-0]

[106] Chang, B-Y.; Park, S-M. Electrochemical impedance spectroscopy. *Annu. Rev. Anal. Chem. (Palo Alto, Calif.),* **2010**, *3*(1), 207-229.
[http://dx.doi.org/10.1146/annurev.anchem.012809.102211] [PMID: 20636040]

[107] Chinaglia, D. L.; Gozzi, G.; Alfaro, R. A. M.; Hessel, R. Espectroscopia de ImpedâNcia No Laboratório de Ensino. *Rev. Bras. Ensino Física,* **2008**, *30*(4), 4504.

[108] Huang, V.M.; Wu, S.L.; Orazem, M.E.; Pébre, N.; Tribollet, B.; Vivier, V. Local Electrochemical Impedance Spectroscopy: A Review and Some Recent Developments.*Electrochimica Acta*; Elsevier Ltd, **2011**, Vol. 56, pp. 8048-8057.
[http://dx.doi.org/10.1016/j.electacta.2011.03.018]

[109] Levie, R.D.; Husovsky, A.A. Instrument for the Automatic Measurement of the Electrode Admittance. *J. Electroanal. Chem. Interfacial Electrochem.,* **1969**, *20*(2), 181-193.
[http://dx.doi.org/10.1016/S0022-0728(69)80119-1]

[110] Macdonald, J.R.; Johnson, W.B. Fundamentals of Impedance Spectroscopy. In: *Impedance Spectroscopy: Theory Experiment, and Applications*; John Wiley and Sons, Inc., **2005**; pp. 1-26.

[111] Silverman, D.C. Rapid Corrosion Screening in Poorly Defined Systems by Electrochemical Impedance Technique. *Corrosion,* **1990**, *46*(7), 589-598.
[http://dx.doi.org/10.5006/1.3585153]

[112] Delgado, A.; García-Sánchez, M. F.; M'Peko, J.-C.; Ruiz-Salvador, a. R.; Rodríguez-Gattorno, G.; Echevarría, Y.; Fernández-Gutierrez, F. An Elementary Picture of Dielectric Spectroscopy in Solids: Physical Basis. *J. Chem. Educ.,* **2003**, *80*(9), 1062.

[113] F.J. Holler; D.A. Skoog; S.R. Crouch.. *Princípios de análise instrumental*; Porto Alegre (RS): Bookman, **2009**.

[114] Shi, Y.; Wang, H.; Cai, N. Simulation of Two-Dimensional Electrochemical Impedance Spectra of Solid Oxide Fuel Cells Using Transient Physical Models. *ECS Trans.,* **2011**, *35*(1), 871-881.

[115] Keddam, M.; Nóvoa, X.R.; Soler, L.; Andrade, C.; Takenouti, H. An Equivalent Electrical Circuit of Macrocell Activity in Facing Electrodes Embedded in Cement Mortar. *Corrosion Science,* **1994**, *36*(7), 1155-1166.

[116] Boukamp, B. A. A Linear Kronig-Kramers Transform Test for Immittance Data Validation. *J. Electrochem. Soc.,* **1995**, *142*(6), 1885.

[117] Yuan, X.; Wang, H.; Colin Sun, J.; Zhang, J. AC Impedance Technique in PEM Fuel Cell Diagnosis-A Review. *Int. J. Hydrogen Energy,* **2007**, *32*(17), 4365-4380.
[http://dx.doi.org/10.1016/j.ijhydene.2007.05.036]

[118] Grilo, J.P.F.; Moura, C.G.; Macedo, D.A.; Rajesh, S.; Figueiredo, F.M.L.; Marques, F.M.B.; Nascimento, R.M. Effect of Composition on the Structural Development and Electrical Conductivity of NiO-GDC Composites Obtained by One-Step Synthesis. *Ceram. Int.,* **2017**, *0–1*(April)
[http://dx.doi.org/10.1016/j.ceramint.2017.04.027]

[119] Araujo, A.J.M.; Sousa, A.R.O.; Grilo, J.P.F.; Campos, L.F.A.; Loureiro, F.J.A.; Fagg, D.P.; Dutra, R.P.S.; Macedo, D.A. Preparation of One-Step NiO/Ni-CGO Composites Using Factorial Design. *Ceram. Int.,* **2016**, *42*(16), 18166-18172.
[http://dx.doi.org/10.1016/j.ceramint.2016.08.131]

[120] Wandekar, R.V. Ali (Basu) M., M.; Wani, B. N.; Bharadwaj, S. R. Physicochemical Studies of NiO-GDC Composites. *Mater. Chem. Phys.,* **2006**, *99*, 2-3, 289-294.
[http://dx.doi.org/10.1016/j.matchemphys.2005.10.025]

[121] Chavan, A.U.; Jadhav, L.D.; Jamale, A.P.; Patil, S.P.; Bhosale, C.H.; Bharadwaj, S.R.; Patil, P.S. Effect of Variation of NiO on Properties of NiO/GDC (Gadolinium Doped Ceria) Nano-Composites.

Ceram. Int., **2012**, *38*(4), 3191-3196.
[http://dx.doi.org/10.1016/j.ceramint.2011.12.023]

[122]   Pyda, J. W. W. A New Method of Preparing Ni / YSZ Cermet Materials. *J. Mat. Sci.,* **2012**, *47*(6), 2807-2817.
[http://dx.doi.org/10.1007/s10853-011-6109-0]

[123]   Pinheiro, L. B.; Martinelli, A. E.; Fonseca, F. C. Effects of Microwave Processing on the Properties of Nickel Oxide / Zirconia / Ceria Composites. *Adv. Mat. Res.,* **2014**, *975*(3), 154-159.
[http://dx.doi.org/10.4028/www.scientific.net/AMR.975.154]

[124]   Grilo, J.P.F.; Macedo, D.A.; Nascimento, R.M.; Marques, F.M.B. Assessment of NiO-CGO Composites as Cermet Precursors. *Solid State Ion,* **2018**, *321*(February), 115-121.
[http://dx.doi.org/10.1016/j.ssi.2018.04.014]

CHAPTER 6

Perovskite-Based Anode Materials for Solid Oxide Fuel Cells

Vladislav A. Kolotygin[*, 1], **Irina E. Kuritsyna**[1] and **Nikolay V. Lyskov**[2]

[1] *Institute of Solid State Physics RAS, Chernogolovka, 142432, Russia*

[2] *Department of Chemical Engineering, Institute of Problems of Chemical Physics RAS, Chernogolovka, 142432, Russia*

Abstract: The present chapter is devoted to comparison of oxide-based anode materials developed during the last 10-15 years, with a particular focus on relationships between their functional characteristics, such as phase and structural stability, electronic and ionic conductivity, thermal and chemical expansion, and the electrochemical activity of the corresponding anodes. In most studies, the strategy of selection of the anode material composition is unclear while no obvious correlations can be revealed between the anode activity and other properties of the material. The situation is complicated by the presence of catalytically active phases in the anode layers, such as CeO_2-based compounds or metallic phases (Pt, Ag, Ni, *etc*). The latter are frequently introduced as current collecting coatings for optimizing the interface contact between the anode and interconnect; however, their influence on the catalytic activity cannot *a priori* be considered to be negligible. For this reason, the electrochemical characteristics of different anode materials studied by the same research group and, consequently, prepared and modified by the same route, frequently appear to be similar. The purpose of this review is to critically discuss the results where the origins of the observed anode performance are arguable, and to emphasize the studies where a reliable analysis of the performance-determining factors has been done.

Keywords: Alternative Anode, Chemical Compatibility, Chemical Expansion, Current Collector, Electrochemical Activity, Fuel Cell, Perovskite, Polarization Resistance, Surface Modification, Thermal Expansion.

INTRODUCTION

Solid oxide fuel cells (SOFCs) provide advantages over other energy conversion engines due to high power output, fuel flexibility, environmental safety, capability to utilize exhaust heat, *etc*. One of the actual concerns related to SOFCs is performance degradation attributable to high-temperature interaction between the

[*] **Corresponding author Vladislav A. Kolotygin:** Department of Chemical Engineering Institute of Solid State Physics RAS, Chernogolovka, 142432, Russia; Tel.: +74965228471; +79153624350; Fax: +7(496) 522 8160; E-mail: kolotygin@issp.ac.ru

Moisés R. Cesário, Cédric Gennequin, Edmond Abi-Aad & Daniel A. de Macedo (Eds.)

cell components or microstructural instability of the electrode layers. In terms of anode materials, one of the strategies to prolongate the durability of the cell is to diminish the content of metallic Ni phase by partial or complete substitution with other components satisfying the requirements to anodes. In the last few years, selected perovskite compositions have been proposed as promising candidates capable of operating as individual anode materials. However, significant performance losses originate from insufficient electrochemical activity of the alternative anodes, which is especially pronounced at low temperatures due to high activation energies of rate-determining processes. On the other hand, the information on relationships between the anode material composition, its functional properties and electrochemical activity is scarce and contradictive. The present chapter is devoted to comparative analysis of perovskite-based anode materials, their short- and long-term durability and discussion of the basic factors determining their electrochemical properties.

Non-Perovskite Oxide-Based Anode Materials

The present part considers the functional and electrochemical properties of materials which have been considered as an alternative to substitute nickel in SOFC anodes. Since metal-ceramic composites containing metals other than nickel (*i.e.*, Cu-, Co-, Fe-, Ru- or alloy-containing cermets) have been discussed in numerous studies [1 - 5], these anodes are not included in the present chapter. Among binary oxides, promising characteristics have been reported for CeO_2-, ZrO_2-, NbO_x-, TiO_x-, MoO_x-, VO_x-based solid solutions and composites, due to enhanced conductivity under anode conditions and, in some cases, a promotion of the catalytic processes [2, 5 - 11]. However, the number of studies on these materials is by far too scarce to discuss their applicability for fuel cell applications and suggest possible routes of their optimization. Moreover, some results require further verification and explanation. In particular, extremely high power density was reported for a cell with MoO_2-based anode (>3 W/cm^2 at 1023 K in $n\text{-}C_{12}H_{26}$ fuel) [11]; however, no information on further testing of this material has been found.

Among the possible combinations of mixed oxides composed of large-size cations in so-called A-sublattice (mainly alkali-earth or rare-arth cations) and small cations in B-sublattice (transition metal cations), one should briefly consider Ti-, Mn-, Fe-, Ni-, Co- Ti-, Nb- and W-containing composition where the molar ratio A:B differs from 1. Although the reduction stability of manganites, ferrites, *etc.* $(Ln,A)_{n+1}B_nO_{3n+1}$ (where Ln is rare-earth metal, A is an alkali-earth metal and B is a transition metal element) with Ruddlesden-Popper (RP) structure as well as $Sr_4Fe_6O_x$-based materials is improved as compared to perovskites analogues, the level of the electronic and ionic conductivity is higher in the perovskites [12 - 16].

$(La,Ca,Sr,Ba)_{0.6}NbO_3$- or $(Na,K,Rb,Cs)WO_3$-based phases with the structure of tungsten bronze generally exhibit sufficient level of the *n*-type electronic conductivity which may even be superior in comparison with that of perovskites. In particular, $Na_{0.8}WO_{3-\delta}$-YSZ composite anodes demonstrated the polarization resistance of ~1 Ohm×cm^2 without additional modifications [17]. However, the performance of the bronze anodes is poor, while the problems associated with stability, interaction with YSZ, oxygen diffusion limitations and especially extremely low thermal expansion coefficients (TECs), significantly limit their utilization [18, 19]. Moreover, the bronze structure is barely tolerant towards introduction of foreign cations into the crystal lattice, which restricts their possible modifications by introduction of catalytically active dopants. Regardless of the disadvantages indicated, some studies of the electrochemical characteristics of perovskite-related layered phases and tetragonal bronzes have been carried out [15, 17, 20, 21]. Pylochlore-type materials $(Gd,Ca)_2Ti_2O_{7-\delta}$ might be considered as an alternative to titanate perovskites due to their high stability in a wide $p(O_2)$ range and better tolerance to oxidation. The materials possess dominant oxygen ionic conductivity while the comparatively low electronic contribution and catalytic activity necessitate the distribution of metallic phases, ceria-based compositions or surface modification, as shown in Table **1** [22 - 25].

Table 1. Anode overpotential values at i=100 mA/cm^2 of selected anodes with perovskite-like structure in hydrogen-containing fuels.

Composition	Electrolyte	T, K	Atmosphere	η, mV	Ref.
$Gd_{1.86}Ca_{0.14}Ti_2O_{7-\delta}$ - Ni - GDC (30-50-20 wt.%) *c.c.: Pt gauze*	LSGM	1073	wet 10% H_2 - 90% N_2	56	[24]
$Gd_2Ti_{0.6}Mo_{1.2}Sc_{0.2}O_{7-\delta}$ *c.c.: Au paste*	YSZ-Al$_2$O$_3$	1205	wet 60% H_2	~30	[25]
$La_{0.6}Sr_{0.4}Fe_{0.8}Mn_{0.2}O_{3-\delta}$ *c.c.: Pt mesh*	LSGMCo	1073	wet H_2	~30	[28]
$SrCo_{0.8}Fe_{0.2}O_{3-\delta}$ *c.c.: Au grid*	YSZ	1223	H_2	410	[21]
$La_{0.6}Sr_{0.4}Co_{0.2}Fe_{0.8}O_{3-\delta}$ *c.c.: Au grid*	YSZ	1223	H_2	390	[21]
$SrFe_{0.7}Mo_{0.3}O_{3-\delta}$ *c.c.: Pt mesh*	LSGM	1073	wet 10% H_2 - 90% N_2	70	[29]
$SrFeCo_{0.5}O_x$ *c.c.: Au grid*	YSZ	1223	H_2	570	[21]

The perovskite lattice is flexible towards cation substitution while the transport properties are generally better compared to other families, like spinels, pyrochlores or garnets. Due to insufficient stability under anode conditions, the choice of the perovskite-type anode materials is restricted by chromites, titanates, molybdates and vanadates. Moreover, appropriate doping of manganites with stabilizing cations without substantial deterioration of the transport and catalytic properties enables to consider these compounds as another group of promising anode materials. Some authors reported application of thermodynamically unstable perovskites such as $(Ln,A)(Co,Fe)O_{3-\delta}$ or $(Ln,A)(Fe,Ni)O_{3-\delta}$, as SOFC anodes (Table **1**) or catalysts for hydrocarbon oxidation [21, 26, 27]. However, one should note that the requirements for catalytic tests differ from those to anode materials, in particular, in terms of the electrocatalytic activity, stability, *etc*. Due to poor stability and large volume changes occurring upon thermal or redox cycling, the presented groups will be omitted from our consideration.

The electrochemical activity of perovskite-based anodes strongly depends on the processing conditions. Moreover, comparison of experimental data is complicated by the fact that some authors provide no information on such important details as the composition of the solid electrolyte, fuel gas atmosphere, fractions of the components in composite electrode, *etc*. Evaluation of the literature data is often obscured by the fact that in most part of works, the anodes are covered with a layer of dispersed metallic particles, sputtered or coated in the form of pastes or inks. Most authors suggest that these particles serve only as an electron-conductive component, without any effect on the activity. However, the contribution of metal-containing components can only be neglected when no surface modification or metal penetration into the electrochemically active zone occurs. As shown by previous studies [30 - 32], Pt, Ag, Au and other noble metals possess non-negligible catalytic activity and may substantially affect the porous microstructure of the anode layer (for example, *via* forming catalytically-active centers inside the electrode pores) which cannot be achieved by introduction of less expensive current collectors made of stainless steels or $LaCrO_{3-\delta}$-based ceramics. At the same time, any wide utilization of noble metals in industrial scales is impossible owing to economic reasons while conventional current collectors have significantly higher bulk and contact resistance. Therefore, the results obtained on laboratory electrochemical cell (the active electrode area below 10 cm^2) covered with catalytically-active pastes are unrelevant to electrochemical batteries of the scale of ~100 cm^2 and more. For the reason discussed, it is more correct to take into account the nature and form of current collecting additives when considering the electrochemical behaviour of the oxide-based anodes.

Chromite-based Perovskite Solid Solutions

A large group of perovskite-based materials is presented by $Ln_{1-x}A_xCrO_{3-\delta}$ (Ln is a rare-earth cation, A = Ca, Sr, Ba) oxides and their B-site substituted derivatives. These materials exhibit negligible variations of oxygen content with temperature or p(O$_2$) which ensure high stability towards reduction and enable to preserve the phase under anode conditions [33 - 37]. The electronic conductivity of chromites is determined by the concentration of *p*-type electronic charge carriers which is proportional to the content of alkali-earth dopants at low concentrations of oxygen vacancies [35]. Strong reduction (for example, at p(O$_2$)<10^{-15} atm, T=1273 K for La$_{0.9}$Ca$_{0.1}$CrO$_{3-\delta}$ [38]) decreases the concentration of electron holes inducing the conductivity drop down to 0.3 - 10 S/cm (Fig. **1**), (Table **2**).

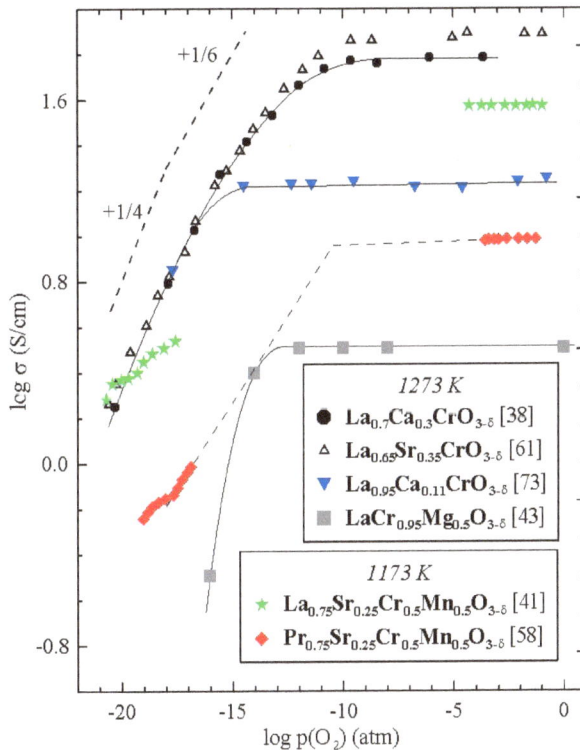

Fig. (1). Isothermal p(O$_2$) dependencies of the total conductivity for selected chromites at 1173 and 1273 K. The data are taken from [38, 41, 43, 58, 61, 73].

Introduction of reducible cations (Mn, Fe, Co, *etc*) into the B-sublattice of chromites generally increases the concentration and/or mobility of the charge carriers; however, the oxygen losses and resultant conductivity drop upon reduction become more pronounced [35, 38 - 43]. This leads to 2 opposite effects on the conductivity; generally, a moderate doping with Mn, Fe, *etc* promotes the electron transfer under the anode conditions provided that the perovskite phase stability is preserved (Table **2**).

Table 2. Total conductivity at 1073 K in reducing atmospheres and TECs of selected chromite-based materials.

Composition	$\sigma_{1073\,K}$, S/cm *(Atmosphere)*	Thermal Expansion		Ref.
		T, K	**TEC×10⁻⁶, K⁻¹** *(Atmosphere)*	
$LaCrO_{3-\delta}$	5.4×10^{-3} *(10% H_2 - 90% N_2)*	320-1270	8.0 *(wet air)*	[45, 46]
$La_{0.9}Ca_{0.1}CrO_{3-\delta}$	2.3 (H_2)	320-1270	8.5 *(wet H_2)*	[45]
$La_{0.7}Ca_{0.3}CrO_{3-\delta}$	4.3 *(H_2)*	570-1270	11.5 *(air)*	[45]
$La_{0.9}Sr_{0.1}CrO_{3-\delta}$	1.7 *(H_2)*	320-1270	9.8 *(wet H_2)*	[45]
$La_{0.7}Sr_{0.3}CrO_{3-\delta}$	4.7 *(H_2)*			[45]
$LaCr_{0.9}Mg_{0.1}O_{3-\delta}$	0.12 *(H_2)*	320-1270	8.4 *(wet H_2)*	[45]
$La_{0.9}Sr_{0.1}Cr_{0.95}Al_{0.05}O_{3-\delta}$		320-1270	10.8 *(wet H_2)*	[47]
$LaCr_{0.9}Ti_{0.1}O_{3-\delta}$	4.6×10^{-2} *(p(O_2)= 4.5×10^{-2} atm)*			[48]
$La_{0.95}Ba_{0.05}Cr_{0.9}Ti_{0.1}O_{3-\delta}$	7.5×10^{-2} *(p(O_2)= 1.1×10^{-22} atm)*			[48]
$La_{0.7}Ca_{0.3}Cr_{0.5}Ti_{0.5}O_{3-\delta}$	2.1×10^{-2} *(4% H_2 - Ar)*	300-1270	10.1 *(air)*	[49, 50]
$La_{0.8}Sr_{0.2}Cr_{0.5}Mn_{0.5}O_{3-\delta}$	5.1 *(wet 5% H_2 - Ar)*			[51]
$La_{0.75}Sr_{0.25}Cr_{0.5}Mn_{0.5}O_{3-\delta}$	0.93 *(5% H_2 - Ar)*	570-1270	12.0 *(air)*	[41, 46]
$La_{0.8}Sr_{0.2}Cr_{0.5}Fe_{0.5}O_{3-\delta}$	6.3 *(5% H_2 - Ar)*	not indicated	11.8 *(air)*	[52]
$La_{0.33}Sr_{0.67}Cr_{0.33}Fe_{0.67}O_{3-\delta}$	0.14 *(p(O_2)= 8.4×10^{-21} atm)*	970-1120	20.7 *(He)*	[40]
$La_{0.83}Ba_{0.17}Cr_{0.5}Fe_{0.5}O_{3-\delta}$	0.12 *(wet 5% H_2 - Ar)*			[53]
$Ce_{0.9}Sr_{0.1}Cr_{0.5}V_{0.5}O_{3-\delta}$	44 *(10% H_2S - N_2)*			[54]
$Ce_{0.3}Sr_{0.7}Cr_{0.5}Mn_{0.5}O_{3-\delta}$		950-1370	15.8 *(air)*	[55]
$Ce_{0.9}Sr_{0.1}Cr_{0.5}Fe_{0.5}O_{3-\delta}$	3.0 *(5% H_2S - N_2)*			[56]
$Pr_{0.7}Ca_{0.3}Cr_{0.6}Mn_{0.4}O_{3-\delta}$	0.23 *(wet 5% H_2 - Ar)*			[57]
$Pr_{0.75}Sr_{0.25}Cr_{0.5}Mn_{0.5}O_{3-\delta}$	0.28 *(5% H_2 - Ar)*	820-1170	9.5 *(air)*	[58]
$Pr_{0.7}Ca_{0.3}Cr_{0.9}Ni_{0.1}O_{3-\delta}$	0.98 *(5% H_2)*			[59]

The low and nearly constant level of the oxygen stoichiometry leads to a moderate chemical expansivity of chromites (<0.25%) while rather strict correlations between the oxygen stoichiometry variations in chromites and corresponding volume changes on redox cycling were reported [36, 44]. Increasing the reducibility enhances the chemical expansion of chromites and results in a slight enhancement of the TEC at the temperatures above 900 K. Nevertheless, the level of the thermal and chemical expansion of chromites is moderate even at 50% substitution degree (Table **2**).

Due to low concentrations of mobile oxygen vacancies, the ionic conductivity of chromites is negligible even after introduction of reducible cations [39, 60, 61]. In particular, for $(La_{0.75}Sr_{0.25})_{0.95}Cr_{0.5}Mn_{0.5}O_{3-\delta}$, σ_O was found to be 3.5×10^{-4} S/cm at $p(O_2) \approx 10^{-15}$ atm, T=1223 K [39]. This results in poor intrinsic electrochemical activity of $(La,A)CrO_{3-\delta}$-like oxides (Table **3**); for example, the polarization resistance of $La_{0.7}Ca_{0.32}CrO_{3-\delta}$ anode is reported to be ~86 Ohm×cm^2 at 1123 K [62]. Therefore, chromites are basically considered as a chemically and mechanically stable electron-conductive matrix where ion-conductive components (CeO_2-, ZrO_2-based oxides) or catalytically-active phases (Ni, Pd, VO_x) are distributed. The presence of large fraction of the stable perovskite phase enables to diminish volume or microstructural changes of the composite anode.

The performance of chromite-based anodes can to some extent be improved by introducing A- or B-site dopants, especially those forming catalytically-active centres (Mn, Fe, *etc*) or dispersed metallic particles (Ni, Ru) on the perovskite surface. Extensive doping leads to destabilization and enhances the chemical expansion of the perovskite phase. On the other hand, the decomposition may be hindered kinetically [39, 41, 57, 60, 63, 64].

Mn-doped chromites $(Ln,A)(Cr,Mn)O_{3-\delta}$ (LnACM) represent the most widely studied perovskite group for potential applications as SOFC anodes. Depending on the composition, the material is characterized by a structural transition on heating or reduction which, however, has no detrimental effects on the transport or mechanical properties [51, 65]. On reduction, small quantities of MnO, Ln_2O_3 or RP-phases may be separated [55, 66]. The TECs of LnACM are slightly higher than those of SOFC electrolytes or interconnects (Table **2**). The chemical expansion for Ln = La is 0.2 - 0.3% at 1073 - 1273 K, $p(O_2) = 10^{-18} - 10^{-13}$ atm [39]. Despite the moderate expansivity, poor adhesion of $La_{0.75}Sr_{0.25}Cr_{0.5}Mn_{0.5}O_{3-\delta}$ to YSZ was reported [67]; this drawback can be solved by fabrication of composites.

The presence of Mn species somewhat improves the oxygen diffusivity and surface exchange kinetics compared to undoped chromites, especially under reducing conditions [39, 68], due to a high tendency of Mn species towards variations of the oxygen environement and oxidation state [41, 69 - 72]. The optimum combination of the chemical stability, thermomechanical properties, electronic transport and catalytic activity was found for $La_{0.75}Sr_{0.25}Cr_{0.5}Mn_{0.5}O_{3-\delta}$;

examples of the anode performance of the cells with the corresponding anodes are shown in Fig. (**2**) and Table **3**.

Table 3. Power density of selected cells with chromite-based anodes.

Anode	Current Collector	Electrolyte (Thickness, μm)	Fuel	P_{max}, mW/cm^2 *(T, K)*	Ref.
LaCrO$_{3-\delta}$ - VO$_x$ - YSZ *(40-40-20 wt.%)*	Au mesh	YSZ (300)	pure H$_2$	80 *(1073)*	[92]
La$_{0.8}$Sr$_{0.2}$CrO$_{3-\delta}$ - GDC *(50 - 50 wt.%)*	Au grids	LSGM (400)	wet H$_2$	200 *(1073)*	[79]
La$_{0.8}$Sr$_{0.2}$CrO$_{3-\delta}$ -YSZ *(45-55 wt.%) impr. with CeO$_2$ (5 wt.%), Pd (0.5 wt.%)*	Ag paste	LSGM (75)	wet H$_2$	380 *(973)*	[80]
La$_{0.7}$Ca$_{0.3}$CrO$_{3-\delta}$ *impr. with GDC*	Ag paste	LSGM (300)	dry H$_2$	390 *(1073)*	[93]
La$_{0.7}$Sr$_{0.3}$CrO$_{3-\delta}$ - GDC *(80 - 20 wt.%)*	Ag paste	LSGM (300)	H$_2$	570 *(1173)* 210 *(1073)*	[94]
La$_{0.7}$Sr$_{0.3}$CrO$_{3-\delta}$ - CeO$_2$ - Ni *(25 - 37.5 - 37.5 mol.%)*	not indicated	YSZ (350)	3% H$_2$O - H$_2$	130 *(1073)*	[95]
La$_{0.75}$Sr$_{0.25}$Cr$_{0.5}$Al$_{0.5}$O$_{3-\delta}$	Pt paste	LSGM (1500)	wet 5% H$_2$ - Ar	80 *(1073)*	[96]
La$_{0.75}$Sr$_{0.25}$Cr$_{0.5}$Mn$_{0.5}$O$_{3-\delta}$	Pt mesh	YSZ (250)	H$_2$	270 *(1223)*	[97]
La$_{0.75}$Sr$_{0.25}$Cr$_{0.5}$Mn$_{0.5}$O$_{3-\delta}$*impr. with Pt and Cu*	Pt mesh	LSGM (250)	dry H$_2$	520 *(1073)*	[75]
La$_{0.6}$Sr$_{0.4}$Cr$_{0.6}$Fe$_{0.4}$O$_{3-\delta}$	Au grids	LSGM (400)	wet H$_2$	460 *(1073)*	[98]
La$_{0.8}$Sr$_{0.2}$Cr$_{0.5}$Fe$_{0.5}$O$_{3-\delta}$	Au paint	YSZ (300)	H$_2$	210 *(1173)*	[52]
La$_{0.8}$Sr$_{0.2}$Cr$_{0.5}$Fe$_{0.5}$O$_{3-\delta}$ - GDC *(50 - 50 wt.%)*	Au paint	YSZ (300)	H$_2$	250 *(1123)*	[52]
La$_{0.8}$Sr$_{0.2}$Cr$_{0.5}$Fe$_{0.5}$O$_{3-\delta}$*-impr. into LSGM scaffold*	Ag ink	LSGM (15)	3% H$_2$O - H$_2$	850 *(1073)*	[85]
La$_{0.33}$Sr$_{0.67}$Cr$_{0.33}$Fe$_{0.67}$O$_{3-\delta}$ **-GDC** *(50 - 50 wt.%)*	Au, Pt paste	LSGM	3% H$_2$O - 97% H$_2$	340 *(1073)*	[89]
La$_{0.33}$Sr$_{0.67}$Cr$_{0.33}$Fe$_{0.67}$O$_{3-\delta}$ - GDC *(50 - 50 wt.%)*	Au grids	GDC (300)	3% H$_2$O - 97% H$_2$	300 *(1073)*	[77]
La$_{0.33}$Sr$_{0.67}$Cr$_{0.33}$Fe$_{0.67}$O$_{3-\delta}$ - GDC *(50 - 50 wt. %)*	Au grids	LSGM (400)	3% H$_2$O - 97% H$_2$	360 *(1073)*	[77]
La$_{0.8}$Sr$_{0.2}$Cr$_{0.48}$Co$_{0.52}$O$_{3-\delta}$ - GDC *(50 - 50 wt.%)*	Au paint	YSZ (300)	H$_2$	140 *(1123)*	[52]
La$_{0.2}$Sr$_{0.2}$Cr$_{0.9}$Ni$_{0.1}$O$_{3-\delta}$ -YSZ *(45-55 wt.%)*	Ag paste	YSZ (75)	wet H$_2$	130 *(973)*	[80]

(Table 2) cont.....

Anode	Current Collector	Electrolyte (Thickness, μm)	Fuel	P_{max}, mW/cm^2 (T, K)	Ref.
$La_{0.7}Ca_{0.3}Cr_{0.8}Ni_{0.2}O_{3-\delta}$	Pt mesh	LSGM (400)	3% H_2O - H_2	250 *(1073)*	[64]
$La_{0.8}Sr_{0.2}Cr_{0.82}Ru_{0.18}O_{3-\delta}$ - GDC *(50 - 50 wt. %)*	Au grids	LSGM (400)	wet H_2	460 *(1073)*	[79]

Protective layers ($Ce_{0.6}La_{0.4}O_{2-\delta}$ [89, 98], $Ce_{0.9}Gd_{0.1}O_{2-\delta}$ [77]) were deposited between the anode and electrolyte. Cathode: LSM [94, 97], LSM-YSZ [95], SCF [75], BSCF [96], LSCF-GDC [77, 79, 89, 98], LSF-YSZ [80], $LaNi_{0.6}Fe_{0.4}O_{3-\delta}$-LSGM [85], Pt [52, 92]. Symmetrical cells were studied in [64, 93].

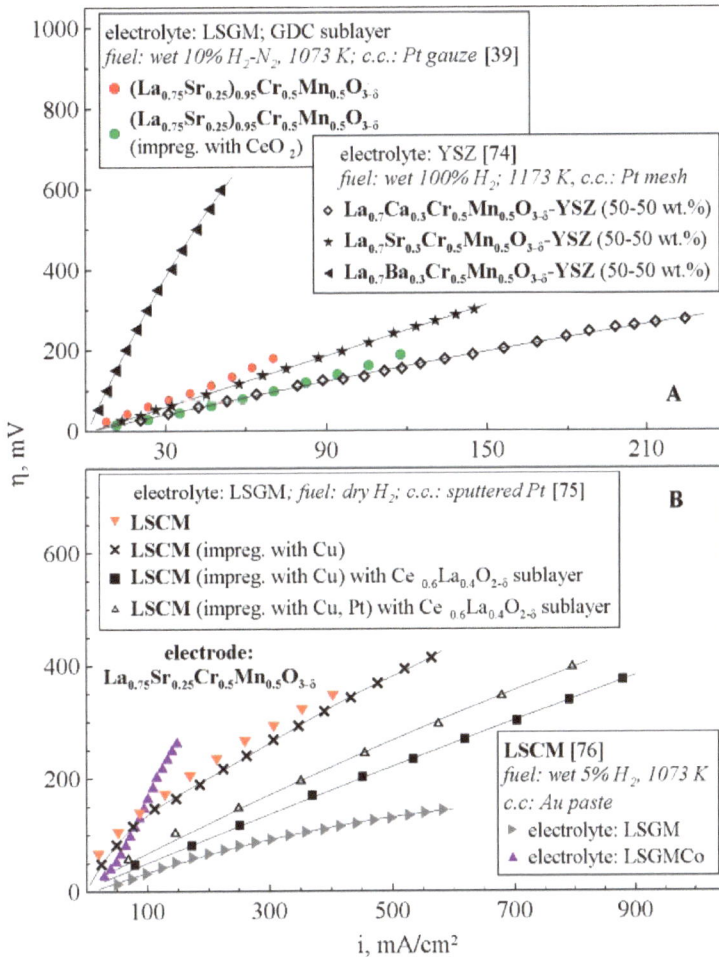

Fig. (2). Polarization curves of selected (La,A)(Cr,Mn)O$_3$-based anodes. The data are taken from [39, 74 - 76].

Promising electrochemical properties were reported for $(La,Sr)(Cr,Fe)O_{3-\delta}$-based perovskite anodes. The effect of Fe content on the anode activity is unobvious (Table **3**) in particular, increasing Fe content above 30% was reported to deteriorate the performance [60]. Although the stability limits of Sr- and Fe-enriched chromites are close to the anode conditions [40, 60], the material and its comparatively high conductivity may be stabilized at temperatures below 1073 K [60] or by partial oxidation of the perovskite phase under anodic polarization [77].

A prospective approach towards optimization of chromite anodes was proposed in [63, 78 - 81] where dissolution of Fe, Co, Ni, Cu or Ru species in the perovskite matrix with their subsequent segregation under reducing conditions allows to produce catalytically-active submicron metallic particles. Exsolution of the metallic phase from the perovskite structure was shown to provide better electrochemical performance than preparation of mechanical mixtures with similar fractions of the components [79]. In terms of microstructural stability, Ru-doping is more preferable compared to Ni due to higher melting point and suppressed sinterability and volatility of Ru particles [81, 82]. In $La_{1-x}Sr_xCr_{1-y}Ru_yO_{3-\delta}$, the compromise between the optimum anode activity and stability was achieved for x = 0.3 and y = 0.08 [79]; this composition is close to $La_{0.7}Sr_{0.3}Cr_{0.95}Ni_{0.05}O_{3-\delta}$ which exhibited the maximum CH_4 conversion compared to other $(La,Sr)(Cr,Ni)O_{3-\delta}$ catalysts [83]. Nevertheless, the procedure of the anode fabrication and operation should be optimized in order to achieve high and stable electrode performance.

Nearly all the groups of chromite-based anodes demonstrated a higher activity after additions of Pd, Rh, Pt, Fe, Ni, VO_x or ceria, due to improved electronic/ionic conductivity, surface kinetics or microstructure-related factors. Selected examples are shown in Table **3**. In particular, the cell with porous YSZ anode layer impregnated with $La_{0.8}Sr_{0.2}CrO_{3-\delta}$, Pd and ceria, yielded the power density 380mW/cm^2 at 973 K in wet H_2 [80]. Stable performance (at least for 24 hours) of 320 - 550 mW/cm^2 and R_η of 0.18 - 0.29 Ohm×cm^2 at 1223 K was observed for $La_{0.75}Sr_{0.25}Cr_{0.5}Mn_{0.5}O_{3-\delta}$ - YSZ - Pt composite electrode in wet H_2 atmospheres, while in wet CH_4, the power density and R_η were 350 mW/cm^2 and 0.26 Ohm×cm^2, respectively [84]. Composite anodes based on $(La,Sr)(Cr,Mn)O_{3-\delta}$ or $(La,Sr)(Cr,Fe)O_{3-\delta}$ showed a noticeable activity towards conversion of hydrocarbons [85 - 87], ethanol [88] while their degradation in H_2S- or PH_3-contaminated fuels was found to be slower compared to Ni-containing cermets [52, 89, 90]. The acceptable stability and functional properties in both oxidizing and reducing atmospheres makes chromites attractive for utilizing in symmetrical SOFCs where both electrodes are made of the same material, although due to low electronic conductivity of Cr-containing cathodes this factor becomes the most critical parameter for the efficiency of such cells [39, 57, 74, 64, 91].

Manganite-Based Perovskites

Manganite-based perovskites have been widely studied as cathode materials due to their high electronic conductivity, moderate thermal expansion, acceptable catalytic activity and suppressed (in comparison with other cathodes) interaction with SOFC components. Reduction of manganites, although being attractive in terms of improvement of the ionic transport [99], results in destabilization of the perovskite phase [100 - 103]. While chromite-, titanate- and molybdate-based perovskites are frequently refered to as alternative anode materials [2, 5, 100], no overviews on manganites stabilized under low-p(O$_2$) conditions have been found so far. On the other hand, the number of studies devoted to assessment of manganites as anode materials or evaluation of their low-p(O$_2$) characteristics has been growing in recent years. The present section is focused on assessment of manganite perovskite as potential anode materials, with primary attention on their tolerance towards reducing atmospheres.

Manganites are characterized by larger oxygen content variations compared to chromites while Mn species are stable in multiple oxidation states (+4/+3/+2) and may adopt the octahedral, pyramidal and tetrahedral coordination with respect to oxygen atoms [69 - 72]. Larger oxygen losses lead to destabilization of the perovskite phase under reducing conditions; the decomposition may be suppressed kinetically or by introducing donor-like or other stabilizing cations into A- or B-sublattice (Ln^{3+}, Ti^{4+}, Cr^{3+}, Nb^{5+}, Al^{3+}). In particular, the onset of formation of the binary oxides and RP-phases from La$_{0.8}$Sr$_{0.2}$MnO$_{3-\delta}$ and La$_{0.8}$Sr$_{0.2}$Mn$_{0.5}$Al$_{0.5}$O$_{3-\delta}$ was observed at p(O$_2$)≈10^{-17} and ~10^{-18} atm (T =1023 K), respectively, indicating a partial stabilizing effect of Al^{3+} cations [102, 104]. The conductivity of (La,Sr)MnO$_{3-\delta}$ perovskites in air is basically higher in comparison with that of chromites and exhibits a similar drop on strong reduction. Despite the higher conductivity even under anode conditions (Table **4**), this factor remains critical for manganite-based anodes [78, 105].

Due to larger variations of the oxygen content as well as formation of oxygen-excessive structures under oxidizing conditions [101, 103, 106 - 108], the chemically-induced expansivity of manganites may exceed 0.2-0.5% [55, 105, 109 - 111]. In particular, reduction of La$_{0.8}$Sr$_{0.2}$Mn$_{0.5}$Al$_{0.5}$O$_{3-\delta}$ perovskite in H$_2$-Ar atmosphere results in ~0.7% expansion [104], unacceptable for the anode applications. Depending on the dopant nature and its content, the TECs of manganites is (11-15)×10^{-6} K^{-1} (Table **4**). Manganites generally exhibit a superior ionic conductivity in comparison with chromites due to a larger concentration of oxygen vacancies [39, 55, 99, 112], although the information on the transport properties of manganites under reducing conditions is scarce due to insufficient stability. A good adhesion of the manganite layers to ZrO$_2$-based electrolytes attributable to a high dissolution of Mn in zirconia phases was reported [113 - 116].

A satisfactory electrochemical activity was reported for $La_{1-x}Sr_xMnO_{3-\delta}$ anodes; increasing Sr content up to x=0.5 leads to an improvement of the anode properties [15], possibly due to increasing electronic/ionic conductivity of the perovskite phase, despite the instability of Sr-enriched manganites under the anode conditions [55, 102, 106, 124]. Introducing of stabilizing cations (Al, Sc, Ti, Ga, Nb) into manganite lattice generally deteriorates the anode performance, as demonstrated in Table **5** and Fig. (**3**).

Table 4. Total conductivity at 1073 K in reducing atmospheres and TEC in air of selected manganite-based materials.

Composition	σ_{1073}, S/cm *(Atmosphere)*	$TEC_{air} \times 10^6$, K^{-1} *(T range, K)*	Ref.
$LaMnO_{3-\delta}$	57 *(p(O$_2$)≈ 4.1×10^{-18} atm)*	11.8 *(280-1100)*	[101, 117]
$La_{0.9}Sr_{0.1}MnO_{3-\delta}$	81 *(p(O$_2$)≈ 3.3×10^{-19} atm)*	9.9 *(300-1270)*	[101, 118]
$La_{0.8}Sr_{0.2}Mn_{0.3}Al_{0.7}O_{3-\delta}$	6.4×10^{-2} *(4% H$_2$ - 9.5% H$_2$O - Ar)*		[119]
$La_{0.8}Sr_{0.2}Mn_{0.5}Al_{0.5}O_{3-\delta}$	1.5 *(4% H$_2$ - 9.5% H$_2$O - Ar)*	11.5 *(300-1270)*	[104, 119]
$La_{0.8}Sr_{0.2}Mn_{0.8}Sc_{0.2}O_{3-\delta}$	5.1 *(wet H$_2$)*		[120]
$La_{0.8}Sr_{0.2}Mn_{0.8}Sc_{0.2}O_{3-\delta}$	3.8 *(dry H$_2$)*		[120]
$La_{0.4}Sr_{0.6}Mn_{0.2}Ti_{0.8}O_{3-\delta}$	9.5×10^{-2}*(wet 4% H$_2$-Ar)*		[111]
$La_{0.4}Sr_{0.6}Mn_{0.4}Ti_{0.6}O_{3-\delta}$	8.7×10^{-2}*(wet 4% H$_2$-Ar)*		[111]
$La_{0.4}Sr_{0.6}Mn_{0.6}Ti_{0.4}O_{3-\delta}$	1.3 *(wet 4% H$_2$-Ar)*	11.9 *(300-1270)*	[111]
$La_{0.5}Sr_{0.5}Mn_{0.5}Ti_{0.5}O_{3-\delta}$		12.8 *(923-1223)*	[105]
$(La_{0.55}Sr_{0.45})_{0.95}Mn_{0.5}Ti_{0.5}O_{3-\delta}$	1.7 *(p(O$_2$) ≈ 10^{-19} atm)*	12.5 *(923-1223)*	[105]
$La_{0.9}Sr_{0.1}Mn_{0.2}Ga_{0.8}O_{3-\delta}$	0.10 *(H$_2$)*		[121]
$La_{0.9}Sr_{0.1}Mn_{0.2}Ga_{0.8}O_{3-\delta}$	6.5×10^{-2}*(H$_2$)*		[122]
$La_{0.9}Sr_{0.1}Mn_{0.43}Ga_{0.57}O_{3-\delta}$	0.38 *(H$_2$)*		[122]
$SrMn_{0.5}Nb_{0.5}O_{3-\delta}$	4.4×10^{-2}*(wet 5% H$_2$-Ar)*	12.2 *(923-1223)*	[78, 123]
$La_{0.75}Sr_{0.25}Cr_{0.5}Mn_{0.5}O_{3-\delta}$	0.93 *(5% H$_2$ - Ar)*	12.0 *(570-1270)*	[41, 46]
$Sr_{0.7}Ce_{0.3}Mn_{0.8}Cr_{0.2}O_{3-\delta}$	4.8 *(p(O$_2$) ≈ 6.7×10^{-19} atm)*		[55]
$La_{0.65}Ce_{0.1}Sr_{0.25}Cr_{0.5}Mn_{0.5}O_{3-\delta}$	0.73 *(3% H$_2$O - H$_2$)*	11.5 *(300-1790)*	[66]
$Pr_{0.7}Ca_{0.3}Cr_{0.4}Mn_{0.6}O_{3-\delta}$	~0.70 *(wet 5% H$_2$-Ar)*		[57]

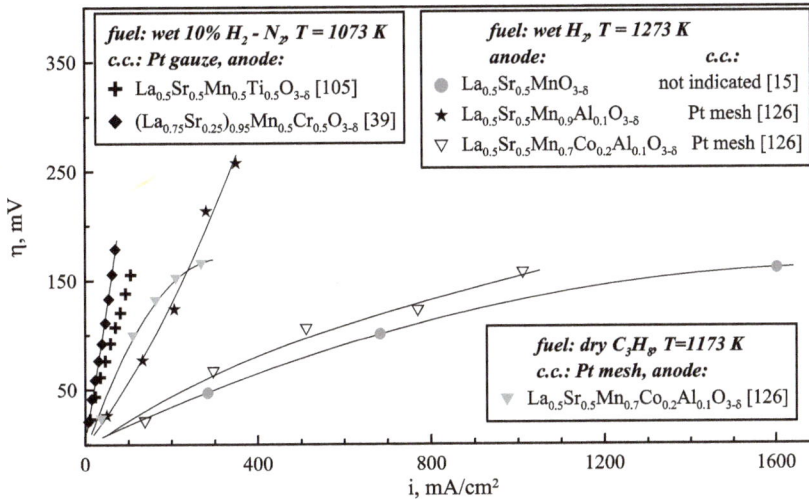

Fig. (3). Polarization curves of selected manganite-based anodes. The data were obtained on cells with LSGM [39, 105] or LSGMCo [15, 126] electrolytes.

High power density was produced for $La_{0.8}Sr_{0.2}Mn_{0.9}Sc_{0.1}O_{3-\delta}$ perovskite anode activated by CeO_2, Ag, Pd and impregnated into YSZ scaffolds (~700 mW/cm^2 in 3% H_2O - H_2) [113]. The power density as high as ~360 mW/cm^2 at 1129 K in wet H_2 was observed for a cell with $La_{0.4}Sr_{0.6}Mn_{0.6}Ti_{0.4}O_{3-\delta}$ - YSZ anode. The anode polarization resistance under given conditions is ~0.3 Ohm×cm^2 [111]; similar value was obtained for A-deficient $(La_{0.8}Sr_{0.2})_{0.94}Mn_{0.5}Al_{0.5}O_{3-\delta}$ - YSZ (65 - 35 vol.%) composite anode covered with Au paste [104].

Promising results were obtained for $(La,Sr)(Mn,Fe)O_3$-based materials, which exhibit acceptable phase and structural stability upon reduction [28, 72, 125]. Anode $La_{0.6}Sr_{0.4}Mn_{0.1}Fe_{0.9}O_{3-\delta}$ - $Ce_{0.6}Mn_{0.3}Fe_{0.1}O_{2-\delta}$ demonstrated a power density of ~300 mW/cm^2 in wet H_2 at 1073 K and a remarkable activity towards electrochemical oxidation of hydrocarbons [125].

Summarizing the results observable on manganite-based anodes, one should note that the dominant part of the electrodes studied contains large amounts of additives or current-collecting layers promoting the anode activity which complicate evaluation of the applicability of manganite-based anodes in industrial scale. Moreover, in contrast to chromite-based anodes, high-temperature utilization of manganites may be detrimental in terms of poor chemical stability, electronic conductivity drop, microstructural changes and especially mechanical degradation of the electrode layers. Therefore, utilization of $(La,A)MnO_{3-\delta}$-based anodes without expensive modifications is possible only if reasonable performances may be achieved at comparatively low temperatures (950 - 1100 K).

By far, no appropriate anode composition has been found. At the same time, attractive characteristics of manganite perovskites offer numerous possibilities to improve both activity and stability by compositional modifications.

Table 5. Power density of selected cells with manganite-based anodes.

Anode	Current Collector	Electrolyte (Thickness, μm)	Fuel	P_{max}, mW/cm^2 (T, K)	Ref.
$LaMnO_{3-\delta}$	not indicated	LSGMNi (300)	3% H_2O - H_2	60 (1273)	[15]
$La_{0.5}Sr_{0.5}MnO_{3-\delta}$	not indicated	LSGMNi (300)	3% H_2O - H_2	630 (1273) / 450 (1173)	[15]
$La_{0.8}Sr_{0.2}Mn_{0.7}Co_{0.2}Al_{0.1}O_{3-\delta}$	Pt mesh	LSGM (285)	3% H_2O - H_2	950 (1273)	[126]
$La_{0.8}Sr_{0.2}Mn_{0.9}Sc_{0.1}O_{3-\delta}$ - YSZ** (40-60 wt.%)	Ag paste	YSZ (85)	3% H_2O - H_2	220 (973)	[113]
$La_{0.8}Sr_{0.2}Mn_{0.9}Sc_{0.1}O_{3-\delta}$ -CeO$_2$- Pd -YSZ* (40-10-1-49 wt.%)	Ag paste	YSZ (85)	3% H_2O - H_2	400 (973)	[113]
			3% H_2O - CH_4	450 (1073) / 340 (973)	
$La_{0.8}Sr_{0.2}Mn_{0.8}Sc_{0.2}O_{3-\delta}$	Au	ScSZ (200)	3% H_2O - H_2	170 (1173)	[127]
			CO	110 (1173)	
$La_{0.8}Sr_{0.2}Mn_{0.8}Sc_{0.2}O_{3-\delta}$	Au paste	ScSZ (300)	3% H_2O - H_2	310 (1173)	[120]
$La_{0.8}Sr_{0.2}Mn_{0.8}Sc_{0.2}O_{3-\delta}$ -CeO$_2$ - Pd - YSZ* (40-10-1-49 wt.%)	Ag paste	YSZ (85)	3% H_2O - H_2	550 (1073)	[113]
			3% H_2O - CH_4	420 (1073)	
$La_{0.8}Sr_{0.2}Mn_{0.7}Sc_{0.3}O_{3-\delta}$ -CeO$_2$- Pd - YSZ* (40-10-1-49 wt.%)	Ag paste	YSZ (85)	3% H_2O - H_2	280 (1073)	[113]
$La_{0.4}Sr_{0.6}Mn_{0.6}Ti_{0.4}O_{3-\delta}$ -8YSZ (50-50 wt.%)	Au paste	YSZ (200)	wet H_2	360** (1129)	[111]
			wet CH_4	50** (1129)	
$La_{0.9}S_{0.1}Mn_{0.8}Ga_{0.2}O_{3-\delta}$	porous Pt	LSGM (1650)	2% H_2O - H_2	90 (1123)	[121]
$La_{0.9}S_{0.1}Mn_{0.57}Ga_{0.43}O_{3-\delta}$	not indicated	LSGM (800)	3% H_2O - H_2	180 (1073)	[122]
GDC-Ni (66-34 wt.%) on $La_{0.9}Mn_{0.8}Ni_{0.2}O_{3-\delta}$	Ni gauze	YSZ (90)	50% H_2O - 50% H_2	170** (1123)	[128]
$Pr_{0.6}Sr_{0.4}MnO_{3-\delta}$	not indicated	LSGMNi (300)	3% H_2O - H_2	270 (1273)	[15]
$PrBaMn_2O_{5+\delta}$ - YSZ* (45-55 wt. %)	Ag paste	YSZ (~55)	3% H_2O - H_2	530 (973)	[4]
$PrBaMn_2O_{5+\delta}$ - FeNi -YSZ* (45-15 - 40 wt.%)	Ag paste	YSZ (~55)	3% H_2O - H_2	810 (973)	[4]
			3% H_2O - C_3H_8	~300 (973)	[4]

* anode was infiltrated into porous YSZ scaffold. ** U=0.7 V. Cathode: LSM-YSZ [111, 128], LSCF [122], LSF-YSZ [113], $La_{0.4}Ba_{0.6}CoO_{3-\delta}$ [15], $La_{0.9}Sr_{0.1}Co_{0.8}Ga_{0.2}O_{3-\delta}$ [121], $Sm_{0.5}Sr_{0.5}CoO_{3-\delta}$ [126], (Gd,Ba,Sr)(Co,Fe)O$_{5-\delta}$-YSZ [4]. Symmetrical cells were studied in [120, 127].

Titanates with Perovskite-Like Structure

$(A,Ln)TiO_{3-\delta}$ oxides and their derivatives are characterized by complex relationships between the defect structure and phase composition. Low redox potential of $Ti^{4+/3+}$ couple leads to negligible equilibrium content of Ti^{3+} species at high oxygen chemical potentials and the charge neutrality is achieved by formation of secondary SrO-, Ln_2O_3-phases or their intergrowths in the perovskite lattice [129 - 131]. In contrast, cation deficient $(A,Ln)_{1-x}TiO_3$-like oxides are stable in air while reduction causes their decomposition [132, 133]. Thus, titanates are succeptible to redox cycling, and the major strategies towards development of titanate-based anodes are focused on their stabilization by kinetic factors or by introduction of reducible cations (primarily into B-sublattice), such as Co, Mn, Fe *etc*, which enable to preserve the charge neutrality with a minimum participation of Ti^{4+} in the redox processes. The latter approach also allows to accelerate the equilibration kinetics on redox cycling [134 - 136].

Titanates are characterized by dominating *n*-type electronic conductivity, with a high delocalization degree of the electronic charge carriers which ensures high conductivity (>100 - 1000 S/cm) [137, 138]. However, generation of large amounts of mobile electrons requires strongly reducing conditions and appropriate amounts of donor-like cations (Ln^{3+}, Nb^{5+}, *etc*). Thus, both the electron concentration and the total conductivity decrease on oxidation; this effect may be accompanied by microstructural degradation or kinetic stabilization of the oxidized phase which leads to a pratically irreversible conductivity drop [139, 140]; examples are shown in Fig. (**4**).

The influence of the intrinsic properties of perovskite phase on the electrochemical properties is currently unobvious. For most anodes, the electrode properties are affected by microstructure, equilibration kinetics, poor catalytic activity, especially without substituting cations in B-sublattice, and peculiarities of the anode processing or testing conditions [138 - 142]. Despite the high electronic conductivity, donor-doped titanates demonstrate poor performances [139, 143, 144]; such a behaviour may be attributed to decreasing content of oxygen vacancies at higher dopant amounts [132, 145]. In particular, individual anodes $Sr_{0.94}Ti_{0.9}Nb_{0.1}O_{3-\delta}$ deposited onto YSZ and covered with Pt paste, showed the polarization resistance in the range of 0.5 - 160 Ohm×cm^2 at 1123 K, depending on the sequence of reduction or oxidation runs [139, 147]. These data suggest a significantly weaker effect of the electronic conductivity on the anode properties of titanates as compared to chromites and manganites.

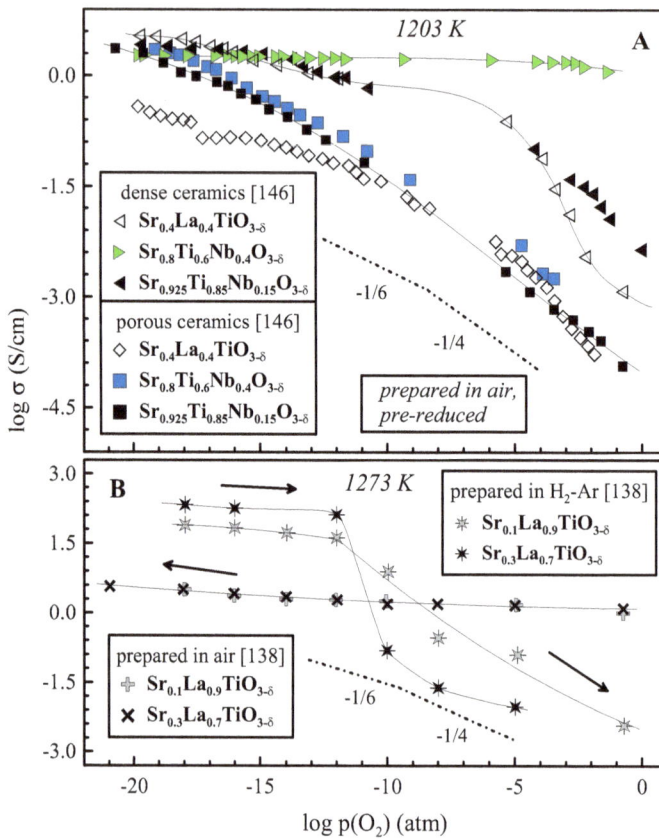

Fig. (4). Isothermal dependencies of the total conductivity *vs.* p(O$_2$) for selected titanates with various microstructure (A) and preparation conditions (B). The data are taken from [138, 146].

Most electrochemical studies were carried on (Sr,La)TiO$_{3-\delta}$ or (Sr,Y)TiO$_{3-\delta}$ perovskites. A positive impact of a partial replacement of Sr with Ca [148 - 150] or Ba [151], as well as introduction of Mn, Fe, Co, Ni into B-sublattice [105, 145, 152], on the electrochemical behaviour of titanates, was reported.

Quite promising results were obtained for Mn-substituted titanate anodes [105, 111, 153, 154]; the polarization resistance of Sr$_{0.4}$La$_{0.6}$Ti$_{1-x}$Mn$_x$O$_{3-\delta}$ (0.2≤x≤0.8) was reported to be as low as 0.01 - 0.1 Ohm×cm^2 at 1073 K in wet H$_2$ [154]. Increasing Mn content improves the anode performance attributable to the improvement of the electronic conductivity [111, 154]. Further optimization may be achieved by a partial replacement of Mn with Ga which enables to attain a promising combination of the electronic and ionic conductivity as well as the catalytic properties [155]. The polarization resistance as low as 0.3 Ohm×cm^2 at

1073 K was observed for $(Sr,La)Ti_{0.5}Mn_{0.5}O_{3-\delta}$ anodes while the power density yielded by a cell with a similar anode achieved ~200 mW/cm^2 in wet hydrogen [105, 156].

A positive effect of introduction of 20-50% Fe into $(Sr,La)TiO_{3-\delta}$ or $(Sr,Y)TiO_{3-\delta}$ on the anode properties has been reported [157 - 159]. For example, the polarization resistance of $Sr_{0.92}Y_{0.08}Ti_{0.75}Fe_{0.25}O_{3-\delta}$ anodes is ~0.7 Ohm×cm^2 at 1073 K [157]. Acceptable electrode activity was also achieved upon doping with Sc [160], Co [161, 162], Ni [163], consistently with higher ionic conductivity and/or catalytic activity towards fuel oxidation processes.

Introduction of ion-conductive ZrO_2- or CeO_2-based materials or catalytically-active metallic species (Ni, Pd, Ru) favours the electrochemical performance of titanate-based anodes; examples are presented in Table **6**. Impregnation of porous YSZ scaffolds with solutions containing dissolved precursors for synthesis of $Sr_{0.6}La_{0.4}TiO_{3-\delta}$, $CeO_{2-\delta}$ and Pd, allowed to achieve the power density of 1.1 W/cm^2 [164]. Promising results were reported for anodes composed of YSZ infiltrated with $Sr_{0.5}Ba_{0.1}La_{0.4}TiO_{3-\delta}$ [165] or $Sr_{0.92}Y_{0.08}Ti_{0.8}Fe_{0.2}O_{3-\delta}$ [159]. Poor adhesion of titanates to zirconia-based substrate may be improved by utilization of titanates substituted in B-sublattice [114, 159] or by preliminary introduction of CeO_2-based phases into YSZ scaffolds [144].

Table 6. Power densities of selected cells with titanate-based anodes in H_2-containing fuel.

Anode	Current Collector	Electrolyte (Thickness, μm)	P_{max}, mW/cm^2 (T, K)	Ref.
$Sr_{0.8}La_{0.2}TiO_{3-\delta}$ - GDC (50:50 wt.%), *impr. with Ni*	Pt paste	LSGM (500)	270 *(1073)*	[169]
$Sr_{0.8}La_{0.2}TiO_{3-\delta}$*nanofibers, impr. with GDC (0.8 wt.%)*	Au mesh	ScCeSZ (50)	380 *(1223)*	[170]
$Sr_{0.8}La_{0.2}TiO_{3-\delta}$*impr. with Ni (5.1 vol.%) on porous LSGM layer*	Au grid	LSGM (18)	510 *(923)*	[171]
Support:$Sr_{0.8}La_{0.2}TiO_{3-\delta}$, *Protect. layer:* Ni-SDC (50 - 50 wt.%) *Active layer:* Ni-YSZ (50 - 50 wt.%)	-	YSZ (~10)	850 *(1073)*	[172]
$Sr_{0.7}La_{0.2}TiO_{3-\delta}$*impr. with GDC+Cu*	Ag paste	YSZ (50-75)	550 *(1023)*	[173]
$Sr_{0.7}La_{0.3}TiO_{3-\delta}$ - YSZ (50 - 50%) *impr. with CeO_2 +Pd (10+1 wt.%)*	Ag paste	YSZ (75)	650 *(1173)*	[174]
YSZ impr. with $Sr_{0.7}La_{0.3}TiO_{3-\delta}$ + CeO_2+ *Pd (45+5+0.5 wt.%)*	not indicated	YSZ	1100 *(1173)*	[164]
YSZ impr. with CeO_2, Pd	$Sr_{0.6}La_{0.4}TiO_{3-\delta}$	YSZ (75)	650 *(1073)*	[175]

(Table 6) cont.....

Anode	Current Collector	Electrolyte (Thickness, μm)	P_{max}, mW/cm² (T, K)	Ref.
$Sr_{0.25}Ca_{0.45}La_{0.2}TiO_{3-\delta}$ *impr. with CeO_2 + Ni (6 +3.1 wt%)*	Ag paste	YSZ (~40)	960 *(1073)*	[150]
$Sr_{0.88}Y_{0.08}TiO_{3-\delta}$ - YSZ (50 - 50 wt.%) *impr. with Ni*	$Sr_{0.88}Y_{0.08}TiO_{3-\delta}$	YSZ (500)	340 *(1273)*	[176]
$Sr_{0.88}Y_{0.08}TiO_{3-\delta}$-YSZ (50-50 wt.%) *impr. with CeO_2, Ru*	Pt mesh	YSZ (~10)	510 *(1073)*	[177]
$Sr_{0.88}Y_{0.08}TiO_{3-\delta}$ - LDC (50 - 50 wt.%)	Au paste	LSGM (300)	460 *(1123)*	[178]
$Sr_{0.88}Y_{0.08}TiO_{3-\delta}$ - $Ce_{0.6}La_{0.4}O_{2-\delta}$- Pd (49.25 - 49.25 - 1.5 wt.%)		interlayer LDC	1000 *(1123)*	
$Sr_{0.67}La_{0.33}Ti_{0.95}Sc_{0.05}O_{3-\delta}$-YSZ *(graded)*	Au paste	YSZ (2000)	130 *(1173)*	[160]
$(Sr_{0.4}La_{0.6})_{0.97}Ti_{0.5}Mn_{0.5}O_{3-\delta}$	not indicated	LSGMCo (500)	200 *(1173)*	[156]
$Sr_{0.4}La_{0.6}Ti_{0.6}Mn_{0.4}O_{3-\delta}$	not indicated	GDC (500)	270 *(1073)*	[154]
YSZ impr. with $Sr_{0.92}Y_{0.08}Ti_{0.8}Fe_{0.2}O_{3-\delta}$ + CeO_2 +Pd (40+10+1 wt.%)	Ag paste	YSZ (85)	420 *(1073)*	[159]
$Sr_{0.8}La_{0.2}Ti_{0.98}Co_{0.02}O_{3-\delta}$ - GDC (30-70%), *impr. with Ni (15 wt.%)*	not indicated	LSGM (250)	920 *(1073)*	[179]
$Sr_{0.8}La_{0.2}Ti_{0.9}Ni_{0.1}O_{3-\delta}$	Pt mesh	ScCeSZ (300)	150 *(1073)*	[167]
$Sr_{0.67}La_{0.33}Ti_{0.92}Mn_{0.04}Ga_{0.04}O_{3-\delta}$ - YSZ *(graded)*	Au paste	YSZ (330)	500 *(1223)*	[155]
$Sr_{0.99}Ti_{0.9}Nb_{0.1}O_{3-\delta}$ - FeCr steel (70 - 30 vol.%) *impr. with GDC+Ni*	Pt paste	ScYSZ (not indicated)	1400* *(1023)* *(U=650 mV)*	[180]

Cathodes: LSM [150, 155], LSM - YSZ [172, 176, 177], LSF - YSZ [159, 164, 174, 175], LSCo [180], LSCo - YSZ [173], LSCF [179], LSCF-GDC [167, 171], BSCF [169], BSCF-GDC [154], $Sm_{0.5}Sr_{0.5}CoO_{3-\delta}$ [156, 178], Pt [160, 170].

Similar to chromite-based anodes, *3d-* or *4f-*cations which form secondary catalytically-active phases under reducing (Fe, Co, Ni, MnO_x) or oxidizing ($CeO_{2-\delta}$) conditions may be introduced into the titanate perovskite lattice. Microscopic studies demonstrated that formation of these phases takes place preferentially at the "defects" on the material surface (borders, edges, cavities, grain boundaries); thus, adjusting the cation ratios and processing conditions allows to tune the particle size and their distribution over the surface [142, 161, 163, 166, 167]. Structured anodes, consisting of nanoparticles with $(La_{0.75}Sr_{0.2}Ba_{0.05})_{0.175}Ce_{0.825}O_{2-\delta}$ as the shell adhered to $Sr_{0.7}La_{0.3}Ti_{1-x}Nb_xO_{3-\delta}$ core were proposed in [168]; however, the power density of the corresponding cell achieved only 30 mW/cm² at 973 K.

Basically, the activity of titanate-based anodes in natural gas-derived fuel is poor [181, 182] but may be improved by addition of catalytically-active phases into the perovskite anode or by wet impregnation of titanate-precursor solution into porous ceramic scaffolds. In particular, the power density of the cell comprising a $(Sr,La)(Ti,Mn,Ga)O_{3-\delta}$ - YSZ graded anode was reported to achieve 350 mW/cm² in wet CH_4. No substantial carbon deposition was found after electrochemical studies while the polarization resistance gradually improved during the test [155]. Attractive results were achieved on multilayered anodes where the electrochemical process takes place on a cermet-based functional layer covered with a porous titanate support. The latter provides the mechanical strength, electronic current collection and ensures inhibited gas diffusion which increases the ratio $(H_2O+CO_2)/C_xH_y$ in the vicinity of Ni-contained phase making the process of carbon deposition thermodynamically unfavourable. Application of this approach allowed to achieve the power densities 300 - 500 mW/cm² [172, 175, 183].

As noted above, the characteristics of titanates are strongly influenced by the preparation atmosphere. In terms of the anode activity, preliminary high-temperature reduction or cathodic polarization is considered to be beneficial [138, 146]. However, most titanate-based electrodes showed a specific feature, associated with fast and in some cases irreversible performance deterioration observable on oxidation; examples will be demonstrated in the corresponding section.

SrMoO₃-based Oxides with Disordered and Double Perovskite Structure

Among molybdate-based compounds, the most promising in terms of potential anode application are $AMoO_{3-\delta}$ and $(A,Ln)(Mo,B)O_{6-\delta}$ (A=Ca, Sr, Ba, B = Mg, Fe, Co, Ni, Mn); the latter group is characterizied by a partial ordering in B-sublattice [184, 185]. For $AMoO_{3-\delta}$ perovskites, the largest stability region was observed for $SrMoO_{3-\delta}$ ($10^{-15} < p(O_2) < 10^{-12}$ atm at 1473 K [186]); oxidation or reduction results in formation of insulating $SrMoO_4$ scheelite phase or SrO+Mo mixture, respectively [186]. The scheelite-like phases are frequently produced during synthesis of Mo-containing oxides and may affect their functional characteristics or result in discrepancies between the experimental data. Another factor limiting applications of Mo-doped perovskites is large volatility of $MoO_x(OH)_y$ compounds [187, 188]. $AMoO_{3-\delta}$-like materials exhibit an excellent *n*-type electronic conductivity (up to 10 kS/cm); however, utilization of pure $SrMoO_3$-based anodes is problematic due to large volume variations on redox cycling [189 - 191]. On the other hand, formation of $SrMoO_3$ traces under reducing conditions is suggested to promote the electrochemical properties of molybdate-based anodes [192, 193].

The stability of molybdates may to some extent be improved by substitution with Cr, Ti, Nb or V species [191, 194 - 196]. In particular, V-doped $SrMoO_{3-\delta}$ show better tolerance towards oxidation while their conductivity varies in the range of 1 - 5 kS/cm under SOFC operation conditions and exhibits a metallic-like behavior [194]. However, large volume changes on redox cycling [191] still significantly limit the applicability of the material.

Among $(A,Ln)(Mo,B)O_6$-like series, where B relates to a 2-fold charged cation (Mg^{2+}, Fe^{2+}, Co^{2+}, Ni^{2+}), the most studied representative is $Sr_2MgMoO_{6-\delta}$. The material is fairly stable under anode environments, but is susceptible towards oxidation [190, 192], strong reduction [197, 198] or the presence of CO_2 or CH_4 in the atmosphere [192, 198]. The conductivity of Mg-substituted molybdates is 3-10 S/cm under anodic conditions (Table **7**), although this value may be affected by the presence of high-conductive $SrMoO_{3-\delta}$ impurity [190, 195]. Introducing Ln^{3+} into the A-sublattice of $Sr_2MgMoO_{6-\delta}$ increases the conductivity up to ~30 S/cm due to formation of additional *n*-type charge carriers; on the other hand, Ln-doped materials are less tolerant towards oxidation [199, 200].

Table 7. Total conductivity at 1073 K in reducing atmospheres and TEC of selected molybdate-based perovskites.

Composition	σ_{1073}, S/cm *(Atmosphere)*	Thermal Expansion		Ref.
		T, K	**TEC×10⁶, K⁻¹*(Atmosphere)***	
$CaMoO_{3-\delta}$	760 *(5% H₂ - N₂)*	300 - 1273	11.0 *(air, N₂, H₂)*	[202, 203]
$CaMo_{0.5}V_{0.5}O_{3-\delta}$	~500 *(10% H₂ form. gas)*	973 - 1173	10.0 *(10% H₂ form. gas)*	[191]
$SrMo_{0.5}V_{0.5}O_{3-\delta}$		373 - 773	13.3 *(air)*	[204]
		773 - 1173	15.4 *(air)*	
		373 - 1173	12.6 *(10% H₂ - N₂)*	
$Sr_2MgMoO_{6-\delta}$	8.6 *(p(O₂)= 7×10⁻²¹ atm)*	691 - 1074	12.7 *(air)*	[184]
$Sr_2MgMoO_{6-\delta}$		573 - 1203	12.7 *(5% H₂ - N₂)*	[205]
$Sr_{1.4}La_{0.6}MgMoO_{6-\delta}$	5.0 *(5% H₂-Ar)*			[206]
$Sr_{1.6}Sm_{0.4}MgMoO_{6-\delta}$	7.3 *(H₂)*	573 - 1273	13.5 *(N₂)*	[207]
$Sr_2MgMo_{0.8}Nb_{0.2}O_{6-\delta}$	0.13 *(p(O₂)=10⁻²¹ atm)*	573 - 1273	13.7 *(air)*	[208]
		573 - 1273	14.6 *(5% H₂ - N₂)*	
$SrMo_{0.9}Cr_{0.1}O_{3-\delta}$	167 *(5% H₂ - N₂)*	573 - 1123	10.9 *(air)*	[196]
		673 - 1123	12.8 *(5% H₂ - N₂)*	
$Sr_2MnMoO_{6-\delta}$		873 - 1203	13.0 *(5% H₂ - N₂)*	[205]
$Ba_2MnMoO_{6-\delta}$	1.7 *(dry H₂)*	873 - 1173	15.6 *(H₂)*	[209]

(Table 7) cont.....

Composition	σ_{1073}, S/cm *(Atmosphere)*	T, K	TEC×10^6, K^{-1}*(Atmosphere)*	Ref.
$Ca_2FeMoO_{6-\delta}$	380 *(H_2)*			[210]
$SrMo_{0.9}Fe_{0.1}O_{3-\delta}$	184 *(5% H_2 - N_2)*	573 - 1123	13.8 *(air)*	[211]
		573 - 1123	12.8 *(5% H_2 - N_2)*	
$SrMo_{0.8}Fe_{0.2}O_{3-\delta}$	129 *(5% H_2 - N_2)*	573 - 1123	13.4 *(5% H_2 - N_2)*	
$Sr_2FeMoO_{6-\delta}$	220 *(H_2)*	300 - 1273	13.9 *(N_2)*	[210]
		300 - 1173	14.0 *(N_2)*	[201]
$Sr_2Fe_{1.5}Mo_{0.5}O_{6-\delta}$	860 *(H_2)*	473 - 1473	18.1 *(air)*	[212]
		473 - 983	14.1 *(air)*	[213]
$Ba_2FeMoO_{6-\delta}$	30 *(H_2)*	300 - 1273	13.3 *(N_2)*	[210]
$Ba_2FeMoO_{6-\delta}$	190 *(dry H_2)*	873 - 1173	14.3 *(H_2)*	[209]
$SrMo_{0.9}Co_{0.1}O_{3-\delta}$		573 - 1123	13.4 *(air)*	[214]
		573 - 1123	12.8 *(5% H_2 - N_2)*	
$Sr_2CoMoO_{6-\delta}$		543 - 1123	11.2 *(5% H_2 - N_2)*	[215]
$Sr_2Co_{1.2}Mo_{0.8}O_{6-\delta}$	3.8 *(H_2)*	873 - 1173	15.8 *(air)*	[216]
$Sr_2Co_{0.7}Mg_{0.3}MoO_{6-\delta}$	6.7 (5% H_2 - N_2)	323 - 1573	13.9 *(air)*	[217]
$Sr_2NiMoO_{6-\delta}$	1.9 (5% H_2 - Ar)			[218]
$Sr_2NiMoO_{6-\delta}$	53 (H_2)	300 - 1253	12.1 *(air)*	[219]
$Ba_2NiMoO_{6-\delta}$	2.3 (dry H_2)			[209]

Due to flexible oxidation state and oxygen coordination of Fe species, $SrFe_{1-x}Mo_xO_{3-\delta}$-based compositions are more stable towards redox cycling. The materials possess a comparatively high electronic conductivity, originating from a contribution of both $Mo^{6+/5+/4}$ and $Fe^{3+/2+}$ species into the generation of the electronic structure as well as larger overlapping of the *d*-cations with 2*p*-oxygen orbitals and exhibit the *p-n* transition on redox cycling [29, 185, 195, 201].

It is unobvious if the electrochemical behaviour of molybdates is associated with the intrinsic properties of the base material or is governed by the presence of the secondary phases generated during the processing of molybdates, *i.e.* $SrMoO_3$, MoO_x, Co- or Ni-containing compounds, metallic particles, *etc.* One should note that most molybdate-based anodes were essentially studied in contact with LSGM-based electrolytes and $Sr(Co,Fe)O_3$-based cathodes (Table **8**). Therefore, to provide an adequate electrochemical assessment of molybdates, optimization of the corresponding electrolyte and cathode materials is necessary.

Depending on other functional materials, the power density of the cells with $Sr_2MgMoO_{6-\delta}$-based anodes varies in the range of 500 - 1000 mW/cm^2 in hydrogen fuels (Table **8**). A somewhat improvement is observed on substitution of rare-earth cations for Sr [206, 207]; the effect of introduction of transition metals (Mn, Cr, Nb) into B-sublattice of $Sr_2MgMoO_{6-\delta}$ is less obvious [208, 220], suggesting that Mo^{n+} cations and/or Mo-containing impurities are primarily responsible for the catalytic properties.

Table 8. Power densities of selected cells with molybdate-based anodes and LSGM electrolyte in hydrogen-containing fuels.

Anode	Current Collector	Fuel	T, K	P_{max}, mW/cm^2 *(Electrolyte Thickness, μm)*	Ref.
$SrMoO_{3-\delta}$ - GDC (40-60 wt%) [a]	Ag paste	3% H_2O - H_2	1173	580 *(400)*	[223]
$Sr_{0.5}Ca_{0.5}MoO_{3-\delta}$ - GDC	Ag paste	H_2	1073	330 *(~350)*	[224]
$SrMo_{0.9}Mg_{0.1}O_{3-\delta}$ [a]	Pt paste	pure H_2	1073	680 *(300)*	[225]
$Sr_2MgMoO_{6-\delta}$ [a]	Pt mesh	dry H_2	1073	710 *(300)*	[184, 220]
$Sr_2MgMoO_{6-\delta}$ [a]	Pt paste	dry H_2	1073	810* *(300)*	[184, 220]
		3% H_2O - H_2	1073	760* *(300)*	
$Sr_2MgMoO_{6-\delta}$ [a]	Pt paste	H_2	1073	1040 *(250)*	[32]
$Sr_2MgMoO_{6-\delta}$ - func. layer	Au paste ($Sr_{0.7}La_{0.3}TiO_{3-\delta}$ - c.c. layer [a])	H_2	1073	620 *(250)*	[32]
$Sr_{1.6}Sm_{0.4}MgMoO_{6-\delta}$ [a]	not indicated	dry H_2	1073	500 *(300)*	[207]
$Sr_2Mg_{0.8}Mn_{0.2}MoO_{6-\delta}$ [a]	Pt paste	H_2	1073	690* *(300)*	[220]
$Sr_2Mg_{0.9}Cr_{0.1}MoO_{6-\delta}$ [a]	Pt paste	H_2	1073	730* *(300)*	[220]
$SrMo_{0.9}Cr_{0.1}O_{3-\delta}$ [a]	Au paste	100% H_2	1123	750 *(300)*	[196]
$Sr_2MnMoO_{6-\delta}$ [a]	Pt paste	H_2	1073	610* *(300)*	[220]
$Ba_2MnMoO_{6-\delta}$ [b]	Ag paste	dry H_2	1073	450 *(300)*	[209]
$SrMo_{0.9}Fe_{0.1}O_{3-\delta}$ [a]	Pt paste	100% H_2	1073	580 *(300)*	[211]
$SrMo_{0.75}Fe_{0.25}O_{3-\delta}$ - SDC (60 - 40 wt.%)	Ag paste	3% H_2O - H_2	1073	600 *(600)*	[226]
$Ca_2FeMoO_{6-\delta}$ [a]	not indicated	dry H_2	1123	190 *(300)*	[210]
$Sr_2FeMoO_{6-\delta}$	Ag paste	H_2	1073	600 *(300)*	[201]
$SrFe_{1.5}Mo_{0.5}O_{3-\delta}$	Au paste	3% H_2O - H_2	1073	610 *(265)*	[212]

(Table 8) cont.....

Anode	Current Collector	Fuel	T, K	P_{max}, mW/cm^2 (Electrolyte Thickness, μm)	Ref.
$SrFe_{1.5}Mo_{0.5}O_{3-\delta}$	not indicated	3% H_2O - H_2	1073	970 (~20)	[221]
$Sr_2Fe_{1.5}Mo_{0.5}O_{6-\delta}$	Au paste	3% H_2O - H_2	1073	720 (250)	[222]
$Sr_2Fe_{1.5}Mo_{0.5}O_{6-\delta}$ - Ni (98 - 2 wt.%)	Au paste	3% H_2O - H_2	1073	1170 (250)	[222]
$Ba_2FeMoO_{6-\delta}$[b]	Ag paste	dry H_2	1073	520 (300)	[209]
$Sr_2CoMoO_{6-\delta}$[a]	Pt paste	H_2	1073	730 (300)	[218]
$Sr_2CoMoO_{6-\delta}$[a]	Pt paste	H_2	1073	1020 (300)	[227]
$Sr_2Co_{1.1}Mo_{0.9}O_{6-\delta}$[c]	Ag paste	100% H_2	1123	660 (300)	[216]
$SrMo_{0.9}Co_{0.1}O_{3-\delta}$[a]	Pt paste	100% H_2	1073	540 (300)	[214]
$Ba_2CoMoO_{6-\delta}$[b]	Ag paste	dry H_2	1073	490 (300)	[209]
$Sr_2NiMoO_{6-\delta}$	Ag paste	H_2	1073	590 (300)	[219]
$Ba_2NiMoO_{6-\delta}$[b]	Ag paste	dry H_2	1123	360 (300)	[209]

Cathodes: SCF [32, 184, 196, 209, 211, 214, 216, 218, 220, 225, 227], LSCF [222 - 224], BSCF [201, 219], SmBaCo$_2$O$_{5+\delta}$ [207, 210], Sm$_{0.5}$Sr$_{0.5}$CoO$_{3-\delta}$ - SDC [226]. In [212, 221] symmetrical cells were studied. LSGM was used as solid electrolyte. [a] Ce$_{0.6}$La$_{0.4}$O$_{2-\delta}$, [b] Ce$_{0.6}$Sm$_{0.4}$O$_{2-\delta}$, [c] Ce$_{0.8}$Sm$_{0.2}$O$_{2-\delta}$ sublayer was coated between the anode and electrolyte. *after 50 cycles

The performance as high as 970 mW/cm^2 at 1073 K was demonstrated for Sr$_2$Fe$_{1.5}$Mo$_{0.5}$O$_{6-\delta}$ anode [221] while addition of Ni (~2 wt.%) increases the power density to 1200 mW/cm^2 [222]. A reasonable activity was also reported for SrNiMoO$_6$- and SrCoMoO$_6$-based anodes where formation of Ni- and Co-enriched phases may promote the electrochemical processes or catalytic reforming of hydrocarbons [209, 218, 219].

Vanadates with Perovskite-Like Structure

Compared to titanate- and molybdate-based materials, AVO$_3$-based oxides (A = Ca, Sr) show a better tolerance towards high contents of donor cations and are characterized by a somewhat lower temperature of single phase formation. The latter factor enables to fabricate electrode layers by combustion synthesis techniques inside porous matrix of an ion-conductive material at comparatively low temperatures [228 - 230]. This is an important advantage since vanadates are characterized by somewhat increased TEC in reducing atmospheres (Table **9**) and conventional techniques of anode application may be inappropriate. Vanadates are also susceptible towards oxidation leading to partially irreversible formation of LnVO$_4$-, Sr$_2$V$_2$O$_7$- or Sr$_3$V$_2$O$_8$-like phases [191, 228, 231]. LnVO$_{3-\delta}$ exhibit a semiconducting-like behaviour with dominant *p*-type charge carriers [232, 233]. The highest conductivity was observed for Ln = La while doping with Ca/Sr cations increases the conductivity and gradually leads to a change of the transport mechanism into metallic-like behaviour with domination of *n*-type charge carriers. For $La_{1-x}Sr_xVO_{3-\delta}$ series, the maximum conductivity was reported for x = 0.3 - 0.4 [230, 234]; thus, most electrochemical studies were carried out on the corresponding anodes prepared by infiltration of La-, Sr- and V-containing solutions into a porous matrix of the corresponding electrolyte material. A somewhat optimization of the electrochemical properties and suppression of the reactivity of the anode material with other components may be achieved by application of vanadates with a moderate deficit in B-sublattice [235].

Table 9. Total conductivity and TECs of selected vanadates.

Composition	σ_{1073}, S/cm (Atmosphere)	Thermal Expansion		Ref.
		T, K	TEC×10^6, K^{-1}(Atmosphere)	
$SrVO_{3-\delta}$	730 (10% H$_2$ - N$_2$)	300 - 1223	18.0 (10% H$_2$ - N$_2$)	[239]
$Sr_3V_2O_8$	7.3×10^{-5}(air)	300 - 1273	14.3 (air)	[239]
$Sr_2V_2O_7$	1.6×10^{-3}(air) 6.7×10^{-5} (p(O$_2$)= 5×10^{-5} atm)	300 - 1223	15.0 (air)	[239]
$CeVO_x$	3.8×10^{-2}(air) 2.4 (10% H$_2$ - N$_2$)	400 - 800	5.6 (air)	[233]
$Ce_{0.8}Ca_{0.2}VO_x$	16 (10% H$_2$ - N$_2$)	400 - 800	5.6 (air)	[233]
$La_{0.7}Sr_{0.3}VO_{3-\delta}$	150 (3% H$_2$O - H$_2$)	823 - 1223	11.5 (4% H$_2$ - Ar)	[240]
$Y_{0.2}Sr_{0.8}VO_{3-\delta}$	350 (10% H$_2$ - N$_2$)	300-773	12.8 (10% H$_2$ - N$_2$)	[241]
		773-1223	16.9 (10% H$_2$ - N$_2$)	
$SrV_{0.97}Nb_{0.03}O_{3-\delta}$	470 (10% H$_2$ - N$_2$)	373-773	14.8 (10% H$_2$ - N$_2$)	[231]
$SrV_{0.9}Nb_{0.1}O_{3-\delta}$	210 (10% H$_2$ - N$_2$)	373-773	15.2 (10% H$_2$ - N$_2$)	[231]
		773-1223	20.5 (10% H$_2$ - N$_2$)	
$SrV_{0.7}Nb_{0.3}O_{3-\delta}$	37 (10% H$_2$ - N$_2$)	373-1223	13.3 (10% H$_2$ - N$_2$)	[231]
$La_{0.2}Sr_{0.8}V_{0.9}Nb_{0.1}O_{3-\delta}$	250 (10% H$_2$ - N$_2$)	373 - 1273	14.9(10% H$_2$ - N$_2$)	[242]
$Y_{0.1}Sr_{0.9}V_{0.9}Al_{0.1}O_{3-\delta}$	280 (10% H$_2$ - N$_2$)	300-1223	15.1 (10% H$_2$ - N$_2$)	[241]
$Y_{0.2}Sr_{0.8}V_{0.9}Al_{0.1}O_{3-\delta}$	230 (10% H$_2$ - N$_2$)	300-1223	15.0 (10% H$_2$ - N$_2$)	[241]
$Y_{0.2}Sr_{0.8}V_{0.9}Nb_{0.1}O_{3-\delta}$	140 (10% H$_2$ - N$_2$)	373 - 1273	14.9 (10% H$_2$ - N$_2$)	[242]

The performance of cells with vanadate-based composites anodes without specific modifications is basically poor (Table **10**) and may be improved by anode

Table 10. Power density of selected cells with vanadate-based anodes in hydrogen-containing fuels.

Anode	Fuel	T, K	P_{max}, mW/cm^2 *(Electrolyte Thickness, μm)*	Ref.
LaVO$_{3-\delta}$ - YSZ (10-90 vol.%) [a] + 2.8 vol.% CeO$_{2-\delta}$ + 0.5 vol.% Pd	3% H$_2$O-H$_2$	973	295* *(70-80)*	[235]
LaV$_{0.9}$O$_{3-\delta}$ -YSZ (10-90 vol.%) [a] + 2.8 vol.% CeO$_{2-\delta}$ + 0.5 vol.% Pd	3% H$_2$O-H$_2$	973	290* *(70-80)*	[235]
La$_{0.7}$Sr$_{0.3}$VO$_{3-\delta}$	100% H$_2$	1223	155 *(250)*	[247]
La$_{0.7}$Sr$_{0.3}$VO$_{3-\delta}$ - YSZ (25-75 wt%) [a] + 10wt% CeO$_{2-\delta}$ + 1 wt.% Pd	4.2% H$_2$O-H$_2$	1073	490 *(55)*	[236]
		973	300 *(55)*	
		1073	6 W [b] *(55)*	
La$_{0.7}$Sr$_{0.3}$VO$_{3-\delta}$ - YSZ (10-90 vol.%) [a] + 2.8 vol.% CeO$_{2-\delta}$ + 0.5 vol.% Pd	3% H$_2$O-H$_2$	1073	960 *(80)*	[248]
		973	280 *(80)*	
La$_{0.7}$Sr$_{0.3}$V$_{0.9}$O$_{3-\delta}$ - YSZ (15-85 vol.%) [a] + 2.8 vol.% CeO$_{2-\delta}$ + 0.5 vol.% Pd	3% H$_2$O-H$_2$	1073	1000 *(80)*	[249]
La$_{0.7}$Sr$_{0.3}$VO$_{3-\delta}$ - SDC (11.5-88.5 vol.%) [a] + 1.5 vol.% CeO$_{2-\delta}$ + 0.5 vol.% Pd	wet H$_2$-Ar (20-80%)	973	430** *(120)*	[237]
		873	170** *(120)*	
La$_{0.7}$Sr$_{0.3}$VO$_{3-\delta}$ - SDC (20-80 vol.%) [a] + 2.1vol.% CeO$_{2-\delta}$ ⎮ 0.3 vol.% Pd	3% H$_2$O-H$_2$	873	230 *(130)*	[250]
La$_{0.7}$Sr$_{0.3}$V$_{0.8}$O$_{3-\delta}$ - YSZ (10-90 vol.%) [a] + 2.8 vol.% CeO$_{2-\delta}$ + 0.5 vol.% Pd	3% H$_2$O-H$_2$	973	390* *(70-80)*	[235]
La$_{0.7}$Sr$_{0.3}$V$_{0.95}$O$_{3-\delta}$ - YSZ (10-90 vol.%) [a] + 2.8 vol.% CeO$_{2-\delta}$ + 0.5 vol.% Pd	3% H$_2$O-H$_2$	973	330 *(80)*	[248]
SrVO$_{3-\delta}$ - YSZ (10-90 vol.%) [a] + 2.8 vol.% CeO$_{2-\delta}$ + 0.5 vol.% Pd	3% H$_2$O-H$_2$	973	220* *(70-80)*	[235]
Ce$_{0.7}$Sr$_{0.3}$VO$_{3-\delta}$ - YSZ (30-70 wt.%) [a] + 1wt.% Pd	3% H$_2$O-H$_2$	973	470 *(80)*	[230]
Ce$_{0.8}$Co$_{0.2}$VO$_{3-\delta}$ - YSZ (30-70 wt.%) [a]	3% H$_2$O-H$_2$	973	300 *(80)*	[245]
Ce$_{0.8}$Sr$_{0.1}$Co$_{0.1}$VO$_{3-\delta}$ - YSZ (30-70 wt.%) [a]	3% H$_2$O-H$_2$	973	~450 *(65)*	[246]
Ce$_{0.8}$Sr$_{0.1}$Co$_{0.05}$Cu$_{0.05}$VO$_{3-\delta}$ - YSZ (30-70 wt.%) [a]	3% H$_2$O-H$_2$	973	~420 *(65)*	[246]
Ce$_{0.8}$Ni$_{0.2}$VO$_{3-\delta}$ - YSZ (30-70 wt.%) [a]	3% H$_2$O-H$_2$	973	310 *(80)*	[245]
Ce$_{0.7}$Sr$_{0.1}$Ni$_{0.2}$VO$_{3-\delta}$ - YSZ (30-70 wt.%) [a]	3% H$_2$O-H$_2$	973	310 *(80)*	[245]
Ce$_{0.8}$Sr$_{0.1}$Cu$_{0.1}$VO$_{3-\delta}$ - YSZ (30-70 wt.%) [a]	3% H$_2$O-H$_2$	973	~150 *(65)*	[246]

Electrolytes: SDC [237, 250], YSZ - other stidues. Anode current collectors: Ag paste [230, 235, 237, 245, 246, 248 - 250], Pt paste or paint [236, 247].Cathodes: LSM-YSZ [247], YSZ impregnated with LSF [230, 235, 236, 245, 246, 248, 249], SDC impregnated with LSCo [237, 250]. [a] the composites were prepared by infiltration of the precursors for vanadate synthesis into YSZ scaffolds, [b] obtained on a cell "5×5cm^2. The data were collected after 2 redox cycles (*) or after 200 hours of testing (**). For other values, this parameter was not indicated.

impregnation with Ce-, Pd- or Pt-containing solutions [235 - 238]. For $La_{0.7}Sr_{0.3}VO_{3-\delta}$ - YSZ anode (S = 5×5 cm^2) impregnated with CeO_2 and Pd the power output achieved 6 W; however, no stability issues were reported [236]. Promising results were also obtained on $(Ce,Sr)VO_{3-\delta}$-based anodes. This phenomenon is consistent with the improvement of the catalytic properties of vanadate-based materials after addition of ceria-based phases, possibly due to formation of $CeVO_x$ oxides, as suggested in [228, 243, 244].

The presence of $Sr_2V_2O_7$ and $SrVO_6$ impurities which form eutectic mixture with low melting point (~800 K) may ensure a better adhesion between the anode and electrolyte, although an excess of the liquid phases may cover catalytically-active sites on the surface [235]. The polarization resistance of $Ce_{0.8}Ca_{0.2}VO_{3-\delta}$ - YSZ - Ni composite anode deposited onto LSGM electrolyte is ~4 Ohm×cm^2 at 973 K; the performance losses are attributed to the interaction between the anode and electrolyte layers [24].

Adijanto *et al* [245, 246] demonstrated that the electrochemical activity increases in the sequence $CeVO_{3-\delta} < Ce_{0.8}Cu_{0.2}VO_{3-\delta} < Ce_{0.8}Co_{0.2}VO_{3-\delta} \approx Ce_{0.8}Ni_{0.2}VO_{3-\delta}$. This order correlates with the catalytic activity of the corresponding metallic phase (Cu, Ni, Co) formed on the oxide surface. The activity of $(Ce,Sr)(V,Co,Cu)O_{3-\delta}$ anode under low currents was found to be close to that for $Ce_{0.8}Sr_{0.1}Co_{0.1}VO_{3-\delta}$ while increasing anodic overpotential above a critical level leads to a significant performance drop, possibly due to oxidation of metallic Co particles [246]. The examples demonstrated above confirm that the anode behavior correlates with the activity of the dopant metal or other additives rather than with the intrinsic properties of the vanadate phase.

Short-Term Performance of Perovskite-Based Anode Materials

In order to satisfy the economical demands, the degradation rate of a SOFC assembly should not exceed 1% per 10000 hours [251]. By far, no studies with comparable testing period have been reported for alternative anodes; most tests were limited by 100-500 hours. The present section considers possible origins and degradation mechanisms for oxide-based anodes.

Due to the high phase, structural and thermomechanical stability, the performance stability of chromite-based anodes is primarily affected by the behaviour of additional components (metallic phases, VO_x, doped $CeO_{2-\delta}$, *etc*) which suffer from catalytic deactivation, vaporization, microstructural or volume changes leading to cracking or delamination of the electrode layer [52, 54, 252 - 254]. For example, the performance loss as high as 20% per 200 hours was detected for $La_{0.8}Sr_{0.2}Cr_{0.98}V_{0.02}O_{3-\delta}$ - GDC - Ni composite anode in C_3H_8 at 1023 K [253], while the anode $La_{0.8}Sr_{0.2}Cr_{0.97}V_{0.03}O_{3-\delta}$ - GDC impregnated with Ru (1 mg/cm^2) showed no degradation in wet CH_4 [252], which confirms that the stability is primarily governed by the nature and amount of microstructurally-unstable phases.

Another factor which influences the durability of chromite electrodes is thermal or redox cycling; the latter may be induced by changing the gas environment or by oscillating polarization. In [255], the performance of $La_{0.75}Sr_{0.25}Cr_{0.5}Mn_{0.5}O_{3-\delta}$ - YSZ anode was deteriorated on cycling "wet H_2 - reformate gas - wet O_2" at 1173 K by 15% per 4 cycles; the initial performance drop was followed by stabilization. $La_{0.8}Sr_{0.2}Cr_{0.98}V_{0.02}O_{3-\delta}$ - GDC - Ni showed a loss of activity on oxidation in air, which was rapidly recovered after switching back to H_2 or C_3H_8. This composite anode could withstand up to 11 redox cycles of 1 hour each [253]. The performance of $La_{0.75}Sr_{0.25}Cr_{0.5}Mn_{0.5}O_{3-\delta}$ - GDC (33 - 67 wt. %) anode showed a slight (3%) decrease after cycling in wet CH_4 at 1123 K for 50 hours under currents oscillating in the range 200 - 400 mA/cm^2 [256].

Concerning the stability in hydrocarbon atmospheres, the deposition of carbon at high ratios C:O is predicted thermodynamically [253, 254], but for undoped chromites the process may be hampered kinetically due to their low activity towards hydrocarbon conversion and carbon deposition [254, 257]. Formation of carbonacious deposits may be, however, promoted by introduction of Mn [41, 88], Fe [258], Co [259] or Ni [83, 259] into the anode composition, by the presence of VO_x [260] or other catalitically-active components. High currents densities may ensure electrochemical oxidation of deposited carbon or even prevent its precipitation [41, 86].

As noted above, manganite-based anodes are characterized by insufficient chemical and dimensional instability under reducing conditions, as well as poor chemical compatibility with zirconia-based electrolytes. While the anode $La_{0.4}Sr_{0.6}Mn_{0.6}Ti_{0.4}O_{3-\delta}$ - YSZ showed only a slight initial increase in polarization and ohmic resistance in H_2 at 1068 K (~10% per 200 hours) [111], the polarization resistance of $La_{0.8}Sr_{0.2}Mn_{0.5}Al_{0.5}O_{3-\delta}$ - YSZ under similar condition irreversibly showed a ~2-fold raise after 100 operating hours [104]. $La_{0.8}Sr_{0.2}Mn_{0.8}Sc_{0.2}O_{3-\delta}$ exhibited rapid performance drop (70% per 3 hours) in CO under OCV while anodic polarization or addition of CO_2 into the fuel allowed to suppress the dergadation suggesting that the anode conditions are close to the stability limit of the perovskite phase [127]. The same anode material showed a satisfactory stability in CH_4 [120], possibly due to inhibited methane reforming over the manganite which enables to preserve the local oxygen pressure high enough to ensure the phase stability of the perovskite.

Preliminary reduced titanate-based anodes are extremely susceptible to oxidation, accompanied by the conductivity drop, large volume variations and/or microstructural changes. For example, $Sr_{0.895}Y_{0.07}TiO_{3-\delta}$ - YSZ - Ni anode showed a satisfactory tolerance towards cycling between H_2, N_2 and air at 1073 K while cooling to 973 K induced a ~35% performance loss after 200 cycles [140]. The average degradation rate of $Sr_{0.99}Ti_{0.9}Nb_{0.1}O_{3-\delta}$ - FeCr - $Ce_{0.8}Gd_{0.2}O_{2-\delta}$ - Ni composite anode at 923 K was ~9% per 1000 hours (Fig. 5). The $Sr_{0.94}Ti_{0.9}Nb_{0.1}O_{3-\delta}$ - YSZ anode showed a twofold increase in the polarization resistance after 2 exposures to oxidizing atmosphere and subsequent switch to H_2 [139]. On the other hand, Ni-cermet anode with $Sr_{0.8}La_{0.2}TiO_{3-\delta}$-based support was shown to be capable to endure >5 cycles "dry H_2 - air"; considering a poor tolerance of Ni-based anodes towards oxidation, introduction of the titanate layer seems beneficial [172]. The anode $Sr_{0.6}La_{0.4}TiO_{3-\delta}$ showed a rapid performance decay in wet H_2; afterwards, the activity was stable both in reducing atmospheres and upon redox cycling, although remained rather low [138].

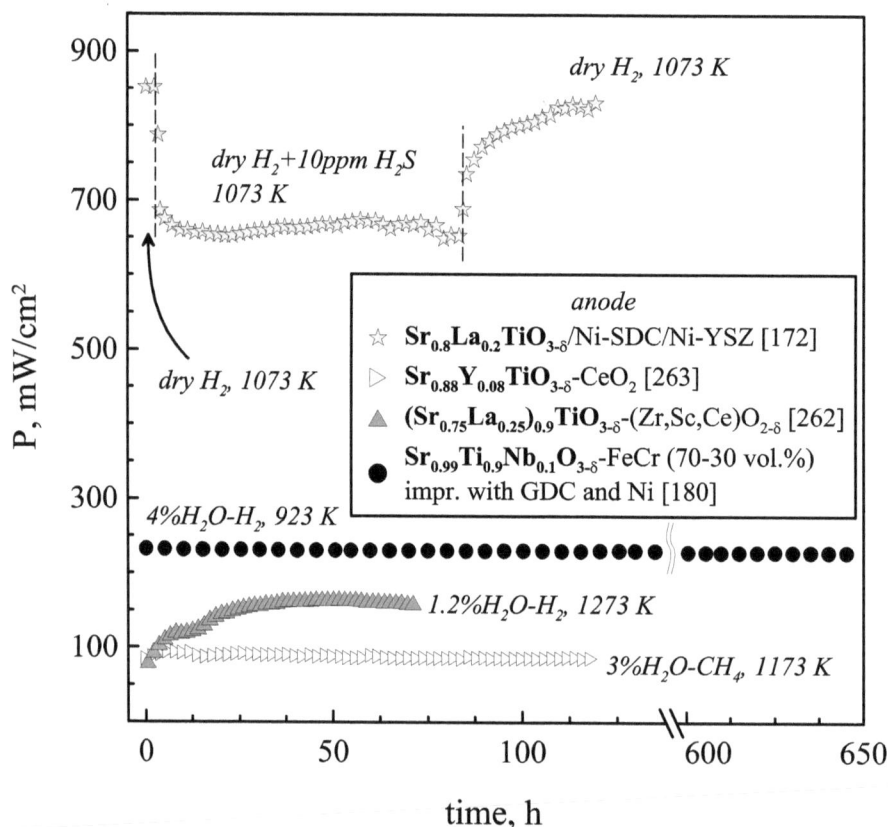

Fig. (5). Selected results of short-term stability testing of cells with titanate-based anodes in H_2-, CH_4- and H_2S-containing fuels. The data are taken from [172, 180, 262, 263].

The information on the stability of molybdate- or vanadate-based anodes is scarce. Similar to titanates, the origins of degradation are generally associated with the instability of the perovskite phase or catalytically-active impurities towards oxidation [190 - 192, 228, 231]. Nevertheless, the $SrFe_{0.75}Mo_{0.25}O_{3-\delta}$ anode endured 5 - 16 cycles "$H_2 \rightarrow CH_4 \rightarrow$ air" or "$H_2 + 100$ ppm $H_2S \rightarrow$ air" without a significant drop of the produced power [261]. In some studies, the electrode activation during operation was observed originating from increasing conductivity, co-sintering of the anode layer with the current collector or formation of catalytically-active phases under operation conditions [32, 211, 221, 223]; however, the testing time was limited by 5 - 100 hours. Testing $La_{0.7}Sr_{0.3}VO_{3-\delta}$ - $Ce_{0.85}Sm_{0.15}O_{2-\delta}$ anodes in H_2 - Ar atmosphere showed a 6 - 8% degradation level after 200 operation hours while nearly stable performance for 300 hours was demonstrated by $La_{0.7}Sr_{0.3}VO_{3-\delta}$ - YSZ composite anode in CH_4 and H_2S-containing fuel, although the power output achieved only ~125 mW/cm^2 (Fig. **6**).

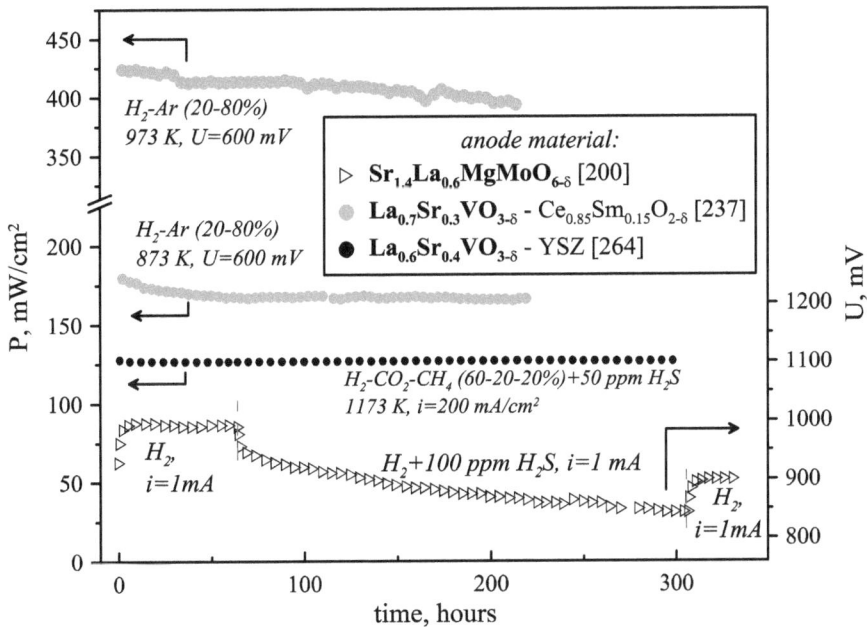

Fig. (**6**). Selected results of short-term stability testing of molybdate- and vanadate-based anodes. The data are taken from [200, 237, 264].

Relationships between the Intrinsic Properties of the Anode Material and Electrochemical Behavior: Selected Examples

Appropriate selection of the anode materials requires understanding their functional properties and their possible influence on the electrochemical performance. Unfortunately, to date the number of oxide-based anodes tested under appropriate experimental conditions is insufficient to reveal certain correlations between the material intrinsic properties and electrochemical activity. In order to achieve the acceptable level of performance, the electrode material should satisfy a list of requirements, including sufficient electronic and ionic transport, electrocatalytic activity, chemical and mechanical stability [2, 5, 100]. Compositional optimization of electrode materials or components of composite electrodes in most cases affects the electrode activity; however, evaluation of the effect of any individual parameter on the electrochemical properties is obscured by simultaneous effects of other characteristics. Consequently, the combined influence of the functional properties should be considered rather than the individual effect of each of those.

The multiplicity of performance-determining factors has been noted for cathode materials [2, 265]. Regardless of the complex compositional influence on the electrode performance, the current constriction effects have been tracked for numerous composition groups with the conductivity below 100 S/cm while the polarization behavior for materials with higher conductivity is primarily governed by other factors such as ionic transport-, surface exchange kinetics-related issues or insufficient chemical/mechanical compatibility between the cell components [100, 265 - 267].

For anodes, the effect of the electronic transport is even more difficult to detect, while comparison between various representatives of the same anode group is complicated by the low availability of the information on the transport properties of the anode materials under appropriate conditions. In particular, for Ni-YSZ cermets, the acceptable level of the total conductivity is achieved only when the volume fraction of the metal exceeds the percolation threshold (20 - 50 vol. %) [5, 100]. Increasing Ni content in Ni-YSZ above 30 vol.% gradually decreases the ohmic losses, consistently with improving electronic conductivity of the anode layer. The polarization resistance, however, exhibits a minimum at ~40 - 50 vol.% Ni associated with substantial microstructural changes leading to a drop of the cermet activity [268]. Similar behavior was noticed for Cu-based cermets [269].

Comparative studies on various perovskite, fluorite, pyrochlore, tungsten bronze and other structural types of anode composition revealed no obvious correlation between the electronic transport and polarization behaviour [24, 270]. On the other hand, the performance of $La_{0.7}A_{0.3}Cr_{0.5}Mn_{0.5}O_{3-\delta}$ anodes increased in the order Ba<Sr<Ca, consistently with the electronic conduction and stability of the materials [74]. Similarly, increasing Mn content in $Pr_{0.7}Ca_{0.3}Cr_{1-x}Mn_xO_{3-\delta}$ from x = 0.2 to 0.8 leads to ~1 order enhancement of the conductivity and anode performance [57]. Comparison of the electrochemical properties of

(La,Sr)(Cr,Mn,Ti)O$_3$- and (La,Sr)(Cr,Fe)O$_3$-based anodes, without additional microstructural modifications, showed that the anode activity correlates with the electronic conductivity under the anode conditions (Fig. **7**) rather than with the ionic transport or phase stability of the materials. At the same time, some correlation was also found between the anode activity and dimensional instability of the anode material suggesting microstructural failure or anode delamination from the electrolyte surface upon reduction [60, 78, 105].

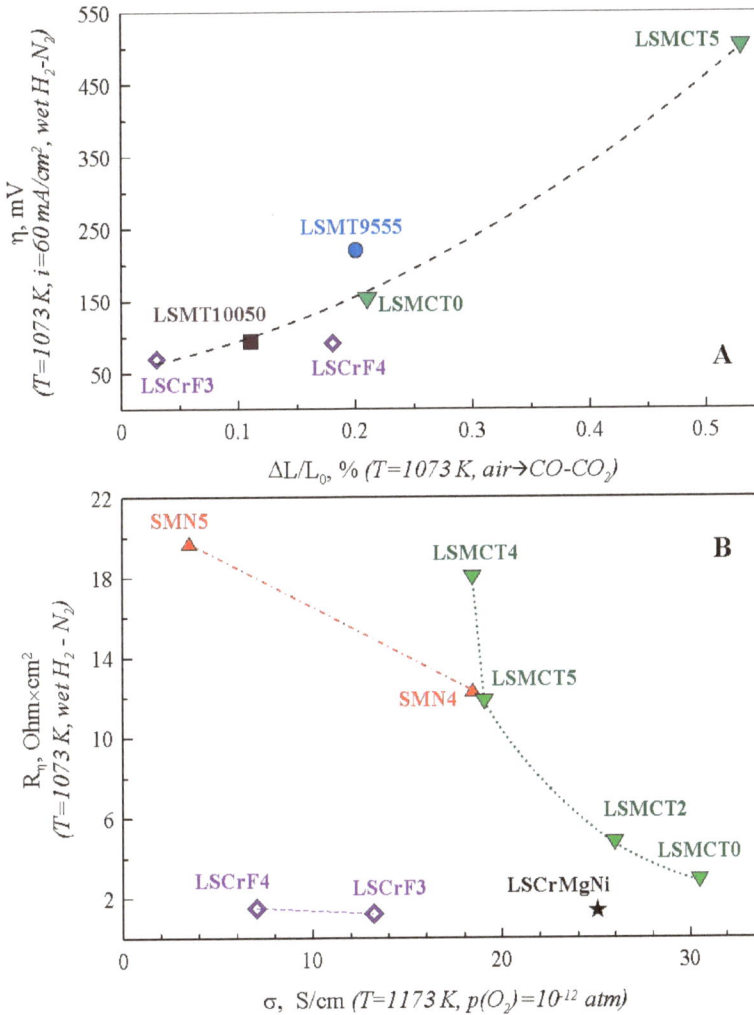

Fig. (7). Correlation between the anode overpotential (A) or zero-current polarization resistance (B) and relative linear dimensional changes on reduction (A) or total conductivity for selected groups of perovskite-like anode materials (B). The data are taken from [60, 78, 105].

The effect of the ionic transport on the electrochemical activity of cermet or oxide-based anodes is less obvious. This is primarily originating from a strong influence of the catalytic properties of the anode constituents or microstructure on the electrochemical activity. In particular, comparison of the polarization behaviour of Ni - $Zr_{0.85}Y_{0.15}O_{2-\delta}$, Ni - $Ce_{0.9}Gd_{0.1}O_{2-\delta}$, $Ce_{0.6}Gd_{0.4}O_{2-\delta}$, $La_{0.75}Sr_{0.25}Cr_{0.97}V_{0.03}O_{3-\delta}$, $Zr_{0.71}Y_{0.12}Ti_{0.17}O_{2-\delta}$ [271] or Ni - $Sr_{0.88}Y_{0.08}Ti_{0.95}B_{0.05}O_{3-\delta}$ (B = Ti, Cr, Co) anodes [162] indeed revealed a non-stringent correlation between the oxide-ionic or protonic transport and electrochemical properties. At the same time, no effect of the ionic conduction of mixed-conducting component has been reported for cermet anodes based on $Ce_{0.8}Ca_{0.2}VO_{3-\delta}$, $Gd_{1.86}Ca_{0.14}Ti_2O_{7-\delta}$, $TbZrO_{2-\delta}$, $Ce_{0.8}Gd_{0.2}O_{2-\delta}$, $Zr_{0.85}Y_{0.15}O_{2-\delta}$ and $La_{0.9}Sr_{0.1}Al_{0.65}Fe_{0.2}Mg_{0.15}O_{3-\delta}$ [24].

More obvious correlation was found between the anode performance and the catalytic activity of the electrode components towards processes occurring at the gas/solid interface, such as gas adsorption, dissociation, surface diffusion of intermediate species towards electrochemically active sites, their "spillover" between the phases, *etc.* Most simulations and experimental studies carried out on metallic surfaces demonstrated a superior catalytic activity towards oxidation processes for Co, Ni and some noble metals [5, 31, 272, 273], while the presence of metallic Cu or Au only provides the electronic percolation in the anode layer and modifies the microstructure [269, 274, 275]. These data are confirmed by comparatively high electrochemical activity of perovskite-based anodes modified by additions of Ni, Ru, Pt, *etc* or coated with Pt paste as current collector; examples are shown in Tables 3, 5, 6, 8 and 10.

Among other effects related to the electrode compositions, one should note a strong influence of the material prehistory which involves the temperature and atmosphere of preparation of the electrode layers, pre-reduction or polarization prior to electrochemical studies, *etc.* This phenomenon is especially important for electrode materials which activity is associated with kinetically-hindered processes, such as segregation of catalytically-active surface phases, generation of additional charge carriers, *etc* [79, 83, 135, 138]. For example, $Sr(Ti,Nb)O_{3-\delta}$-based anode prepared in reducing atmosphere at 1523 K yielded the polarization resistance by ~40% lower than that for the analogue fired under oxidation conditions [139]. The anode performance of $(La,Sr)(Cr,B)O_{3-\delta}$-based anodes (B = Ni, Ru) showed a gradual improvement upon operation under anodic conditions which might indicate the necessity of preliminary high-temperature reduction of the electrode material in order to achieve a sufficient surface coverage with metallic particles [79, 81].

Chemical Compatibility between Electrode/Electrolyte

Exposure of electrochemical cells to temperatures above 1273 K during the preparation procedure or maintenance under operating conditions for a long period may cause the interdiffusion between the cell components or lead to formation of interfacial compounds. This imposes additional restrictions on the choice of the electrode and electrolyte materials.

One of the most critical and widely studied issues in this field is the formation of $Ln_2Zr_2O_7$ or $AZrO_3$ (A = Ca, Sr, Ba) phases with poor ionic transport at the interface of ZrO_2-based electrolytes with oxide-based electrodes. Among the perovskite anode groups, the most pronounced interaction was reported for manganite-based electrodes, although their chemical compatibility was essentially studied in oxidizing atmospheres [100, 103]. The optimum composition in terms of the chemical compatibility with YSZ is $La_{0.6}Sr_{0.4}MnO_{3\pm\delta}$: no interaction was observed up to ~1400 K, while increasing La or Sr content in the perovskite reduces the reaction temperature below 1150 K [100, 103, 276]. Replacement of Sr with Ca was reported to decrease Mn diffusivity and reactivity with the electrolyte [277] while reducing the ionic radius of the rare-earth cation decreases the stability of both the perovskite and $Ln_2Zr_2O_7$ pyrochlore phase; generally, this results in suppressed reactivity [118, 277]. The compatibility may be substantially improved by diminishing (La+Sr)/Mn ratio below 1 [278]. Another problem relates to a high Mn solubility in the fluorite phase, which achieves 5 - 12 wt.% at 1273 - 1573 K and promotes the formation of zirconates [116, 279]. More information on the compatibility of $(Ln,A)MnO_{3\pm\delta}$ with zirconia-based phases may be found in [100, 103, 116].

The compatibility of manganites substituted in B-sublattice generally correlates with their chemical stability. In particular, no interaction was detected between $La_{0.8}Sr_{0.2}Mn_{0.5}Al_{0.5}O_{3\pm\delta}$ [104] or $La_{0.4}Sr_{0.6}Mn_{0.6}Ti_{0.4}O_{3\pm\delta}$ [111] perovskites with YSZ at least up to 1673 K, while for Mn-rich $La_{0.8}Sr_{0.2}Mn_{0.9}Sc_{0.1}O_{3\pm\delta}$ the interaction was observed at ~1420 K [280].

Titanates are also susceptible towards interaction with zirconia-based electrolytes, although to a less extent in comparison with manganites. The chemical compatibility was reported to be better for titanates understoichiometric in A-sublattice [139, 145, 262]; however, most tests have been carried out after annealing the reaction mixture in air, where the cation deficiency is favorable in terms of stability. In contrast, studies by Burnat *et al* demonstrated a worse compatibility of $Sr_{0.7}La_{0.2}TiO_{3-\delta}$ as compared to the stoichiometric analogue at temperatures above 1473 K [281]; the effect was attributed to a substantial diffusion of excessive Ti species along the grain boundaries, partial replacement of Y from the fluorite lattice and formation of interfacial layer with the tetragonal zirconia structure [262, 281]. Ti diffusion may be accompanied by backward migration of Zr, Y or Sc species into the B-sublattice of the perovskite structure [262, 281, 282]. The compatibility between titanates and zirconia was found to be improved under reducing conditions [281], attributed to hindered diffusivity of Ti^{3+} and its lower solubility in the fluorite structure.

No secondary phases were detected at the interface between $(Sr,Ln)TiO_{3-\delta}$ or $(Sr,Y)TiO_{3-\delta}$ with ZrO_2-based compounds up to 1623 - 1823 K [132, 135, 143, 283], while Chen *et al* [262, 284] observed generation of $La_2Zr_2O_7$ and cation interdiffusion at the interface of $(La_{0.25}Sr_{0.75})_{1-x}TiO_{3-\delta}$ (x = 0 - 0.1) with $Zr_{0.79}Sc_{0.2}Ce_{0.01}O_{2-\delta}$ in spite of lower annealing temperatures (1473 K); in the latter study the effect was more pronounced under oxidizing conditions. On the other hand, for $Sr_{0.7}La_{0.2}TiO_{3-\delta}$ a somewhat better compatibility with $Zr_{0.79}Sc_{0.2}Ce_{0.01}O_{2-\delta}$ in comparison with that for $Zr_{0.85}Y_{0.15}O_{2-\delta}$ was demonstrated [281].

The chemical compatibility of titanates is affected by the presence of transition metal cations, such as Fe, in B-sublattice [145, 157]; in particular, interaction of $SrTi_{1-x}Fe_xO_{3-\delta}$ with YSZ was detected at temperatures above 1273 K, irrespective to Ti:Fe ratio [145]. Promising results were obtained for $Sr_{0.94}Ti_{0.9}Nb_{0.1}O_{3-\delta}$ perovskite, chemically compatible with YSZ in air and 9% H_2 - Ar at 1523 K [139]. Annealing at higher temperatures (1723 K at $p(O_2)=10^{-21}$ atm) results in cation interdiffusion and separation of impurities at the grain boundaries, presumably of metallic Nb [282].

The reactivity of $(La,A)CrO_{3-\delta}$ (A = Ca, Sr) with zirconia is primarily associated with cation demixing and formation of Ca(Sr)-Cr-O secondary phases. $(La,Sr)CrO_{3-\delta}$ are basically more compatible with YSZ electrolytes [35, 285, 286]. Interaction of $(La,Ca)CrO_{3-\delta}$ with YSZ originates from formation of CaO, $Ca_m(CrO_4)_n$ phases [287], which may be segregated during synthesis of the chromite perovskites, or from a high diffusivity and solubility of Ca in the fluorite structure [288]. The cation demixing may be promoted by exposure of the electrode to hydrocarbon- or CO/CO_2-containing atmospheres or by polarization [289, 290]. No interaction between $(La,Ca)(Cr,Ti)O_{3-\delta}$ [291], $La_{0.75}Sr_{0.25}Cr_{0.5}Mn_{0.5}O_{3-\delta}$ [41], $Pr_{0.7}Ca_{0.3}(Cr,Mn)O_{3-\delta}$ [57] or $(La,Sr)(Cr,Ru)O_{3-\delta}$ [292] with YSZ was observed at least up to 1573, 1573, 1473 and 1373 K, respectively.

A high reactivity is basically typical for Mo-containing oxide materials [293, 294]. The interaction of $Sr(Mo,B)O_{6-\delta}$ (B = Mg, Fe, Co) with YSZ was found at 1273 - 1673 K, while the amount of the $Sr_2MgMoO_{6-\delta}$ reduces from 50 to 6.6% on heating. Another problem relates to separation of $SrMoO_4$ impurity [193, 198, 212]. The information on the compatibility between vanadate-based compounds with other SOFC components is scarce. Basically, no interaction between $(La,Sr)VO_{3-\delta}$ or $Sr(Mo,V)O_{3-\delta}$ with ZrO_2- or CeO_2-based electrolytes was found in the temperature range of 1173 - 1573 K [204, 234, 240], although the anode powder may contain some amounts of $Sr_3V_2O_8$ phase [238]. It is unclear if the latter compound as well as $SrMoO_4$ impurity in molybdate anodes is separated due to interaction between the anode and electrolyte materials or forms during the processing of the electrode powder irrespective to the presence of the electrolyte.

Application of gallate-based electrolytes necessitates to avoid the direct contact with Fe-, Co- and Ni-containing electrodes, since the corresponding cations may substitute Ga in the perovskite lattice and cause segregation of $LaSrGa_3O_7$, $LaSrGaO_4$, $LaNiO_{3-\delta}$ or other interfacial phases with worse transport characteristics [295, 296]. Another degradation origin is associated with a gradient of La chemical potential between LSGM and anode phase leading to diffusion and phase separation [76, 200, 297]. Obviously, these phenomena are affected by cations ratios in $(La,Sr)(Ga,Mg)O_{3-\delta}$ perovskite, cation deficiency and the presence of other dopant species, such as Co. Unfortunately, this information is often ignored in articles. $CeO_{2-\delta}$-based oxides basically do not exhibit any critical interaction with the electrode materials, although the penetration of mobile cations (Co, Mn) into the fluorite lattice is typical for ceria-based materials [298 - 300].

The onset of interaction between $(La,Sr)(Cr,Fe)O_{3-\delta}$ and LSGM is located at relatively low temperatures (1073 - 1173 K) resulting in cation exchange [297] or formation of $LaSrGa_3O_7$ [77]. The interaction of $La_{0.75}Sr_{0.25}Cr_{0.5}Al_{0.5}O_{3-\delta}$ with LSGM starts at 1473 K producing $SrCrO_4$, $LaCrO_{3-\delta}$ and $LaGaO_3$-based phases and causing the electrode delamination [96]. $La_{0.75}Sr_{0.25}Cr_{0.5}Mn_{0.5}O_{3-\delta}$ reacts with LSGM at 1623 - 1673 K with formation of $LaSrGaO_4$ and $LaSrGa_3O_7$ [76]. No secondary phases or diffusional layers were detected at the interface between $(La_{0.5-x}Sr_{0.5+x})_{1-y}Mn_{0.5}Ti_{0.5}O_{3-\delta}$ (x = 0 - 0.25, y = 0 - 0.3) and $La_{0.8}Sr_{0.2}Ga_{0.8}Mg_{0.15}Co_{0.05}O_{3-\delta}$ after firing at 1573 - 1623 K in air [156] while Ti diffusion from $Sr_{0.8}La_{0.2}TiO_{3-\delta}$ into LSGM was observed at 1673 K [171].

For molybdates, formation of $SrMoO_4$ at the interface with LSGM, CGO or $La_{10}Si_{5.5}Al_{0.5}O_{27}$ at 1073 - 1273 K was found [198]; increasing co-firing temperature leads to higher content of the impurity indicating that the phase is formed during the interaction, irrespective to its initial presence. Increasing the temperature up to 1273 K leads to formation of $LaSrGa_3O_7$ and MgO at the interfaces with LSGM electrolyte. Similar phase relationships were reported for B-site-doped $Sr_2MgMoO_{6-\delta}$, $Sr_2CoMoO_{6-\delta}$ [193, 208, 301], while for Ni-doped molybdates a somewhat better compatibility was reported [193, 219]. Due to increased stability, Fe-rich (up to 75% Fe) $Sr(Fe,Mo)O_{3-\delta}$ with disordered or partially ordered B-site cations demonstrate an appropriate compatibility with LSGM and $(Ce,A)O_{2-\delta}$ [193, 201, 212, 226] which suggests that the reactivity is primarily associated with Mo species. One should note that in most studies the testing temperature was rather low (973 - 1123 K) [226, 302]. An appropriate compatibility was found between $Sr_{1.2}La_{0.8}MgMoO_{6-\delta}$ and $La_{0.8}Sr_{0.2}Ga_{0.8}Mg_{0.2}O_{3-\delta}$ after co-firing at 1573 K in air, at 1073 K in H_2 or 100 ppm H_2S-H_2. Increasing or decreasing La content in the double perovskite causes La migration towards the

region of lower La chemical potential [200]. For $CeVO_3$-based anodes, formation of interfacial layers at the interface with LSGM electrolyte was found [24].

CONCLUDING REMARKS

In the present chapter, the basic non-perovskite and perovskite-like oxide groups attractive as potential SOFC anodes were considered. Unfortunately, a large amount of experimental data, including those which have been published during the preparation of our work, could not be included in the review. We attempted to discuss the main benefits and disadvantages of each group, reveal the major factors responsible for their electrochemical behavior and demonstrate the novel approaches towards optimization of their anode properties. In the conclusion, we would like to emphasize the following aspects.

1. Every year, 30-50 papers focused on development of oxide-based anode materials are published. In accordance with most results, the power density of the cells with the alternative anodes may exceed 1-2 W/cm^2. Such results are extremely optimistic and, unfortunately, do not seem to be reliable. In particular, such a level of the performance is impossible for cells with electrolytes thicker than 200-400 μm. The overall cell resistance calculated from the impedance spectra presented in several papers is also inconsistent with the power density reported.
2. Despite a large number of the high-active anode materials published by far, the amount of studies where the stability for >100 hours was tested is limited. In following works, this subject should be emphasized.
3. Most results obtained by the same research group on various anode compositions are very close. This might indicate that the composition of the anode material has a negligible effect on the electrochemical activity while the latter is governed by other factors. Among those, the most pronounced effect is expected from the nature and form of the current collecting material. In our review, we attempted to consider this effect separately. As follows from Tables **3, 5, 6, 8, 10**, the cells where Pt paste is used as a currect collecting layer exhibit an enhanced power density. It is unobvious if application of less active metal layers affects the electrochemical properties. In the future, we

recommend to the authors of such studies to evaluate the economical benefits of such modifications: namely, to calculate the difference between the income obtained from the power density produced by the cell, and consumes originating from utilization of expensive materials. It would also be desirable to explain the origins of high performance of oxides enriched with Mn, Fe, Co and Ni which are expected to decompose or loss their high conductivity under the anode conditions.

CONSENT FOR PUBLICATION

Not applicable.

CONFLICT OF INTEREST

The authors declare no conflict of interest, financial or otherwise.

ACKNOWLEDGEMENT

V.A. Kolotygin and I.E. Kuritsyna are grateful to the Russian Science Foundation (grant 17-79-30071) for the possibility of searching and analyzing the data on the functional properties and electrochemical behavior of the perovskite-based anodes. The information on the chemical compatibility between the anode and electrolyte materials was collected by N.V. Lyskov under financial support by the Russian Foundation for Basic Research (grant 17-08-00831).

REFERENCES

[1]   McIntosh, S.; Gorte, R.J. Direct hydrocarbon solid oxide fuel cells. *Chem. Rev.,* **2004**, *104*, 4845-4865.
        [http://dx.doi.org/10.1021/cr020725g]

[2]   Tsipis, E.V.; Kharton, V.V. Electrode materials and reaction mechanisms in solid oxide fuel cells: a brief review. III. Recent trends and selected methodological aspects. *J. Solid State Electrochem.,* **2011**, *15*, 1007-1040.
        [http://dx.doi.org/10.1007/s10008-011-1341-8]

[3]   Park, H.C.; Virkar, A.V. Bimetallic (Ni-Fe) anode-supported solid oxide fuel cells with gadolinia-doped ceria electrolyte. *J. Power Sources,* **2009**, *186*, 133-137.
        [http://dx.doi.org/10.1016/j.jpowsour.2008.09.080]

[4]   Kim, S.; Kim, C.; Lee, J.H.; Shin, J.B.; Lim, T-H.; Kim, G.G. Tailoring Ni-based catalyst by alloying with transition metals (M=Ni, Co, Cu, and Fe) for direct hydrocarbon utilization of energy conversion devices. *Electrochim. Acta,* **2017**, *225*, 399-406.
        [http://dx.doi.org/10.1016/j.electacta.2016.12.178]

[5]   Jiang, S.P.; Chan, S.H. A review of anode materials development in solid oxide fuel cells. *J. Mater. Sci.,* **2004**, *39*, 4405-4439.
        [http://dx.doi.org/10.1023/B:JMSC.0000034135.52164.6b]

[6]   Kofstad, P. *Nonstoichiometry, Diffusion and Electrical Conductivity in Binary Metal Oxides*; Wiley-Interscience: New York, **1972**.

[7]   Tu, H.; Apfel, H.; Stimming, U. Performance of alternative oxide anodes for the electrochemical oxidation of hydrogen and methane in solid oxide fuel cells. *Fuel Cells (Weinh.),* **2006**, *6*, 303-306.

[http://dx.doi.org/10.1002/fuce.200500216]

[8]   Boulfrad, S.; Cassidy, M.; Irvine, J.T.S. $NbTi_{0.5}Ni_{0.5}O_4$ as anode compound material for SOFCs. *Solid State Ionics,* **2011**, *197*, 37-41.
[http://dx.doi.org/10.1016/j.ssi.2011.05.007]

[9]   Lashtabeg, A.; Canales-Vázquez, J.; Irvine, J.T.S.; Bradley, J.L. Structure, conductivity, and thermal expansion studies of redox stable rutile niobium chromium titanates in oxidizing and reducing conditions. *Chem. Mater.,* **2009**, *21*, 3549-3561.
[http://dx.doi.org/10.1021/cm900281r]

[10]   Madeira, L.M.; Portela, M.F.; Mazzocchia, C. Nickel molybdate catalysts and their use in the selective oxidation of hydrocarbons. *Catal. Rev.,* **2004**, *46*, 53-110.
[http://dx.doi.org/10.1081/CR-120030053]

[11]   Kwon, B.W.; Ellefson, C.; Breit, J.; Kim, J.; Grant Norton, M.; Ha, S. Molybdenum dioxide-based anode for solid oxide fuel cell applications. *J. Power Sources,* **2013**, *243*, 203-210.
[http://dx.doi.org/10.1016/j.jpowsour.2013.05.133]

[12]   Patrakeev, M.V.; Leonidov, I.A.; Kozhevnikov, V.L.; Kharton, V.V. Ion-electron transport in strontium ferrites: rlationships with structural features and stability. *Solid State Sci.,* **2004**, *6*, 907-913.
[http://dx.doi.org/10.1016/j.solidstatesciences.2004.05.002]

[13]   Mogni, L.; Prado, F.; Caneiro, A.; Manthiram, A. High temperature properties of the n=2 Ruddlesden-Popper phases $(La,Sr)_3(Fe,Ni)_2O_{7-\delta}$. *Solid State Ionics,* **2006**, *177*, 1807-1810.
[http://dx.doi.org/10.1016/j.ssi.2006.03.050]

[14]   Al Daroukh, M.; Vashook, V.V.; Ullmann, H.; Tietz, F.; Arual Raj, I. Oxides of the AMO_3 and A_2MO4-type: structural stability, electrical conductivity and thermal expansion. *Solid State Ionics,* **2003**, *158*, 141-150.
[http://dx.doi.org/10.1016/S0167-2738(02)00773-7]

[15]   Ishihara, T.; Fukui, S.; Enoki, M.; Matsumoto, H. Oxide anode derived from Sr-doped $LaMnO_3$ perovskite oxide for SOFCs using $LaGaO_3$ electrolyte. *J. Electrochem. Soc.,* **2006**, *153*, A2085-A2090.
[http://dx.doi.org/10.1149/1.2344830]

[16]   Kharton, V.V.; Patrakeev, M.V.; Tsipis, E.V.; Avdeev, M.; Naumovich, E.N.; Anikina, P.V.; Waerenborgh, J.C. Oxygen nonstoichiometry, chemical expansion, mixed conductivity, and anodic behavior of Mo-substituted $Sr_3Fe_2O_{7-\delta}$. *Solid State Ionics,* **2010**, *181*, 1052-1063.
[http://dx.doi.org/10.1016/j.ssi.2010.06.004]

[17]   Adijanto, L.; Küngas, R.; Park, J.; Vohs, J.M.; Gorte, R.J. SOFC anodes based on infiltration of tungsten bronzes. *Int. J. Hydrogen Energy,* **2011**, *36*, 15722-15730.
[http://dx.doi.org/10.1016/j.ijhydene.2011.09.059]

[18]   Slater, P.R.; Irvine, J.T.S. Niobium based tetragonal tungsten bronzes as potential anodes for solid oxide fuel cells: synthesis and electrical characterisation. *Solid State Ionics,* **1999**, *120*, 125-134.
[http://dx.doi.org/10.1016/S0167-2738(99)00020-X]

[19]   Kaiser, J.L.; Bradley, P.R.; Irvine, J.T.S. Tetragonal tungsten bronze type phases $(Sr_{1-x}Ba_x)_{0.6}Ti_{0.2}Nb_{0.8}O_{3-\delta}$: material characterisation and performance as SOFC anodes. *Solid State Ionics,* **2000**, *135*, 519-524.
[http://dx.doi.org/10.1016/S0167-2738(00)00432-X]

[20]   Jin, C.; Yang, Z.; Zheng, H.; Yang, C.; Chen, F. $La_{0.6}Sr_{1.4}MnO_4$ layered perovskite anode material for intermediate temperature solid oxide fuel cells. *Electrochem. Commun.,* **2012**, *14*, 75-77.
[http://dx.doi.org/10.1016/j.elecom.2011.11.008]

[21]   Wang, S.; Jiang, Y.; Zhang, Y.; Li, W.; Yan, J.; Lu, Z. Electrochemical performance of mixed ionic-electronic conducting oxides as anodes for solid oxide fuel cell. *Solid State Ionics,* **1999**, *120*, 75-84.
[http://dx.doi.org/10.1016/S0167-2738(98)00558-X]

[22]   Kramer, S.A.; Tuller, H.L. A novel titanate-based oxygen ion conductor: $Gd_2Ti_2O_7$. *Solid State Ionics,*

1995, *82*, 15-23.
[http://dx.doi.org/10.1016/0167-2738(95)00156-Z]

[23] Zakharchuk, K.; Kravchenko, E.; Fagg, D.P.; Frade, J.R.; Yaremchenko, A.A. Mixed ionic-electronic conductivity and thermochemical expansion of Ca and Mo co-substituted pyrochlore-type $Gd_2Ti_2O_7$. *RCS Adv.,* **2016**, *6*, 70186-70196.

[24] Tsipis, E.V.; Kharton, V.V.; Frade, J.R. Mixed conducting components of solid oxide fuel cell anodes. *J. Eur. Ceram. Soc.,* **2005**, *25*, 2623-2626.
[http://dx.doi.org/10.1016/j.jeurceramsoc.2005.03.114]

[25] Mailley, S.C.; Kelaidopoulou, A.; Siddle, A.; Dicks, A.L.; Holtappels, P.; Hatchwell, C.E.; Mogensen, M. Electrocatalytic activity of a $Gd_2Ti_{0.6}Mo_{1.2}Sc_{0.2}O_{7-\delta}$ anode towards hydrogen and methane electrooxidation in a solid oxide fuel cell. *Ionics,* **2000**, *6*, 331-339.
[http://dx.doi.org/10.1007/BF02374149]

[26] Taheri, Z.; Seyed-Matin, N.; Safekordi, A.A.; Nazari, K.; Zarrin Pashne, S. A comparative kinetic study on the oxidative coupling of methane over LSCF perovskite-type catalyst. *Appl. Catal. A,* **2009**, *354*, 143-152.
[http://dx.doi.org/10.1016/j.apcata.2008.11.017]

[27] Fischer, II, J.C.; Chuang, S.S.C. Investigating the CH_4 reaction pathway on a novel LSCF anode catalyst in the SOFC. *Catal. Commun.,* **2009**, *10*, 772-776.
[http://dx.doi.org/10.1016/j.catcom.2008.11.035]

[28] Shin, T.H.; Vanalabhpatana, P.; Ishihara, T. Oxide composite of $Ce(Mn,Fe)O_2$ and $La(Sr)Fe(Mn)O_3$ for anode of intermediate solid oxide fuel cells using $LaGaO_3$ electrolyte. *J. Electrochem. Soc.,* **2010**, *157*, B1896-B1901.
[http://dx.doi.org/10.1149/1.3500979]

[29] Kolotygin, V.A.; Tsipis, E.V.; Markov, A.A.; Patrakeev, M.V.; Waerenborgh, J.C.; Shaula, A.L.; Kharton, V.V. Transport and electrochemical properties of $SrFe(Al,Mo)O3-\delta$ *Russ. J. Electrochem,* **2018**, *54*, 514-526.
[http://dx.doi.org/10.1134/S1023193518060083]

[30] Marwood, M.; Vayenas, C.G. Electrochemical promotion of a dispersed platinum catalyst. *J. Catal.,* **1998**, *178*, 429-440.
[http://dx.doi.org/10.1006/jcat.1998.2156]

[31] Rossmeisl, J.; Bessler, W.G. Trends in catalytic activity for SOFC anode materials. *Solid State Ionics,* **2008**, *178*, 1694-1700.
[http://dx.doi.org/10.1016/j.ssi.2007.10.016]

[32] Bi, Z.H.; Zhu, J.H. Effect of current collecting materials on the performance of the double-perovskite $Sr_2MgMoO_{6-\delta}$ anode. *J. Electrochem. Soc.,* **2011**, *158*, B605-B613.
[http://dx.doi.org/10.1149/1.3569754]

[33] Nakamura, T.; Petzow, G.; Gauckler, L.J. Stability of the perovskite phase $LaBO_3$ (B = V, Cr, Mn, Fe, Co, Ni) in reducing atmosphere. I. Experimental results. *Mater. Res. Bull.,* **1979**, *14*, 649-659.
[http://dx.doi.org/10.1016/0025-5408(79)90048-5]

[34] Mizusaki, J.; Yamauchi, S.; Fueki, K.; Ishikawa, A. Nonstoichiometry of the perovskite-type oxide $La_{1-x}Sr_xCrO_{3-\delta}$. *Solid State Ionics,* **1984**, *12*, 119-124.
[http://dx.doi.org/10.1016/0167-2738(84)90138-3]

[35] Fergus, J.W. Lanthanum chromite-based materials for solid oxide fuel cell interconnects. *Solid State Ionics,* **2004**, *171*, 1-15.
[http://dx.doi.org/10.1016/j.ssi.2004.04.010]

[36] Hilpert, K.; Steibrech, R.W.; Boroomand, F.; Wessel, E.; Meschke, F.; Zuev, A.; Teller, O.; Nickel, H.; Singheiser, L. Defect formation and mechanical stability of perovskites based on $LaCrO_3$ for solid oxide fuel cells (SOFC). *J. Eur. Ceram. Soc.,* **2003**, *23*, 3009-3020.

[http://dx.doi.org/10.1016/S0955-2219(03)00097-9]

[37]   Oishi, M.; Yashiro, K.; Hong, J-O.; Nigara, Y.; Kawada, T.; Mizusaki, M. Oxygen nonstoichiometry of B-site doped $LaCrO_3$. *Solid State Ionics*, **2007**, *178*, 307-312.
[http://dx.doi.org/10.1016/j.ssi.2006.12.018]

[38]   Yasuda, I.; Hikita, T. Electrical conductivity and defect structure of calcium-doped lanthanum chromitres. *J. Electrochem. Soc.*, **1993**, *140*, 1699-1704.
[http://dx.doi.org/10.1149/1.2221626]

[39]   Kharton, V.V.; Tsipis, E.V.; Marozau, I.P.; Viskup, A.P.; Frade, J.R.; Irvine, J.T.S. Mixed conductivity and electrochemical behavior of $(La_{0.75}Sr_{0.25})_{0.95}Cr_{0.5}Mn_{0.5}O_{3-\delta}$. *Solid State Ionics*, **2007**, *178*, 101-113.
[http://dx.doi.org/10.1016/j.ssi.2006.11.012]

[40]   Kozhevnikov, V.L.; Leonidov, I.A.; Bahteeva, J.A.; Patrakeev, M.V.; Mitberg, E.B.; Poeppelmeier, K.R. Disordering and mixed conductivity in the solid solution $LaSr_2Fe_{3-y}Cr_yO_{8+\delta}$. *Chem. Mater.*, **2004**, *16*, 5014-5020.
[http://dx.doi.org/10.1021/cm031084o]

[41]   Tao, S.; Irvine, J.T.S. Synthesis and characterization of $La_{0.75}Sr_{0.25}Cr_{0.5}Mn_{0.5}O_{3-\delta}$, a redox-stable, efficient perovskite anode for SOFCs. *J. Electrochem. Soc.*, **2004**, *151*, A252-A259.
[http://dx.doi.org/10.1149/1.1639161]

[42]   Höfer, H.; Schmidberger, R. Electronic conductivity in the $La(Cr,Ni)O_3$ perovskite system. *J. Electrochem. Soc.*, **1994**, *141*, 782-786.
[http://dx.doi.org/10.1149/1.2054811]

[43]   Koc, R.; Anderson, H.U. Investigation of strontium-doped $La(Cr,Mn)O_3$ for solid oxide fuel cells. *J. Mater. Sci.*, **1992**, *27*, 5837-5843.
[http://dx.doi.org/10.1007/BF01119747]

[44]   Yaremchenko, A.A.; Kharton, V.V.; Kolotygin, V.A.; Patrakeev, M.V.; Tsipis, E.V.; Waerenborgh, J.C. Mixed conductivity, thermochemical expansion and electrochemical activity of Fe-substituted $(La,Sr)(Cr,Mg)O_{3-\delta}$ for solid oxide fuel cell anodes. *J. Power Sources*, **2014**, *249*, 483-496.
[http://dx.doi.org/10.1016/j.jpowsour.2013.10.129]

[45]   Mori, M.; Yamamoto, T.; Itoh, H.; Watanabe, T. Compatibility of alkaline earth metal (Mg, Ca, Sr)-doped lanthanum chromites as separators in planar-type high-temperature solid oxide fuel cells. *J. Mater. Sci.*, **1997**, *32*, 2423-2431.
[http://dx.doi.org/10.1023/A:1018565409603]

[46]   Jiang, S.P.; Liu, L.; Ong, K.P.; Wu, P.; Li, J.; Pu, J. Electrical conductivity and performance of doped $LaCrO_3$ perovskite oxides for solid oxide fuel cells. *J. Power Sources*, **2008**, *176*, 82-89.
[http://dx.doi.org/10.1016/j.jpowsour.2007.10.053]

[47]   Mori, M.; Sammes, N.M. Sintering and thermal expansion characterization of Al-doped and Co-doped lanthanum strontium chromites synthesized by the Pechini method. *Solid State Ionics*, **2002**, *164*, 301-312.
[http://dx.doi.org/10.1016/S0167-2738(01)01020-7]

[48]   Karen, P.; Norby, T. $La_{1-x}Ba_xCr_{1-y}Ti_yO_3$ with varied oxygen content. *J. Electrochem. Soc.*, **1998**, *145*, 264-269.
[http://dx.doi.org/10.1149/1.1838245]

[49]   Pudmich, G.; Boukamp, B.A.; Gonzalez-Cuenca, M.; Jungen, W.; Zipprich, W.; Tietz, F. Chromite/titanate based perovskites for application as anodes in solid oxide fuel cells. *Solid State Ionics*, **2000**, *135*, 433-438.
[http://dx.doi.org/10.1016/S0167-2738(00)00391-X]

[50]   Tietz, F. Thermal expansion of SOFC materials. *Ionics*, **1999**, *5*, 129-139.
[http://dx.doi.org/10.1007/BF02375916]

[51] Plint, S.M.; Connor, P.A.; Tao, S.; Irvine, J.T.S. Electronic transport in the novel SOFC anode material $La_{1-x}Sr_xCr_{0.5}Mn_{0.5}O_{3\pm\delta}$. *Solid State Ionics,* **2006**, *177*, 2005-2008.
[http://dx.doi.org/10.1016/j.ssi.2006.02.025]

[52] Danilovic, N.; Vincent, A.; Luo, J.L.; Chuang, K.T.; Hui, R.; Sanger, A.R. Correlation of fuel cell anode electrocatalytic and *ex situ* catalytic activity of perovskites $La_{0.75}Sr_{0.25}Cr_{0.5}X_{0.5}O_{3-\delta}$ (X=Ti, Mn, Fe, Co). *Chem. Mater.,* **2010**, *22*, 957-965.
[http://dx.doi.org/10.1021/cm901875u]

[53] Azad, A.K.; Eriksson, G.G.; Irvine, J.T.S. Structural, magnetic and electrochmical characterization of $La_{0.83}A_{0.17}Fe_{0.5}Cr_{0.5}O_{3-\delta}$ (A = Ba, Ca) perovskites. *Mater. Res. Bull.,* **2009**, *44*, 1451-1457.
[http://dx.doi.org/10.1016/j.materresbull.2009.03.008]

[54] Danilovic, N.; Luo, J.L.; Chuang, K.T.; Sanger, A.R. Effect of substitution with Cr^{3+} and addition of Ni on the physical and electrochemical properties of $Ce_{0.9}Sr_{0.1}VO_3$ as a H_2S-active anode for solid oxide fuel cells. *J. Power Sources,* **2009**, *194*, 252-262.
[http://dx.doi.org/10.1016/j.jpowsour.2009.04.051]

[55] Yaremchenko, A.A.; Kovalevsky, A.V.; Kharton, V.V. Mixed conductivity, stability and electrochemical behavior of perovskite-type $(Sr_{0.7}Ce_{0.3})_{1-x}Mn_{1-y}Cr_yO_{3-\delta}$. *Solid State Ionics,* **2008**, *179*, 2181-2191.
[http://dx.doi.org/10.1016/j.ssi.2008.07.014]

[56] Zhu, X.; Yan, H.; Zhong, Q.; Zhao, X.; Tan, W. $Ce_{0.9}Sr_{0.1}Cr_{0.5}Fe_{0.5}O_{3-\delta}$ as the anode materials for solid oxide fuel cells running on H_2 and H_2S. *J. Alloys Compd.,* **2011**, *509*, 8360-8364.
[http://dx.doi.org/10.1016/j.jallcom.2011.03.124]

[57] El Himri, A.; Marrero-López, D.; Ruiz-Morales, J.C.; Peña-Martínez, J.; Núñez, P. Structural and electrochemical characterisation of $Pr_{0.7}Ca_{0.3}Cr_{1-y}Mn_yO_{3-\delta}$ as symmetrical solid oxide fuel cell electrodes. *J. Power Sources,* **2009**, *188*, 230-237.
[http://dx.doi.org/10.1016/j.jpowsour.2008.11.050]

[58] Raj, E.S.; Irvine, J.T.S. Synthesis and characterization of $(Pr_{0.75}Sr_{0.25})_{1-x}Cr_{0.5}Mn_{0.5}O_{3-\delta}$ as anode for SOFCs. *Solid State Ionics,* **2010**, *180*, 1683-1689.
[http://dx.doi.org/10.1016/j.ssi.2009.11.006]

[59] Tao, S.; Irvine, J.T.S. Structural and electrochemical properties of the perovskite oxide $Pr_{0.7}Sr_{0.3}Cr_{0.9}Ni_{0.1}O_{3-\delta}$. *Solid State Ionics,* **2008**, *179*, 725-731.
[http://dx.doi.org/10.1016/j.ssi.2008.04.027]

[60] Lü, M.F.; Tsipis, E.V.; Waerenborgh, J.C.; Yaremchenko, A.A.; Kolotygin, V.A.; Bredikhin, S.I.; Kharton, V.V. Thermomechanical, transport and anodic properties of perovskite-type $(La_{0.75}Sr_{0.25})_{0.95}Cr_{1-x}Fe_xO_{3-\delta}$. *J. Power Sources,* **2012**, *206*, 59-69.
[http://dx.doi.org/10.1016/j.jpowsour.2012.01.100]

[61] Yasuda, I.; Hishinuma, M. Electrical conductivity and chemical diffusion coefficient of Sr-doped lanthanum chromites. *Solid State Ionics,* **1995**, *80*, 141-150.
[http://dx.doi.org/10.1016/0167-2738(95)00136-T]

[62] Primdahl, S.; Hansen, J.R.; Grahl-Madsen, L.; Larsen, P.H. Sr-doped $LaCrO_3$ anode for solid oxide fuel cells. *J. Electrochem. Soc.,* **2001**, *148*, A74-A81.
[http://dx.doi.org/10.1149/1.1344519]

[63] Kobsiriphat, W.; Madsen, B.D.; Wang, Y.; Shah, M.; Marks, M.D.; Barnett, S.A. Nickel- and ruthenium-doped lanthanum chromite anodes: effects of nanoscale metal precipitation on solid oxide fuel cell performance. *J. Electrochem. Soc.,* **2010**, *157*, B279-B284.
[http://dx.doi.org/10.1149/1.3269993]

[64] Rath, M.K.; Lee, K.T. Investigation of aliovalent transition metal doped $La_{0.7}Ca_{0.3}Cr_{0.8}X_{0.2}O_{3-\delta}$ (X=Ti, Mn, Fe, Co, and Ni) as electrode materials for symmetric solid oxide fuel cells. *Ceram. Int.,* **2015**, *41*, 10878-10890.

[http://dx.doi.org/10.1016/j.ceramint.2015.05.029]

[65]   Tao, S.W.; Irvine, J.T.S. Phase transition in perovskite oxide $La_{0.75}Sr_{0.25}Cr_{0.5}Mn_{0.5}O_{3-\delta}$ observed by *in situ* high-temperature neutron powder diffraction. *Chem. Mater.*, **2006**, *18*, 5453-5460.
[http://dx.doi.org/10.1021/cm061413n]

[66]   Lay, E.; Gauthier, G.; Dessemond, L. Preliminary studies of the new Ce-doped La/Sr chromo-manganite series as potential SOFC anode or SOEC cathode materials. *Solid State Ionics*, **2011**, *189*, 91-99.
[http://dx.doi.org/10.1016/j.ssi.2011.02.004]

[67]   Jiang, S.P.; Chen, X.J.; Chan, S.H.; Kwok, J.T.; Khor, K.A. $(La_{0.75}Sr_{0.25})(Cr_{0.5}Mn_{0.5})O_3$/YSZ composite anodes for methane oxidation reaction in solid oxide fuel cells. *Solid State Ionics*, **2006**, *177*, 149-157.
[http://dx.doi.org/10.1016/j.ssi.2005.09.010]

[68]   Raj, E.S.; Kilner, J.A.; Irvine, J.T.S. Oxygen diffusion and surface exchange studies on $(La_{0.75}Sr_{0.25})_{0.95}Cr_{0.5}Mn_{0.5}O_{3-\delta}$. *Solid State Ionics*, **2006**, *177*, 1747-1752.
[http://dx.doi.org/10.1016/j.ssi.2006.04.011]

[69]   Caignaert, V. $Sr_2Mn_2O_5$ magnetic structure. *J. Magn. Magn. Mater.*, **1997**, *166*, 117-123.
[http://dx.doi.org/10.1016/S0304-8853(96)00455-6]

[70]   Gélard, I.; Dubourdieu, C.; Pailhes, S.; Petit, S.; Simon, Ch. Neutron diffraction study of hexagonal manganite $YMnO_3$, $HoMnO_3$, and $ErMnO_3$ epitaxial films. *Appl. Phys. Lett.*, **2008**, *92*, 232506.
[http://dx.doi.org/10.1063/1.2943276]

[71]   Alonso, J.M.; Cortés-Gil, R.; Ruiz-González, L.; González-Calbet, J.M.; Hernando, A.; Vallet-Regí, M.; Dávila, M.E.; Asensio, M.C. Influence of the synthetic pathway on the properties of oxygen-deficient manganese-related perovskites. *Eur. J. Inorg. Chem.*, **2007**, 3350-3355.
[http://dx.doi.org/10.1002/ejic.200700108]

[72]   Markov, A.A.; Chesnokov, K.Yu.; Patrakeev, M.V.; Leonidov, I.A.; Chukin, A.V.; Leonidova, O.N.; Kozhevnikov, V.L. Oxygen non-stoichiometry and mixed conductivity of $La_{0.5}Sr_{0.5}Fe_{1-x}Mn_xO_{3-\delta}$. *J. Solid State Electrochem.*, **2016**, *20*, 225-234.
[http://dx.doi.org/10.1007/s10008-015-3027-0]

[73]   Sakai, N.; Kawada, T.; Yokokawa, H.; Dokiya, M.; Iwata, T. Sinterability and electrical conductivity of calcium-doped lanthanum chromites. *J. Mater. Sci.*, **1990**, *25*, 4531-4534.
[http://dx.doi.org/10.1007/BF00581119]

[74]   Zhang, L.; Chen, X.B.; Jiang, S.P.; He, H.Q.; Xiang, Y. Characterization of doped $La_{0.7}A_{0.3}Cr_{0.5}Mn_{0.5}O_{3-\delta}$ (A=Ca, Sr, Ba) electrodes for solid oxide fuel cells. *Solid State Ionics*, **2009**, *180*, 1076-1082.
[http://dx.doi.org/10.1016/j.ssi.2009.05.010]

[75]   Wan, J.; Zhu, J.H.; Goodenough, J.B. $La_{0.75}Sr_{0.25}Cr_{0.5}Mn_{0.5}O_{3-\delta}$+Cu composite anode running on H_2 and CH_4 fuels. *Solid State Ionics*, **2006**, *177*, 1211-1217.
[http://dx.doi.org/10.1016/j.ssi.2006.04.046]

[76]   Peña-Martínez, J.; Marrero-López, D.; Ruiz-Morales, J.C.; Savaniu, C.; Núñez, P.; Irvine, J.T.S. Anodic performance and intermediate temperature fuel cell testing of $La_{0.75}Sr_{0.25}Cr_{0.5}Mn_{0.5}O_{3-\delta}$ at lanthanum gallate electrolytes. *Chem. Mater.*, **2006**, *18*, 1001-1006.
[http://dx.doi.org/10.1021/cm0516659]

[77]   Haag, J.M.; Madsen, B.D.; Barnett, S.A.; Poeppelmeier, K.R. Application of $LaSr_2Fe_2CrO_{9-\delta}$ in Solid Oxide Fuel Cell Anodes. *Electrochem. Solid-State Lett.*, **2008**, *11*, B51-B53.
[http://dx.doi.org/10.1149/1.2836484]

[78]   Kolotygin, V.A.; Tsipis, E.V.; Lü, M.F.; Pivak, Y.V.; Yarmolenko, S.N.; Bredikhin, S.I.; Kharton, V.V. Functional properties of SOFC anode materials based on $LaCrO_3$, $La(Ti,Mn)O_3$ and $Sr(Nb,Mn)O_3$ perovskites: A comparative analysis. *Solid State Ionics*, **2013**, *251*, 28-33.
[http://dx.doi.org/10.1016/j.ssi.2013.01.005]

[79]   Kobsiriphat, W.; Madsen, B.D.; Wang, Y.; Marks, L.D.; Barnett, S.A. $La_{0.8}Sr_{0.2}Cr_{1-x}Ru_xO_{3-\delta}$ - $Gd_{0.1}Ce_{0.9}O_{1.95}$ solid oxide fuel cell anodes: Ru precipitation and electrochemical performance. *Solid State Ionics*, **2009**, *180*, 257-264.
        [http://dx.doi.org/10.1016/j.ssi.2008.12.022]

[80]   Oh, T-S.; Yu, A.S.; Adijanto, L.; Gorte, R.J.; Vohs, J.M. Infiltrated lanthanum strontium chromite anodes for solid oxide fuel cells: structural and catalytic aspects. *J. Power Sources*, **2014**, *262*, 207-212.
        [http://dx.doi.org/10.1016/j.jpowsour.2014.03.141]

[81]   Madsen, B.D.; Kobsiriphat, W.; Wang, Y.; Marks, L.D.; Barnett, S.A. Nucleation of nanometer-scale electrocatalyst particles in solid oxide fuel cell anodes. *J. Power Sources*, **2007**, *166*, 64-67.
        [http://dx.doi.org/10.1016/j.jpowsour.2006.12.080]

[82]   Vernoux, P.; Guindet, J.; Kleitz, M. Gradual internal methane reforming in intermediate-temperature solid-oxide fuel cells. *J. Electrochem. Soc.*, **1998**, *145*, 3487-3492.
        [http://dx.doi.org/10.1149/1.1838832]

[83]   Sauvet, A.L.; Irvine, J.T.S. Catalytic activity for steam methane reforming and physical characterisation of $La_{1-x}Sr_xCr_{1-y}Ni_yO_{3-\delta}$. *Solid State Ionics*, **2004**, *167*, 1-8.
        [http://dx.doi.org/10.1016/j.ssi.2003.11.021]

[84]   Ruiz-Morales, J.C.; Canales-Vázquez, J.; Peña-Martínez, J.; Marrero-López, D.; Núñez, P. On the simultaneous use of $La_{0.75}Sr_{0.25}Cr_{0.5}Mn_{0.5}O_{3-\delta}$ as both anode and cathode material with improved microstructure in solid oxide fuel cells. *Electrochim. Acta*, **2006**, *52*, 278-284.
        [http://dx.doi.org/10.1016/j.electacta.2006.05.006]

[85]   Wei, T.; Zhou, X.; Hu, Q.; Gao, Q.; Han, D.; Lv, X.; Wang, S. A high power density solid oxide fuel cell based on nano-structured $La_{0.8}Sr_{0.2}Cr_{0.5}Fe_{0.5}O_{3-\delta}$ anode. *Electrochim. Acta*, **2014**, *148*, 33-38.
        [http://dx.doi.org/10.1016/j.clcctacta.2014.10.020]

[86]   Liu, J.; Madsen, B.D.; Ji, Z.; Barnett, S.A. A fuel-flexible ceramic-based anode for solid oxide fuel cells. *Electrochem. Solid-State Lett.*, **2002**, *5*, A122-A124.
        [http://dx.doi.org/10.1149/1.1473258]

[87]   Qin, Q.; Ruan, C.; Ye, L.; Gan, L.; Xie, K. Efficient syngas production from methane reforming in solid oxide electrolyser with LSCM cathode loaded with Ni-Cu catalysts. *J. Solid State Electrochem.*, **2015**, *19*, 3389-3399.
        [http://dx.doi.org/10.1007/s10008-015-2966-9]

[88]   Huang, B.; Wang, S.R.; Liu, R.Z.; Ye, X.F.; Nie, H.W.; Sun, X.F.; Wen, T.L. Performance of $La_{0.75}Sr_{0.25}Cr_{0.5}Mn_{0.5}O_{3-\delta}$ perovskite-structure anode material at lanthanum gallate electrolyte for IT-SOFC running on ethanol fuel. *J. Power Sources*, **2007**, *167*, 39-46.
        [http://dx.doi.org/10.1016/j.jpowsour.2007.02.022]

[89]   Gong, M.; Bierschenk, D.; Haag, J.; Poeppelmeier, K.R.; Barnett, S.A.; Xu, C.; Zondlo, J.W.; Liu, X. Degradation of $LaSr_2Fe_2CrO_{9-\delta}$ solid oxide fuel cell anodes in phosphine-containing fuels. *J. Power Sources*, **2010**, *195*, 4013-4021.
        [http://dx.doi.org/10.1016/j.jpowsour.2009.12.117]

[90]   Weniy, T.; Qin, Z.; Han, Y.; Xyufang, Z.; Hongyi, L. Deactivation of anode catalyst $La_{0.75}Sr_{0.25}Cr_{0.5}Mn_{0.5}O_{3\pm\delta}$ in SOFC with fuel containing hydrogen sulfur: The role of lattice oxygen. *Int. J. Hydrogen Energy*, **2012**, *37*, 7398-7404.
        [http://dx.doi.org/10.1016/j.ijhydene.2012.02.008]

[91]   Molero-Sánchez, B.; Prado-Gonjal, J.; Ávila-Brande, D.; Chen, M.; Morán, E.; Birss, V. High performance $La_{0.3}Ca_{0.7}Cr_{0.3}Fe_{0.7}O_{3-\delta}$ air electrode for reversible solid oxide fuel cell applications. *Int. J. Hydrogen Energy*, **2015**, *40*, 1902-1910.
        [http://dx.doi.org/10.1016/j.ijhydene.2014.11.127]

[92]   Peng, C.; Wang, B.; Vincent, A. $LaCrO_3$-VO_x-YSZ anode material for solid oxide fuel cells operating

on H_2S-containing syngas. *J. Mater. Sci.,* **2012**, *47*, 227-233.
[http://dx.doi.org/10.1007/s10853-011-5789-9]

[93]   Zhang, Y.; Shen, Y.; Du, X.; Li, J.; Cao, X.; He, T. Nanostructured GDC-impregnated $La_{0.7}Ca_{0.3}CrO_{3-\delta}$ symmetrical electrodes for solid oxide fuel cells operating on hydrogen and city gas. *Int. J. Hydrogen Energy,* **2011**, *36*, 3673-3680.
[http://dx.doi.org/10.1016/j.ijhydene.2010.12.104]

[94]   Zhang, Y.; Zhou, Q.; He, T. $La_{0.7}Ca_{0.3}CrO_3\text{-}Ce_{0.8}Gd_{0.2}O_{1.9}$ composites as symmetrical electrodes for solid-oxide fuel cells. *J. Power Sources,* **2011**, *196*, 76-83.
[http://dx.doi.org/10.1016/j.jpowsour.2010.07.035]

[95]   Bao, W.; Guan, H.; Cheng, J. A new anode material for intermediate solid oxide fuel cells. *J. Power Sources,* **2008**, *175*, 232-237.
[http://dx.doi.org/10.1016/j.jpowsour.2007.09.026]

[96]   Peña-Martínez, J.; Marrero-López, D.; Pérez-Coll, D.; Ruiz-Morales, J.C.; Núñez, P. Performance of XSCoF (X = Ba, La and Sm) and LSCrX' (X' = Mn, Fe and Al) perovskite-structure materials on LSGM electrolyte for IT-SOFC. *Electrochim. Acta,* **2007**, *52*, 2950-2958.
[http://dx.doi.org/10.1016/j.electacta.2006.09.004]

[97]   Zha, S.; Tsang, P.; Cheng, Z.; Liu, M. Electrical properties and sulfur tolerance of $La_{0.75}Sr_{0.25}Cr_{1-x}Mn_xO_3$ under anodic conditions. *J. Solid State Chem.,* **2005**, *178*, 1844-1850.
[http://dx.doi.org/10.1016/j.jssc.2005.03.027]

[98]   Fowler, D.E.; Haag, J.M.; Boland, C.; Bierschenk, D.M.; Barnett, S.A.; Poeppelmeier, K.R. Stable, low polarization resistance solid oxide fuel cell anodes: $La_{1-x}Sr_xCr_{1-x}Fe_xO_{3-\delta}$ (x = 0.2 - 0.67). *Chem. Mater.,* **2014**, *26*, 3113-3120.
[http://dx.doi.org/10.1021/cm500423n]

[99]   Yasuda, I.; Hishinuma, M. Electrical conductivity and chemical diffusion coefficient of strontium-doped lanthanum manganites. *J. Solid State Chem.,* **1996**, *123*, 382-390.
[http://dx.doi.org/10.1006/jssc.1996.0193]

[100]   Tsipis, E.V.; Kharton, V.V. Electrode materials and reaction mechanisms in solid oxide fuel cells: a brief review. *J. Solid State Electrochem.,* **2008**, *12*, 1367-1391.
[http://dx.doi.org/10.1007/s10008-008-0611-6]

[101]   Mizusaki, J.; Yonemura, Y.; Kamata, H.; Ohyama, K.; Mori, N.; Takai, H.; Tagawa, H.; Dokiya, M.; Naraya, K.; Sasamoto, T.; Inaba, H.; Hashimoto, T. Electronic conductivity, Seebeck coefficient, defect and electronic structure of nonstoichiometric $La_{1-x}Sr_xMnO_3$. *Solid State Ionics,* **2000**, *132*, 167-180.
[http://dx.doi.org/10.1016/S0167-2738(00)00662-7]

[102]   Atsumi, T.; Kamegashira, N. Decomposition oxygen partial pressures of $Ln_{1-x}Sr_xMnO_3$ (Ln=La, Nd and Dy). *J. Alloys Compd.,* **1997**, *257*, 161-167.
[http://dx.doi.org/10.1016/S0925-8388(97)00013-3]

[103]   Jiang, S.P. Development of lanthanum strontium manganite perovskite cathode materials of solid oxide fuel cells: a review. *J. Mater. Sci.,* **2008**, *43*, 6799-6833.
[http://dx.doi.org/10.1007/s10853-008-2966-6]

[104]   Fu, Q.X.; Tietz, F.; Lersch, P.; Stöver, D. Evaluation of Sr- and Mn-substituted $LaAlO_3$ as potential SOFC anode materials. *Solid State Ionics,* **2006**, *177*, 1059-1069.
[http://dx.doi.org/10.1016/j.ssi.2006.02.053]

[105]   Kolotygin, V.A.; Tsipis, E.V.; Ivanov, A.I.; Fedotov, Y.S.; Burmistrov, I.N.; Agarkov, D.A.; Sinitsyn, V.V.; Bredikhin, S.I.; Kharton, V.V. Electrical, electrochemical, and thermomechanical properties of perovskite-type $(La_{1-x}Sr_x)_{1-y}Mn_{0.5}Ti_{0.5}O_{3-\delta}$ (x=0.15-0.75, y=0-0.05). *J. Solid State Electrochem.,* **2012**, *16*, 2335-2348.
[http://dx.doi.org/10.1007/s10008-012-1703-x]

[106] Mizusaki, J.; Mori, N.; Takai, H.; Yonemura, Y.; Minaminue, H.; Tagawa, H.; Dokiya, M.; Inaba, H.; Naraya, K.; Sasamoto, T.; Hashimoto, T. Oxygen nonstoichiometry and defect equilibrium in the perovskite-type oxides $La_{1-x}Sr_xMnO_{3-d}$. *Solid State Ionics*, **2000**, *129*, 163-177. [http://dx.doi.org/10.1016/S0167-2738(99)00323-9]

[107] Mitchell, J.F.; Argyriou, D.N.; Potter, C.D.; Hinks, D.G.; Jorgensen, J.D.; Bader, S.D. Structural phase diagram of $La_{1-x}Sr_xMnO_{3-\delta}$: Relationship to magnetic and transport properties. *Phys. Rev. B*, **1996**, *54*, 6172-6183. [http://dx.doi.org/10.1103/PhysRevB.54.6172]

[108] Tofield, B.C.; Scott, W.R. Oxidative nonstoichiometry in perovskites, an experimental survey; the defect structure of an lanthanum manganite by powder neutron diffraction. *J. Solid State Chem.*, **1974**, *10*, 183-194. [http://dx.doi.org/10.1016/0022-4596(74)90025-5]

[109] Zuev, A.Yu.; Tsvetkov, D.S. Oxygen nonstoichiometry, defect structure and defect-induced expansion of undoped perovskite $LaMnO_{3\pm\delta}$. *Solid State Ionics*, **2010**, *181*, 557-563. [http://dx.doi.org/10.1016/j.ssi.2010.02.024]

[110] Miyoshi, S.; Hong, J-O.; Yashiro, H.; Kaimai, A.; Nigara, Y.; Kawamura, K.; Kawada, T.; Mizusaki, J. Lattice creation and annihilation of $LaMnO_{3+\delta}$ caused by nonstoichiometry change. *Solid State Ionics*, **2002**, *154*, 257-263. [http://dx.doi.org/10.1016/S0167-2738(02)00441-1]

[111] Fu, Q.X.; Tietz, F.; Stöver, D. $La_{0.4}Sr_{0.6}Ti_{1-x}Mn_xO_{3-\delta}$ Perovskites as Anode Materials for Solid Oxide Fuel Cells. *J. Electrochem. Soc.*, **2006**, *153*, D74-D83. [http://dx.doi.org/10.1149/1.2170585]

[112] Marozau, I.P.; Kharton, V.V.; Viskup, A.P.; Frade, J.R.; Samakhval, V.V. Electronic conductivity, oxygen permeability and thermal expansion of $Sr_{0.7}Ce_{0.3}Mn_{1-x}Al_xO_{3-\delta}$. *J. Eur. Ceram. Soc.*, **2006**, *26*, 1371-1378. [http://dx.doi.org/10.1016/j.jeurceramsoc.2005.03.230]

[113] Sengodan, S.; Yeo, H.J.; Shin, J.Y.; Kim, G. Assessment of perovskite-type $La_{0.8}Sr_{0.2}Sc_xMn_{1-x}O_{3-\delta}$ oxides as anodes for intermediate-temperature solid oxide fuel cells using hydrocarbon fuels. *J. Power Sources*, **2011**, *196*, 3083-3088. [http://dx.doi.org/10.1016/j.jpowsour.2010.11.161]

[114] Kim, J.H.; Miller, D.; Schlegl, H.; McGrouther, D.; Irvine, J.T.S. Investigation of microstructural and electrochemical properties of impregnated (La,Sr)(Ti,Mn)$O_{3-\delta}$ as a potential anode material in high-temperature solid oxide fuel cells. *Chem. Mater.*, **2011**, *23*, 3841-3847. [http://dx.doi.org/10.1021/cm2007318]

[115] Corre, G.; Kim, G.; Cassidy, M.; Gorte, R.J.; Irvine, J.T.S. Activation and ripening of impregnated manganese containing perovskite SOFC electrodes under redox cycling. *Chem. Mater.*, **2009**, *21*, 1077-1084. [http://dx.doi.org/10.1021/cm803149v]

[116] Yokokawa, H. Understanding materials compatibility. *Annu. Rev. Mater. Res.*, **2003**, *33*, 581-610. [http://dx.doi.org/10.1146/annurev.matsci.33.022802.093856]

[117] Kharton, V.V.; Yaremchenko, A.A.; Naumovich, E.N. Research on the electrochemistry of oxygen ion conductors in the former Soviet Union. II. Perovskite-related oxides. *J. Solid State Electrochem.*, **1999**, *3*, 303-326. [http://dx.doi.org/10.1007/s100080050161]

[118] Sakaki, Y.; Takeda, Y.; Kato, A.; Imanishi, N.; Yamamoto, O.; Hattori, M.; Iito, M.; Esaki, Y. $Ln_{1-x}Sr_xMnO_3$ (Ln = Pr, Nd, Sm and Gd) as the cathode material for solid oxide fuel cells. *Solid State Ionics*, **1999**, *118*, 187-194. [http://dx.doi.org/10.1016/S0167-2738(98)00440-8]

[119] Fu, Q.X.; Tietz, F.; Stöver, D. Synthesis and electrical conductivity of Sr- and Mn-substituted $LaAlO_3$ as a possible SOFC anode material. *Solid State Ionics,* **2006**, *177*, 1819-1822.
[http://dx.doi.org/10.1016/j.ssi.2006.03.028]

[120] Zheng, Y.; Zhang, C.; Ran, R.; Cai, R.; Shao, Z.; Farrusseng, D. A new symmetric solid-oxide fuel cell with $La_{0.8}Sr_{0.2}Sc_{0.2}Mn_{0.8}O_{3-\delta}$ perovskite oxide as both the anode and cathode. *Acta Mater.,* **2009**, *57*, 1165-1175.
[http://dx.doi.org/10.1016/j.actamat.2008.10.047]

[121] Chen, F.; Liu, M. Study of transition metal oxide doped $LaGaO_3$ as electrode materials for LSGM-based solid oxide fuel cells. *J. Solid State Electrochem.,* **1998**, *3*, 7-14.
[http://dx.doi.org/10.1007/s100080050124]

[122] Fu, Q.; Xu, X.; Peng, D.; Liu, X.; Meng, G. Preparation and electrochemical characterization of Sr- and Mn-doped $LaGaO_3$ as anode materials for LSGM-based SOFCs. *J. Mater. Sci.,* **2003**, *38*, 2901-2906.
[http://dx.doi.org/10.1023/A:1024496805673]

[123] Tao, S.; Irvine, J.T.S. Study on the structural and electrical properties of the double perovskite oxide $SrMn_{0.5}Nb_{0.5}O_{3-\delta}$. *J. Mater. Chem.,* **2002**, *12*, 2356-2360.
[http://dx.doi.org/10.1039/B204248G]

[124] Vashook, V.; Franke, D.; Vasylechko, L.; Zosel, J.; Rebello, J.; Ahlborn, K.; Fichtner, W.; Schmidt, M.; Wen, T-L.; Guth, U. Electrical conductivity and oxygen non-stoichiometry of the double B mixed perovskite series $La_{0.6}Ca_{0.4}Mn_{1-y}Me_yO_{3-\delta}$ with Me = Fe, Co, Ni and x = 0 - 0.6. *Solid State Ionics,* **2008**, *179*, 1101-1107.
[http://dx.doi.org/10.1016/j.ssi.2008.04.025]

[125] Shin, T.H.; Ida, S.; Ishihara, T. Doped CeO_2-$LaFeO_3$ composite oxide as an active anode for direct hydrocarbon-type solid oxide fuel cells. *J. Am. Chem. Soc.,* **2011**, *133*, 19399-19407.
[http://dx.doi.org/10.1021/ja206278f]

[126] Bahrain, A.M.K.; Ida, S.; Ishihara, T. Al-doped $La_{0.5}Sr_{0.5}MnO_3$ as oxide anode for solid oxide fuel cells using dry C_3H_8 fuel. *J. Solid State Electrochem.,* **2017**, *21*, 161-170.
[http://dx.doi.org/10.1007/s10008-016-3356-7]

[127] Su, C.; Wu, Y.; Wang, W.; Zheng, Y.; Ran, R.; Shao, Z. Assessment of nickel cermets and $La_{0.8}Sr_{0.2}Sc_{0.2}Mn_{0.8}O_3$ as solid-oxide fuel cell anodes operating on carbon monoxide fuel. *J. Power Sources,* **2010**, *195*, 1333-1343.
[http://dx.doi.org/10.1016/j.jpowsour.2009.09.015]

[128] Ouweltjes, J.P.; van Tuel, M.; Sillessen, M.; Rietveld, G. Redox tolerant SOFC anodes with high electrochemical performance. *Fuel Cells (Weinh.),* **2009**, *9*, 873-882.
[http://dx.doi.org/10.1002/fuce.200800142]

[129] Horikiri, F.; Iizawa, N.; Han, L.Q.; Sato, K.; Yashiro, K.; Kawada, T.; Mizusaki, J. Defect equilibrium and electron transport in the bulk of single crystal $SrTi_{1-x}Nb_xO_3$ (x = 0.01, 0.001, 0.0002). *Solid State Ionics,* **2008**, *179*, 2335-2344.
[http://dx.doi.org/10.1016/j.ssi.2008.10.001]

[130] Ueda, K.; Yanagi, H.; Hosono, H.; Kawazoe, H. Carrier generation and compensation in Y- and Nb-doped $CaTiO_3$ single crystals. *Phys. Rev. B,* **1997**, *56*, 12998-13005.
[http://dx.doi.org/10.1103/PhysRevB.56.12998]

[131] Meyer, R.; Waser, R.; Helmbold, J.; Borchardt, G. Observation of vacancy defect migration in the cation sublattice of complex oxides by ^{18}O tracer experiments. *Phys. Rev. Lett.,* **2003**, *90*, 105901.
[http://dx.doi.org/10.1103/PhysRevLett.90.105901]

[132] Neagu, D.; Irvine, J.T.S. Structure and properties of $La_{0.4}Sr_{0.4}TiO_3$ ceramics for use as anode materials in solid oxide fuel cells. *Chem. Mater.,* **2010**, *22*, 5042-5053.
[http://dx.doi.org/10.1021/cm101508w]

[133] Blennow, P.; Hansen, K.K.; Wallenberg, L.R.; Mogensen, M. Effects of Sr/Ti-ratio in SrTiO$_3$-based SOFC anodes investigated by the use of cone-shaped electrodes. *Electrochim. Acta,* **2006**, *52*, 1651-1661.
[http://dx.doi.org/10.1016/j.electacta.2006.03.096]

[134] Hui, S.; Petric, A. Electrical conductivity of yttrium-doped SrTiO$_3$: influence of transition metal additives. *Mater. Res. Bull.,* **2002**, *37*, 1215-1231.
[http://dx.doi.org/10.1016/S0025-5408(02)00774-2]

[135] Hui, S.; Petric, A. Evaluation of yttrium-doped SrTiO$_3$ as an anode for solid oxide fuel cells. *J. Eur. Ceram. Soc.,* **2002**, *22*, 1673-1681.
[http://dx.doi.org/10.1016/S0955-2219(01)00485-X]

[136] Steinsvik, S.; Bugge, R.; Gjønnes, J.; Taftø, J.; Norby, T. The defect structure of SrTi$_{1-x}$Fe$_x$O$_{3-y}$ (x = 0 - 0.8) investigated by electrical conductivity measurements and electron energy loss spectroscopy (EELS). *J. Phys. Chem. Solids,* **1997**, *58*, 969-976.
[http://dx.doi.org/10.1016/S0022-3697(96)00200-4]

[137] Moos, R.; Schöllhammer, S.; Härdtl, K.H. Electron mobility of Sr$_{1-x}$La$_x$TiO$_3$ ceramics between 600 °C and 1000 °C. *Appl. Phys., A Mater. Sci. Process.,* **1997**, *65*, 291-294.
[http://dx.doi.org/10.1007/s003390050581]

[138] Marina, O.A.; Canfield, N.L.; Stevenson, J.W. Thermal, electrical, and electrocatalytical properties of lanthanum-doped strontium titanate. *Solid State Ionics,* **2002**, *149*, 21-28.
[http://dx.doi.org/10.1016/S0167-2738(02)00140-6]

[139] Blennow, P.; Hansen, K.K.; Wallenberg, L.R.; Mogensen, M. Electrochemical characterization and redox behavior of Nb-doped SrTiO$_3$. *Solid State Ionics,* **2009**, *180*, 63-70.
[http://dx.doi.org/10.1016/j.ssi.2008.10.011]

[140] Ma, Q.; Tietz, F.; Leonide, A.; Ivers-Tiffée, E. Anode-supported planar SOFC with high performance and redox stability. *Electrochem. Commun.,* **2010**, *12*, 1326-1328.
[http://dx.doi.org/10.1016/j.elecom.2010.07.011]

[141] Périllat-Merceroz, C.; Roussel, P.; Huvé, M.; Capoen, E.; Rosini, S.; Gélin, P.; Vannier, R-N.; Gauthier, G.H. Pure and Mn-doped La$_4$SrTi$_5$O$_{17}$ layered perovskite as potential solid oxide fuel cell material: Structure and anodic performance. *J. Power Sources,* **2015**, *274*, 806-815.
[http://dx.doi.org/10.1016/j.jpowsour.2014.10.131]

[142] Neagu, D.; Tsekouras, G.; Miller, D.N.; Ménard, H.; Irvine, J.T.S. *In situ* growth of nanoparticles through control of microstructue. *Nat. Chem.,* **2013**, *11*, 916-923.
[http://dx.doi.org/10.1038/nchem.1773]

[143] He, H.; Huang, Y.; Vohs, J.M.; Gorte, R.J. Characterization of YSZ-YST composites for SOFC anodes. *Solid State Ionics,* **2004**, *175*, 171-176.
[http://dx.doi.org/10.1016/j.ssi.2004.09.033]

[144] Afshar, M.R.; Yan, N.; Zahiri, B.; Mitlin, D.; Chuang, K.T.; Luo, J-L. Impregnation of La$_{0.4}$Ce$_{0.6}$O$_{1.8}$-La$_{0.4}$Sr$_{0.6}$TiO$_3$ as solid oxide fuel cell anode in H$_2$S-containing fuels. *J. Power Sources,* **2015**, *274*, 211-218.
[http://dx.doi.org/10.1016/j.jpowsour.2014.10.052]

[145] Fagg, D.P.; Kharton, V.V.; Frade, J.R.; Ferreira, A.A.L. Stability and mixed ionic-electronic conductivity of (Sr,La)(Ti,Fe)O$_{3-\delta}$ perovskites. *Solid State Ionics,* **2003**, *156*, 45-57.
[http://dx.doi.org/10.1016/S0167-2738(02)00257-6]

[146] Slater, P.R.; Fagg, D.P.; Irvine, J.T.S. Synthesis and electrical characterisation of doped perovskite titanates as potential anode materials for solid oxide fuel cells. *J. Mater. Chem.,* **1997**, *7*, 2495-2498.
[http://dx.doi.org/10.1039/a702865b]

[147] Bernuy-Lopez, C.; Knibbe, R.; He, Z.; Mao, X.; Hauch, A.; Nielse, K.A. Electrochemical characterisation of solid oxide cell electrodes for hydrogen production. *J. Power Sources,* **2011**, *196*,

4396-4403.
[http://dx.doi.org/10.1016/j.jpowsour.2010.10.102]

[148] Yakub, A.; Savaniu, C.; Janjua, N.K.; Irvine, J.T.S. Preparation *via* a solution method of $La_{0.2}Sr_{0.25}Ca_{0.45}TiO_3$ and its characterization for anode supported solid oxide fuel cells. *J. Mater. Chem.,* **2013**, *1*, 14189-14197.
[http://dx.doi.org/10.1039/c3ta12860a]

[149] Lu, L.Y.; Verbraeken, M.C.; Cassidy, M.; Irvine, J.T.S. A Solid Oxide Fuel Cell with lanthanum and calcium co-doped strontium titanate as support. *ECS Trans.,* **2013**, *57*, 1415-1422.
[http://dx.doi.org/10.1149/05701.1415ecst]

[150] Holtappels, P.; Irvine, J.T.S.; Iwanschitz, B.; Kuhn, L.T.; Lu, L.; Ma, Q.; Malzbender, J.; Mai, A.; Ramos, T.; Rass-Hansen, J.; Sudireddy, B.R.; Tietz, F.; Vasechko, V.; Veltzé, S.; Verbraeken, M. Full ceramic fuel cells based on strontium titanate anodes, a approach towards more robust SOFCs. *ECS Trans.,* **2013**, *57*, 1175-1184.
[http://dx.doi.org/10.1149/05701.1175ecst]

[151] Vincent, A.L.; Hanifi, A.R.; Zazulak, M.; Luo, J-L.; Chuang, K.T.; Sanger, A.R.; Etsell, T.; Sarkar, P. Preparation and characterization of an solid oxide fuel cell tubular cell for direct use with sour gas. *J. Power Sources,* **2013**, *240*, 411-416.
[http://dx.doi.org/10.1016/j.jpowsour.2013.03.104]

[152] Tsekouras, G.; Neagu, D.; Irvine, J.T.S. Step-change in high temperature steam electrolysis performance of perovskite oxide cathodes with exsolution of B-site dopants. *Energy Environ. Sci.,* **2013**, *6*, 256-266.
[http://dx.doi.org/10.1039/C2EE22547F]

[153] Ovalle, A.; Ruiz-Morales, J.C.; Canales-Vázquez, J.; Marrero-López, D.; Irvine, J.T.S. Mn-substituted titanates as efficient anodes for direct methane SOFCs. *Solid State Ionics,* **2006**, *177*, 1997-2003.
[http://dx.doi.org/10.1016/j.ssi.2006.06.014]

[154] Rath, M.T.; Ahn, B-G.; Choi, B-H.; Ji, M-J.; Lee, K-T. Effects of manganese substitution at the B-site of lanthanum-rich strontium titanate anodes on fuel cell performance and catalytic activity. *Ceram. Int.,* **2013**, *39*, 6343-6353.
[http://dx.doi.org/10.1016/j.ceramint.2013.01.060]

[155] Ruiz-Morales, J.C.; Canalez-Vázquez, J.; Savaniu, C.; Marrero-López, D.; Núñez, P.; Zhou, W.; Irvine, J.T.S. A new anode for solid oxide fuel cells with enhanced OCV under methane operation. *Phys. Chem. Chem. Phys.,* **2007**, *9*, 1821-1830.
[http://dx.doi.org/10.1039/B617266K]

[156] Ivanov, A.I.; Agarkov, D.A.; Burmistrov, I.N.; Kudrenko, E.A.; Bredikhin, S.I.; Kharton, V.V. Synthesis and properties of fuel cell anodes based on $(La_{0.5+x}Sr_{0.5-x})_{1-y}Mn_{0.5}Ti_{0.5}O_{3-\delta}$ (x = 0 - 0.25, y = 0 - 0.03). *Russ. J. Electrochem.,* **2014**, *50*, 730-736.
[http://dx.doi.org/10.1134/S1023193514080047]

[157] Yoon, J.S.; Yoon, M.Y.; Kwak, C.; Park, H.J.; Lee, S.M.; Lee, K.H.; Hwang, H.J. $Y_{0.08}Sr_{0.92}Fe_xTi_{1-x}O_{3-\delta}$ perovskite for solid oxide fuel cell anodes. *Mater. Sci. Eng. B,* **2012**, *177*, 151-156.
[http://dx.doi.org/10.1016/j.mseb.2011.10.016]

[158] Canalez-Vázquez, J.; Ruiz-Morales, J.C.; Marrero-López, D.; Peña-Martínez, J.; Núñez, P.; Gómez-Romero, P. Fe-substituted (La,Sr)TiO$_3$ as potential electrodes for symmetrical fuel cells (SFCs). *J. Power Sources,* **2007**, *171*, 552-557.
[http://dx.doi.org/10.1016/j.jpowsour.2007.05.094]

[159] Sengodan, S.; Yoon, J.S.; Yoon, M.Y.; Hwang, H.J.; Shin, J. Kim. G. Electrochemical performance of YST infiltrated and Fe doped YST infltrated YSZ anodes for IT-SOFC. *Electrochem. Solid-State Lett.,* **2013**, *2*, F45-F49.

[160] Canalez-Vázquez, J.; Ruiz-Morales, J.C.; Irvine, J.T.S.; Zhou, W. Sc-substituted oxygen excess titanates as fuel electrodes for SOFCs. *J. Electrochem. Soc.,* **2005**, *152*, A1458-A1465.

[http://dx.doi.org/10.1149/1.1921747]

[161] Cui, S-H.; Li, J-H.; Zhou, X-W.; Wang, G-Y.; Luo, J-L.; Chuang, K.T.; Bai, Y.; Qiao, L-J. Cobalt doped LaSrTiO$_{3-\delta}$ as an anode catalyst: effect of Co nanoparticle precipitation on SOFC operating on H$_2$S-containing hydrogen. *J. Mater. Chem. A Mater. Energy Sustain.*, **2013**, *1*, 9689-9696.
[http://dx.doi.org/10.1039/c3ta11315a]

[162] Yang, L.; De Jonghe, L.C.; Jacobsen, C.P.; Visco, S.J. B-Site doping and catalytic activity of Sr(Y)TiO$_3$. *J. Electrochem. Soc.*, **2007**, *154*, B949-B955.
[http://dx.doi.org/10.1149/1.2752970]

[163] Arrivé, C.; Delahaye, T.; Joubert, O.; Gauthier, G. Exsolution of nickel nanoparticles at the surface of a conducting titanate as potential hydrogen electrode material for solid oxide electrochemical cells. *J. Power Sources*, **2013**, *223*, 341-348.
[http://dx.doi.org/10.1016/j.jpowsour.2012.09.062]

[164] Lee, S.; Kim, G.; Vohs, J.M.; Gorte, R.J. SOFC anodes based on infiltration of La$_{0.3}$Sr$_{0.7}$TiO$_3$. *J. Electrochem. Soc.*, **2008**, *155*, B1179-B1183.
[http://dx.doi.org/10.1149/1.2976775]

[165] Vincent, A.L.; Hanifi, A.R.; Luo, J-L.; Chuang, K.T.; Sanger, A.R.; Etsell, T.H.; Sarkar, P. Porous YSZ impregnated with La$_{0.4}$Sr$_{0.5}$Ba$_{0.1}$TiO$_3$ as a possible composite anode for SOFCs fueled with sour feeds. *J. Power Sources*, **2012**, *215*, 301-306.
[http://dx.doi.org/10.1016/j.jpowsour.2012.04.063]

[166] Périllat-Merceroz, C.; Gauthier, G.; Roussel, P.; Huvé, M.; Capoen, E.; Vannier, R-N.; Gauthier, G. New insights on the structure and reducibility of 3D *versus* 2D La/Sr titanates for SOFC anodes. *Solid State Ionics*, **2013**, *247-248*, 76-85.
[http://dx.doi.org/10.1016/j.ssi.2013.06.007]

[167] Park, B.H.; Choi, G.M. Ex-solution of Ni nanoparticles in a La$_{0.2}$Sr$_{0.8}$Ti$_{1-x}$Ni$_x$O$_{3-\delta}$ alternative anode for solid oxide fuel cell. *Solid State Ionics*, **2014**, *262*, 345-348.
[http://dx.doi.org/10.1016/j.ssi.2013.10.016]

[168] Chang, H-Y.; Wang, S-H.; Wang, Y-M.; Lai, C-W.; Lin, C-H.; Cheng, S-Y. Novel core-shell structure of perovskite anode and characterization. *Int. J. Hydrogen Energy*, **2012**, *37*, 7771-7778.
[http://dx.doi.org/10.1016/j.ijhydene.2012.02.022]

[169] Yoo, K.B.; Choi, G.M. Performance of La-doped strontium titanate (LST) anode on LaGaO3-based SOFC. *Solid State Ionics*, **2009**, *180*, 867-871.
[http://dx.doi.org/10.1016/j.ssi.2009.02.013]

[170] Fan, L.; Xiong, Y.; Liu, L.; Wang, Y.; Kishimoto, H.; Yamaji, K.; Horita, T. Performance of Gd$_{0.2}$Ce$_{0.8}$O$_{1.9}$ infiltrated La$_{0.2}$Sr$_{0.8}$TiO$_3$ nanofiber scaffolds as anodes for solid oxide fuel cells. *J. Power Sources*, **2014**, *264*, 125-131.
[http://dx.doi.org/10.1016/j.jpowsour.2014.04.109]

[171] Miller, E.C.; Gao, Z.; Barnett, S.A. Fabrication of solid oxide fuel cells with a thin (La$_{0.9}$Sr$_{0.1}$)$_{0.98}$(Ga$_{0.8}$Mg$_{0.2}$)O$_{3-\delta}$ electrolyte on a Sr$_{0.8}$La$_{0.2}$TiO$_3$ support. *Fuel Cells (Weinh.)*, **2013**, *6*, 1060-1067.
[http://dx.doi.org/10.1002/fuce.201300155]

[172] Pillai, M.R.; Kim, I.; Bierschenk, D.M.; Barnett, S.A. Fuel-flexible operation of a solid oxide fuel cell with Sr$_{0.8}$La$_{0.2}$TiO$_3$ support. *J. Power Sources*, **2008**, *185*, 1086-1093.
[http://dx.doi.org/10.1016/j.jpowsour.2008.07.063]

[173] Savaniu, C.D.; Irvine, J.T.S. La-doped SrTiO$_3$ as anode material for IT-SOFC. *Solid State Ionics*, **2011**, *192*, 491-493.
[http://dx.doi.org/10.1016/j.ssi.2010.02.010]

[174] Kim, G.; Gross, M.D.; Wang, W.; Vohs, J.M.; Gorte, R.J. SOFC anodes based on LST-YSZ composites and on Y$_{0.04}$Ce$_{0.48}$Zr$_{0.48}$O$_2$. *J. Electrochem. Soc.*, **2008**, *155*, B360-B366.

[http://dx.doi.org/10.1149/1.2840473]

[175] Gross, M.D.; Vohs, J.M.; Gorte, R.J. A strategy for achieving high performance with SOFC ceramic anodes. *Electrochem. Solid-State Lett.,* **2007**, *10*, B65-B69.
[http://dx.doi.org/10.1149/1.2432942]

[176] Ikebe, T.; Muroyama, H.; Matsui, T.; Eguchi, K. Fabrication of redox tolerant anode with an electronic conductive oxide of Y-doped $SrTiO_3$. *J. Electrochem. Soc.,* **2010**, *157*, B970-B974.
[http://dx.doi.org/10.1149/1.3421792]

[177] Kurokawa, H.; Yang, L.; Jacobson, C.P.; De Jonghe, L.C.; Visco, S.J. Y-doped $SrTiO_3$ based sulfur tolerant anode for solid oxide fuel cells. *J. Power Sources,* **2007**, *164*, 510-518.
[http://dx.doi.org/10.1016/j.jpowsour.2006.11.048]

[178] Lu, X.C.; Zhu, J.H.; Yang, Z.; Xia, G.; Stevenson, J.W. Pd-impregnated SYT/LDC composite as sulfur-tolerant anode for solid oxide fuel cells. *J. Power Sources,* **2009**, *192*, 381-384.
[http://dx.doi.org/10.1016/j.jpowsour.2009.03.009]

[179] Yoo, K.B.; Park, B.H.; Choi, G.M. Stability and performance of SOFC with $SrTiO_3$-based anode in CH_4 fuel. *Solid State Ionics,* **2012**, *225*, 104-107.
[http://dx.doi.org/10.1016/j.ssi.2012.05.017]

[180] Blennow, P.; Sudireddy, P.R.; Persson, A.H.; Klemensø, T.; Nielsen, J.; Thydén, K. Infiltrated $SrTiO_3$:FeCr-based anodes for metal-supported SOFC. *Fuel Cells (Weinh.),* **2013**, *13*, 494-505.
[http://dx.doi.org/10.1002/fuce.201200176]

[181] Vincent, A.; Luo, J.L.; Chuang, K.T.; Sanger, A.R. Effect of Ba doping on performance of LST as anode in solid oxide fuel cells. *J. Power Sources,* **2010**, *195*, 769-774.
[http://dx.doi.org/10.1016/j.jpowsour.2009.08.018]

[182] Buccheri, M.A.; Hill, J.M. Methane electrochemical oxidation pathway over a Ni/YSZ and $La_{0.3}Sr_{0.7}TiO_3$ bi-layer SOFC anode. *J. Electrochem. Soc.,* **2012**, *159*, B361-B367.
[http://dx.doi.org/10.1149/2.001204jes]

[183] Ma, Q.; Tietz, F.; Leonide, A.; Ivers-Tiffée, E. Electrochemical perfomances of solid oxide fuel cells based on Y-substituted $SrTiO_3$ ceramic anode mterials. *J. Power Sources,* **2011**, *196*, 7308-7312.
[http://dx.doi.org/10.1016/j.jpowsour.2010.07.094]

[184] Huang, Y-H.; Dass, R.I.; Denyszyn, J.C.; Goodenough, J.B. Synthesis and characterization of $Sr_2MgMoO_{6-\delta}$. An anode material for the solid oxide fuel cell. *J. Electrochem. Soc.,* **2006**, *153*, A1266-A1272.
[http://dx.doi.org/10.1149/1.2195882]

[185] Zheng, K.; Świerczek, K. Physicochemical properties of rock-salt-type ordered Sr_2MMoO_6 (M = Mg, Mn, Fe, Co, Ni) double perovskites. *J. Eur. Ceram. Soc.,* **2014**, *34*, 4273-4284.
[http://dx.doi.org/10.1016/j.jeurceramsoc.2014.06.032]

[186] Kamata, K.; Nakamura, T.; Sata, T. Valence stability of molybdenum in alkaline earth molybdates. *Mater. Res. Bull.,* **1975**, *10*, 373-378.
[http://dx.doi.org/10.1016/0025-5408(75)90007-0]

[187] Lin, S.S. Volatile products from interaction between steel and H_2O/CO_2 at high temperatures. *J. Electrochem. Soc.,* **1980**, *127*, 1108-1111.
[http://dx.doi.org/10.1149/1.2129827]

[188] Fryburg, G.C.; Miller, R.A.; Kohl, F.J.; Stearns, C.A. Volatile products in the corrosion of Cr, Mo, Ti, and four superalloys exposed to O_2 containing H_2O and gaseous NaCI. *J. Electrochem. Soc.,* **1977**, *124*, 1738-1743.
[http://dx.doi.org/10.1149/1.2133147]

[189] Hayashi, S.; Aoki, R.; Nakamura, T. Metallic conductivity in perovskite-type compounds $AMoO_3$ (A = Ba, Sr, Ca) down to 2.5 K. *Mater. Res. Bull.,* **1979**, *14*, 409-413.
[http://dx.doi.org/10.1016/0025-5408(79)90107-7]

[190] Vasala, S.; Yamauchi, H.; Karppinen, M. Role of $SrMoO_4$ in Sr_2MgMoO_6 synthesis. *J. Solid State Chem.*, **2011**, *184*, 1312-1317.
[http://dx.doi.org/10.1016/j.jssc.2011.03.045]

[191] Aguadero, A.; de la Calle, C.; Alonso, J.A.; Pérez-Coll, D.; Escudero, M.J.; Daza, L. Structure, thermal stability and electrical properties of $Ca(V_{0.5}Mo_{0.5})O_3$ as solid oxide fuel cell anode. *J. Power Sources*, **2009**, *192*, 78-83.
[http://dx.doi.org/10.1016/j.jpowsour.2008.12.035]

[192] van den Bossche, M.; McIntosh, S. On the methane oxidation activity of $Sr_2(MgMo)_2O_{6-\delta}$: a potential anode material for direct hydrocarbon solid oxide fuel cells. *J. Mater. Chem.*, **2011**, *21*, 7443-7451.
[http://dx.doi.org/10.1039/c1jm10523j]

[193] dos Santos-Gómez, L.; León-Reina, L.; Porras-Vázquez, J.M.; Losilla, E.R.; Marrero-López, D. Chemical stability and compatibility of double perovskite anode materials for SOFCs. *Solid State Ionics*, **2013**, *293*, 1-7.
[http://dx.doi.org/10.1016/j.ssi.2013.03.005]

[194] Fotiev, V.; Bazuev, G.; Zubkov, V. Synthesis and electrical properties of $SrV_xMo_{1-x}O_3$ solid solutions with perovskite structure. *Неорган. матер.*, **1987**, *23*, 1005-1008. [in Russian].

[195] Graves, C.; Sudireddy, B.R.; Mogensen, M. Molybdate based ceramic negative-electrode materials for solid oxide cells. *ECS Trans.*, **2010**, *28*, 173-192.

[196] Martínez-Coronado, R.; Alonso, J.A.; Aguadero, A.; Fernández-Díaz, M.T. New $SrMo_{1-x}Cr_xO_{3-\delta}$ perovskites as anodes in solid-oxide fuel cells. *Int. J. Hydrogen Energy*, **2014**, *39*, 4067-4073.
[http://dx.doi.org/10.1016/j.ijhydene.2013.04.149]

[197] Bernuy-Lopez, C.; Pelloquin, D.; Raveau, B.; Allix, M.; Claridge, J.B.; Rosseinsky, M.J.; Wang, P.; Blcloch, A. Phasoid intergrowth between the double perovskite Sr_2MgMoO_6 and the n=2 R-P phase $Sr_3Mo_2O_7$. *Solid State Ionics*, **2010**, *181*, 889-893.
[http://dx.doi.org/10.1016/j.ssi.2010.04.011]

[198] Marrero-López, D.; Peña-Martínez, J.; Ruiz-Morales, J.C.; Gabás, M.; Núñez, P.; Aranda, M.A.G.; Ramos-Barrado, J.R. Redox behaviour, chemical compatibility and electrochemical performance of $Sr_2MgMoO_{6-\delta}$ as SOFC anode. *Solid State Ionics*, **2010**, *180*, 1672-1682.
[http://dx.doi.org/10.1016/j.ssi.2009.11.005]

[199] Marrero-López, D.; Peña-Martínez, J.; Ruiz-Morales, J.C.; Martín-Sedeño, M.C.; Núñez, P. High temperature phase transition in SOFC anodes based on $Sr_2MgMoO_{6-\delta}$. *J. Solid State Chem.*, **2009**, *182*, 1027-1034.
[http://dx.doi.org/10.1016/j.jssc.2009.01.018]

[200] Howell, T.G.; Kuhnell, C.P.K.; Reitz, T.L.; Eigenbrodt, B.C.; Singh, R.N. $Sr_{2-x}La_xMgMoO_6$ and $Sr_{2-x}La_xMgNbO_6$ for use as sulfur-tolerant anodes without a buffer layer. *J. Am. Ceram. Soc.*, **2014**, *97*, 3636-3642.
[http://dx.doi.org/10.1111/jace.13208]

[201] Wang, Z.; Tian, Y.; Li, Y. Direct CH_4 fuel cell using Sr_2FeMoO_6 as an anode material. *J. Power Sources*, **2011**, *196*, 6104-6109.
[http://dx.doi.org/10.1016/j.jpowsour.2011.03.053]

[202] Im, H.N.; Jeon, S.Y.; Choi, M.B.; Kim, H.S.; Song, S.J. Chemical stability and electrochemical properties of $CaMoO_{3-\delta}$ for SOFC anode. *Ceram. Int.*, **2012**, *38*, 153-158.
[http://dx.doi.org/10.1016/j.ceramint.2011.05.155]

[203] Im, H.N.; Choi, M.B.; Jeon, S.Y.; Song, S.J. Structure, thermal stability and electrical conductivity of $CaMoO_{4+\delta}$. *Ceram. Int.*, **2011**, *37*, 49-53.
[http://dx.doi.org/10.1016/j.ceramint.2010.08.004]

[204] Aguadero, A.; de la Calle, C.; Pérez-Coll, D.; Alonso, J.A. Study of the crystal structure, thermal stability and conductivity of $Sr(V_{0.5}Mo_{0.5})O_{3+\delta}$ as SOFC material. *Fuel Cells (Weinh.)*, **2011**, *11*, 44-50.

[http://dx.doi.org/10.1002/fuce.201000070]

[205] Troncoso, L.; Martínez-Lope, M.J.; Alonso, J.A.; Fernández-Díaz, M.T. Evaluation of Sr_2MMoO_6 (M=Mg, Mn) as anode materials in solid-oxide fuel cells: A neutron diffraction study. *J. Appl. Phys.,* **2013**, *113*, 023511.
[http://dx.doi.org/10.1063/1.4774764]

[206] Ji, Y.; Huang, Y-H.; Ying, J-R.; Goodenough, J.B. Electrochemical performance of La-doped $Sr_2MgMoO_{6-\delta}$ in natural gas. *Electrochem. Commun.,* **2007**, *9*, 1881-1885.
[http://dx.doi.org/10.1016/j.elecom.2007.04.006]

[207] Zhang, L.; He, T. Performance of double perovskite $Sr_{2-x}Sm_xMgMoO_{6-\delta}$ as solid oxide fuel cell anodes. *J. Power Sources,* **2011**, *196*, 8352-8359.
[http://dx.doi.org/10.1016/j.jpowsour.2011.06.064]

[208] Escudero, M.J.; Gómez de Parada, I.; Fuerte, A.; Daza, L. Study of $Sr_2Mg(Mo_{0.8}Nb_{0.2})O_{6-\delta}$ as anode material for solid oxide fuel cells using hydrocarbons as fuel. *J. Power Sources,* **2013**, *243*, 654-660.
[http://dx.doi.org/10.1016/j.jpowsour.2013.05.198]

[209] Zhang, Q.; Wei, T.; Huang, Y-H. Electrochemical performance of double-perovskite Ba_2MMoO_6 (M = Fe, Co, Mn, Ni) anode materials for solid oxide fuel cells. *J. Power Sources,* **2012**, *198*, 59-65.
[http://dx.doi.org/10.1016/j.jpowsour.2011.09.092]

[210] Zhang, L.; Zhou, Q.; He, Q.; He, T. Double-perovskites $A_2FeMoO_{6-\delta}$ (A = Ca, Sr, Ba) as anodes for solid oxide fuel cells. *J. Power Sources,* **2010**, *195*, 6356-6366.
[http://dx.doi.org/10.1016/j.jpowsour.2010.04.021]

[211] Martínez-Coronado, R.; Alonso, J.A.; Aguadero, A.; Fernández-Díaz, M.T. Optimized energy conversion efficiency in solid-oxide fuel cells implementing $SrMo_{1-x}Fe_xO_{3-\delta}$ perovskites as anodes. *J. Power Sources,* **2012**, *208*, 153-158.
[http://dx.doi.org/10.1016/j.jpowsour.2012.02.002]

[212] Liu, Q.; Dong, X.; Xiao, G.; Zhao, F.; Chen, F. A novel electrode material for symmetrical SOFCs. *Adv. Mater.,* **2010**, *22*, 5478-5482.
[http://dx.doi.org/10.1002/adma.201001044]

[213] Xiao, G.; Liu, Q.; Zhao, F.; Zhang, L.; Changrong, X.; Chen, F. $Sr_2Fe_{1.5}Mo_{0.5}O_6$ as cathodes for intermediate-temperature solid oxide fuel cells with $La_{0.8}Sr_{0.2}Ga_{0.87}Mg_{0.13}O_3$ electrolyte. *J. Electrochem. Soc.,* **2011**, *158*, B455-B460.
[http://dx.doi.org/10.1149/1.3556085]

[214] Martínez-Coronado, R.; Alonso, J.A.; Fernández-Díaz, M.T. $SrMo_{0.9}Co_{0.1}O_3$: A potential anode for intermediate-temperature solid-oxide fuel cells (IT-SOFC). *J. Power Sources,* **2014**, *258*, 76-82.
[http://dx.doi.org/10.1016/j.jpowsour.2014.02.031]

[215] Aguadero, A.; Alonso, J.A.; Martínez-Coronado, R.; Martínez-Lope, M.J.; Fernández-Díaz, M.T. Evaluation of $Sr_2CoMoO_{6-\delta}$ as anode material in solid-oxide fuel cells: A neutron diffraction study. *J. Appl. Phys.,* **2011**, *109*, 034907.
[http://dx.doi.org/10.1063/1.3544068]

[216] Wei, T.; Zhang, Q.; Huang, Y-H.; Goodenough, J.B. Cobalt-based double-perovskite symmetrical electrodes with low thermal expansion for solid oxide fuel cells. *J. Mater. Chem.,* **2012**, *22*, 225-231.
[http://dx.doi.org/10.1039/C1JM14756K]

[217] Xie, Z.; Zhao, H.; Du, Z.; Chen, T.; Chen, N.; Liu, X.; Skinner, S.J. Effects of Co doping on the electrochemical performance of double perovskite oxide $Sr_2MgMoO_{6-\delta}$ as an anode material for solid oxide fuel cells. *J. Phys. Chem. C,* **2012**, *116*, 9734-9743.
[http://dx.doi.org/10.1021/jp212505c]

[218] Huang, Y-H.; Liang, G.; Croft, M.; Lehtimäki, M.; Karppinen, M.; Goodenough, J.B. Double-perovskite anode materials Sr_2MMoO_6 (M = Co, Ni) for solid oxide fuel cells. *Chem. Mater.,* **2009**, *21*, 2319-2326.

[http://dx.doi.org/10.1021/cm8033643]

[219] Wei, T.; Ji, Y.; Meng, X.; Zhang, Y. $Sr_2NiMoO_{6-\delta}$ as anode material for $LaGaO_3$-based solid oxide fuel cell. *Electrochem. Commun.,* **2008**, *10*, 1369-1372.
[http://dx.doi.org/10.1016/j.elecom.2008.07.005]

[220] Huang, Y-H.; Dass, R.I.; Xing, Z-L.; Goodenough, J.B. Double perovskites as anode materials for solid-oxide fuel cells. *Science,* **2006**, *312*, 254-257.
[http://dx.doi.org/10.1126/science.1125877]

[221] Meng, X.; Liu, X.; Han, D.; Wu, H.; Li, J.; Zhan, Z. Symmetrical solid oxide fuel cells with impregnated $SrFe_{0.75}Mo_{0.25}O_{3-\delta}$ electrodes. *J. Power Sources,* **2014**, *252*, 58-63.
[http://dx.doi.org/10.1016/j.jpowsour.2013.11.049]

[222] Xiao, G.; Jin, C.; Liu, Q.; Heyden, A.; Chen, F. Ni modified ceramic anodes for solid oxide fuel. *J. Power Sources,* **2012**, *201*, 43-48.
[http://dx.doi.org/10.1016/j.jpowsour.2011.10.103]

[223] Du, Z.; Zhao, H.; Yang, C.; Shen, Y.; Yan, C.; Zhang, Y. Optimization of strontium molybdate based composite anode for solid oxide fuel cells. *J. Power Sources,* **2015**, *274*, 568-574.
[http://dx.doi.org/10.1016/j.jpowsour.2014.10.062]

[224] Xiao, P.; Ge, X.; Liu, Z.; Wang, J-Y.; Wang, X. $Sr_{1-x}Ca_xMoO_3$-$Gd_{0.2}Ce_{0.8}O_{1.9}$ as the anode in solid oxide fuel cells: Effects of Mo precipitation. *J. Alloys Compd.,* **2014**, *587*, 326-331.
[http://dx.doi.org/10.1016/j.jallcom.2013.10.187]

[225] Cascos, V.; Alonso, J.A.; Fernández-Díaz, M.T. Novel Mg-doped $SrMoO_3$ perovskites designed as anode materials for Solid Oxide Fuel Cells. *Materials (Basel),* **2016**, *9*, 588.
[http://dx.doi.org/10.3390/ma9070588]

[226] He, B.; Zhao, L.; Song, S.; Liu, T.; Chen, F.; Xia, C. $Sr_2Fe_{1.5}Mo_{0.5}O_{6-\delta}$-$Sm_{0.2}Ce_{0.8}O_{1.9}$ composite anodes for intermediate-temperature solid oxide fuel cells. *J. Electrochem. Soc.,* **2012**, *159*, B619-B626.
[http://dx.doi.org/10.1149/2.020206jes]

[227] Zhang, P.; Huang, Y-H.; Cheng, J-G.; Mao, Z-Q.; Goodenough, J.B. Sr_2CoMoO_6 anode for solid oxide fuel cell running on H_2 and CH_4 fuels. *J. Power Sources,* **2011**, *196*, 1738-1743.
[http://dx.doi.org/10.1016/j.jpowsour.2010.10.007]

[228] Vo, N.M.; Gross, M.D. The effect of vanadium deficiency on the stability of Pd and Pt catalysts in lanthanum strontium vanadate solid oxide fuel cell anodes. *J. Electrochem. Soc.,* **2012**, *159*, B641-B646.
[http://dx.doi.org/10.1149/2.092205jes]

[229] Yoon, S.E.; Song, S.H.; Choi, J.; Ahn, J-Y.; Kim, B-K.; Park, J-S. Coelectrolysis of steam and CO_2 in a solid oxide electrolysis cell with ceramic composite electrodes. *Int. J. Hydrogen Energy,* **2014**, *39*, 5497-5504.
[http://dx.doi.org/10.1016/j.ijhydene.2014.01.124]

[230] Adijanto, L.; Padmanabhan, V.B.; Holmes, K.J.; Gorte, R.J.; Vohs, J.M. Physical and electrochemical properties of alkaline earth doped, rare earth vanadates. *J. Solid State Chem.,* **2012**, *190*, 12-17.
[http://dx.doi.org/10.1016/j.jssc.2012.01.065]

[231] Macías, J.; Yaremchenko, A.A.; Fagg, D.P.; Frade, J.R. Structural and defect chemistry guidelines for $Sr(V,Nb)O_3$-based SOFC anode materials. *Phys. Chem. Chem. Phys.,* **2015**, *17*, 10749-10758.
[http://dx.doi.org/10.1039/C5CP00069F]

[232] Petit, C.T.G.; Lan, R.; Cowin, P.I.; Tao, S. Structure and conductivity of strontium-doped cerium orthovanadates $Ce_{1-x}Sr_xVO_4$ ($0{\leq}x{\leq}0.175$). *J. Solid State Chem.,* **2010**, *183*, 1231-1238.
[http://dx.doi.org/10.1016/j.jssc.2010.03.032]

[233] Tsipis, E.V.; Patrakeev, M.V.; Kharton, V.V.; Vyshatko, N.P.; Frade, J.R. Ionic and p-type electronic transport in zircon-type $Ce_{1-x}A_xVO_{4{\pm}\delta}$ (A = Ca, Sr). *J. Mater. Chem.,* **2002**, *12*, 3738-3745.
[http://dx.doi.org/10.1039/B206004C]

[234] Ge, X.M.; Chan, S.H. Lanthanum strontium vanadate as potential anodes for solid oxide fuel cells. *J. Electrochem. Soc.,* **2009**, *156*, B386-B391.
[http://dx.doi.org/10.1149/1.3058585]

[235] Park, J-S.; Luo, J.; Adijanto, L.; Vohs, J.M.; Gorte, R.J. The stability of lanthanum strontium vanadate for solid oxide fuel cells. *J. Power Sources,* **2013**, *222*, 123-128.
[http://dx.doi.org/10.1016/j.jpowsour.2012.08.084]

[236] Ni, C.S.; Vohs, J.M.; Gorte, R.J.; Irvine, J.T.S. Fabrication and characterization of a large-area solid oxide fuel cell based on dual tape cast YSZ electrode skeleton supported YSZ electrolytes with vanadate and ferrite perovskite-impregnated anodes and cathodes. *J. Mater. Chem. A Mater. Energy Sustain.,* **2014**, *2*, 19150-19155.
[http://dx.doi.org/10.1039/C4TA04789C]

[237] Tamm, K.; Möller, P.; Nurk, G.; Lust, E. Investigation of time stability of Sr-doped lanthanum vanadium oxide anode and Sr-doped lanthanum cobalt oxide cathode based on samaria doped ceria electrolyte using electrochemical and TOF-SIMS methods. *J. Electrochem. Soc.,* **2016**, *163*, F586-F592.
[http://dx.doi.org/10.1149/2.0111607jes]

[238] Song, S-H.; Yoon, S-E.; Choi, J.; Kim, B-K.; Park, J-S. A high-performance ceramic composite anode for protonic ceramic fuel cells based on lanthanum strontium vanadate. *Int. J. Hydrogen Energy,* **2014**, *39*, 16534-16540.
[http://dx.doi.org/10.1016/j.ijhydene.2014.03.219]

[239] Macías, J.; Yaremchenko, A.A.; Frade, J.R. Redox transitions in strontium vanadates: electrical conductivity and dimensional changes. *J. Alloys Compd.,* **2014**, *601*, 186-194.
[http://dx.doi.org/10.1016/j.jallcom.2014.02.148]

[240] Cheng, Z.; Zha, S.; Aguilar, L.; Liu, M. Chemical, electrical, and thermal properties of strontium doped lanthanum vanadate. *Solid State Ionics,* **2005**, *176*, 1921-1928.
[http://dx.doi.org/10.1016/j.ssi.2005.05.009]

[241] Yaremchenko, A.A.; Brinkmann, B.; Janssen, R.; Frade, J.R. Electrical conductivity, thermal expansion and stability of Y- and Al-substituted $SrVO_3$ as prospective SOFC anode material. *Solid State Ionics,* **2013**, *247-248*, 86-93.
[http://dx.doi.org/10.1016/j.ssi.2013.06.002]

[242] Macías, J.; Yaremchenko, A.A.; Frade, J.R. Enhanced stability of perovskite-like $SrVO_3$-based anode materials by donor-type substitutions. *J. Mater. Chem. A Mater. Energy Sustain.,* **2016**, *4*, 10186-10194.
[http://dx.doi.org/10.1039/C6TA02672A]

[243] Cousin, R.; Dourdin, M.; Abi-Aad, E.; Courcot, D.; Capelle, S.; Guelton, M.; Aboukaïs, A. Formation of $CeVO_4$ phase during the preparation of CuVCe oxide catalysts. *J. Chem. Soc., Faraday Trans.,* **1997**, *93*, 3863-3867.
[http://dx.doi.org/10.1039/a702532g]

[244] Daniell, W.; Ponchel, A.; Kuba, S.; Anderle, F.; Weingand, T.; Gregory, D.H.; Knözinger, H. Characterization and catalytic behavior of VO_x-CeO_2 catalysts for the oxidative dehydrogenation of propane. *Top. Catal.,* **2002**, *20*, 65-74.
[http://dx.doi.org/10.1023/A:1016399315511]

[245] Adijanto, L.; Padmanabhan, V.B.; Küngas, R.; Gorte, R.J.; Vohs, J.M. Transition metal-doped rare earth vanadates: a regenerable catalytic material for SOFC anodes. *J. Mater. Chem.,* **2012**, *22*, 11396-11402.
[http://dx.doi.org/10.1039/c2jm31774e]

[246] Adijanto, L.; Padmanabhan, V.B.; Gorte, R.J.; Vohs, J.M. Polarization-induced hysteresis in CuCo-doped rare-earth vanadates SOFC anodes. *J. Electrochem. Soc.,* **2012**, *159*, F751-F756.
[http://dx.doi.org/10.1149/2.042211jes]

[247] Cheng, Z.; Zha, S.; Aguilar, L.; Wang, D.; Winnick, J.; Liu, M. A Solid oxide fuel cell running on H_2S/CH_4 fuel mixtures. *Electrochem. Solid-State Lett.,* **2006**, *9*, A31-A33.
[http://dx.doi.org/10.1149/1.2137467]

[248] Park, J-S.; Hasson, I.D.; Gross, M.D.; Chen, C.; Vohs, J.M.; Gorte, R.J. A high-performance solid oxide fuel cell anode based on lanthanum strontium vanadate. *J. Power Sources,* **2011**, *196*, 7488-7494.
[http://dx.doi.org/10.1016/j.jpowsour.2011.05.028]

[249] Yoon, S-E.; Ahn, J-Y.; Kim, B-K.; Park, J-S. Improvements in co-electrolysis performance and long-term stability of solid oxide electrolysis cells based on ceramic composite cathodes. *Int. J. Hydrogen Energy,* **2015**, *40*, 13558-13565.
[http://dx.doi.org/10.1016/j.ijhydene.2015.08.012]

[250] Tamm, K.; Küngas, R.; Gorte, R.J.; Lust, E. Solid oxide fuel cell anodes prepared by infiltration of strontium doped lanthanum vanadate into doped ceria electrolyte. *Electrochim. Acta,* **2013**, *106*, 398-405.
[http://dx.doi.org/10.1016/j.electacta.2013.05.127]

[251] Zegers, P. Fuel cell commercialization: The key to a hydrogen economy. *J. Power Sources,* **2006**, *154*, 497-502.
[http://dx.doi.org/10.1016/j.jpowsour.2005.10.051]

[252] Vernoux, P.; Guillodo, M.; Fouletier, J.; Hammou, A. Alternative anode material for gradual methane reforming in solid oxide fuel cells. *Solid State Ionics,* **2000**, *135*, 425-431.
[http://dx.doi.org/10.1016/S0167-2738(00)00390-8]

[253] Madsen, B.D.; Barnett, S.A. Effect of fuel composition on the performance of ceramic-based solid oxide fuel cell anodes. *Solid State Ionics,* **2005**, *176*, 2545-2553.
[http://dx.doi.org/10.1016/j.ssi.2005.08.004]

[254] Sfeir, J.; Buffat, P.A.; Möckli, P.; Xanthopoulos, N.; Vasques, R.; Mathieu, J.; Herle, J.V.; Thampi, K.R. Lanthanum chromite based catalysts for oxidation of methane directly on SOFC anodes. *J. Catal.,* **2001**, *202*, 229-244.
[http://dx.doi.org/10.1006/jcat.2001.3286]

[255] Bastidas, D.M.; Tao, S.; Irvine, J.T.S. A symmetrical solid oxide fuel cell demonstrating redox stable perovskite electrodes. *J. Mater. Chem.,* **2006**, *16*, 1603-1605.
[http://dx.doi.org/10.1039/b600532b]

[256] Chen, X.J.; Liu, Q.L.; Khor, K.A.; Chan, S.H. High-performance $(La,Sr)(Cr,Mn)O_3/(Gd,Ce)O_{2-\delta}$ composite anode for direct oxidation of methane. *J. Power Sources,* **2007**, *165*, 34-40.
[http://dx.doi.org/10.1016/j.jpowsour.2006.11.075]

[257] Baker, R.T.; Metcalfe, I.S. Activity and deactivation of $La_{0.8}Ca_{0.2}CrO_3$ in dry methane using temperature-programmed techniques. *Appl. Catal. A,* **1995**, *126*, 297-317.
[http://dx.doi.org/10.1016/0926-860X(95)00018-6]

[258] Tao, S.; Irvine, J.T.S. Catalytic properties of the perovskite oxide $La_{0.75}Sr_{0.25}Cr_{0.5}Fe_{0.5}O_{3-\delta}$ in relation to its potential as a solid oxide fuel cell anode material. *Chem. Mater.,* **2004**, *16*, 4116-4121.
[http://dx.doi.org/10.1021/cm049341s]

[259] Baker, R.T.; Metcalfe, I.S. Activity and deactivation of $La_{0.8}Ca_{0.2}Cr_{0.9}X_{0.1}O_3$ (X = Ni, Co) in dry methane using temperature programmed techniques. *Appl. Catal. A,* **1995**, *126*, 319-332.
[http://dx.doi.org/10.1016/0926-860X(95)00019-4]

[260] Xu, Z.; Luo, J.L.; Chuang, K.T.; Sanger, A.R. $LaCrO_3$-VO_x-YSZ anode catalyst for solid oxide fuel cell using impure hydrogen. *J. Phys. Chem. C,* **2007**, *111*, 16679-16685.
[http://dx.doi.org/10.1021/jp074672o]

[261] Liu, Q.; Bugaris, D.E.; Xiao, G.; Chmara, M.; Ma, S.; zur Loye, H-C.; Amiridis, M.D.; Chen, F. $Sr_2Fe_{1.5}Mo_{0.5}O_{6-\delta}$ as a regenerative anode for solid oxide fuel cells. *J. Power Sources,* **2011**, *196*, 9148-

9153.
[http://dx.doi.org/10.1016/j.jpowsour.2011.06.085]

[262] Chen, G.; Kishimoto, H.; Yamaji, K.; Kuramoto, K.; Horita, T. Effect of interaction between A-site deficient LST and ScSZ on electrochemical properties of SOFC. *J. Electrochem. Soc.,* **2015**, *162,* F223-F228.
[http://dx.doi.org/10.1149/2.0131503jes]

[263] Sun, X.; Wang, S.; Wang, Z.; Qian, J.; Wen, T.; Huang, F. Evaluation of $Sr_{0.88}Y_{0.08}TiO_3$-CeO_2 as composite anode for solid oxide fuel cells running on CH4 fuel. *J. Power Sources,* **2009**, *187,* 85-89.
[http://dx.doi.org/10.1016/j.jpowsour.2008.10.067]

[264] Ge, X.; Zhang, L.; Fang, Y.; Zeng, J.; Chan, S.H. Robust solid oxide cells for alternate power generation and carbon conversion. *RCS Adv.,* **2011**, *1,* 715-724.

[265] Adler, S.B. Factors governing oxygen reduction in solid oxide fuel cell cathodes. *Chem. Rev.,* **2004**, *104,* 4791-4843.
[http://dx.doi.org/10.1021/cr020724o]

[266] Ishihara, T.; Fukui, S.; Nishiguchi, H.; Takita, Y. Mixed electronic-oxide ionic conductor of $BaCoO_3$ doped with La for cathode of intermediate-temperature-operating solid oxide fuel cell. *Solid State Ionics,* **2002**, *152,* 609-613.
[http://dx.doi.org/10.1016/S0167-2738(02)00394-6]

[267] Hansen, K.K.; Mogensen, M. Evaluation of LSF based SOFC cathodes using cone-shaped electrodes. *ECS Trans.,* **2008**, *13,* 153-160.

[268] Koide, H.; Someya, Y.; Yoshida, T.; Maruyama, T. Properties of Ni/YSZ cermet as anode for SOFC. *Solid State Ionics,* **2000**, *132,* 253-260.
[http://dx.doi.org/10.1016/S0167-2738(00)00652-4]

[269] Ruiz-Morales, J.C.; Canales-Vázquez, J.; Marrero-López, D.; Irvine, J.T.S.; Núñez, P. Improvement of the electrochemical properties of novel solid oxide fuel cell anodes, $La_{0.75}Sr_{0.25}Cr_{0.5}Mn_{0.5}O_{3-\delta}$ and $La_4Sr_8Ti_{11}Mn_{0.5}Ga_{0.5}O_{37.5-\delta}$, using Cu-YSZ-based cermets. *Electrochim. Acta,* **2007**, *52,* 7217-7225.
[http://dx.doi.org/10.1016/j.electacta.2007.05.060]

[270] Holtappels, P.; Bradley, J.; Irvine, J.T.S.; Kaiser, A.; Mogensen, M. Electrochemical characterization of ceramic SOFC anodes. *J. Electrochem. Soc.,* **2001**, *148,* A923-A929.
[http://dx.doi.org/10.1149/1.1383774]

[271] Primdahl, S.; Mogensen, M. Mixed conductor anodes: Ni as electrocatalyst for hydrogen conversion. *Solid State Ionics,* **2002**, *152,* 597-608.
[http://dx.doi.org/10.1016/S0167-2738(02)00393-4]

[272] Setoguchi, T.; Okamoto, K.; Eguchi, K.; Arai, H. Effects of anode material and fuel on anodic reaction of solid oxide fuel cells. *J. Electrochem. Soc.,* **1992**, *139,* 2875-2880.
[http://dx.doi.org/10.1149/1.2068998]

[273] Jiang, S.P.; Badwal, S.P.S. Hydrogen oxidation at the nickel and platinum electrodes on yttria-tetragonal zirconia electrolyte. *J. Electrochem. Soc.,* **1997**, *144,* 3777-3784.
[http://dx.doi.org/10.1149/1.1838091]

[274] McIntosh, S.; Gorte, R.J.; Vohs, J.M. An examination of lanthanide additives on the performance of Cu/YSZ cermet anodes. *Electrochim. Acta,* **2002**, *47,* 3815-3821.
[http://dx.doi.org/10.1016/S0013-4686(02)00352-3]

[275] Lu, C.; Worrel, W.L.; Vohs, J.M.; Gorte, R.J. A comparison of Cu-ceria-SDC and Au-ceria-SDC composites for SOFC anodes. *J. Electrochem. Soc.,* **2003**, *150,* A1357-A1359.
[http://dx.doi.org/10.1149/1.1608003]

[276] van Roosmalen, J.A.M.; Cordfunke, E.H.P. Chemical reactivity and interdiffusion of $(La,Sr)MnO_3$ and $(Zr,Y)O_2$ solid oxide fuel cell cathode and electrolyte materials. *Solid State Ionics,* **1992**, *52,* 303-312.
[http://dx.doi.org/10.1016/0167-2738(92)90177-Q]

[277] Faaland, S.; Einarsrud, M-A.; Wiik, K.; Grande, T. Reactions between $La_{1-x}Ca_xMnO_3$ and CaO-stabilized ZrO_2. *J. Mater. Sci.,* **1999**, *34*, 5811-5819.
[http://dx.doi.org/10.1023/A:1004766419921]

[278] Jiang, S.P.; Zhang, J.P.; Raprakash, Y.; Milosevic, D.; Wilshier, K. An investigation of shelf-life of strontium doped $LaMnO_3$ materials. *J. Mater. Sci.,* **2000**, *35*, 2735-2741.
[http://dx.doi.org/10.1023/A:1004766212164]

[279] Kawada, T.; Sakai, N.; Yokokawa, H.; Dokiya, M.; Anzai, I. Reaction between solid oxide fuel cell materials. *Solid State Ionics,* **1992**, *50*, 189-196.
[http://dx.doi.org/10.1016/0167-2738(92)90218-E]

[280] Zheng, Y.; Ran, R.; Gu, H.; Cai, R.; Shao, Z. Characterization and optimization of $La_{0.8}Sr_{0.2}Sc_{0.1}Mn_{0.9}O_{3-\delta}$ based composite electrodes for intermediate-temperature solid-oxide fuel cells. *J. Power Sources,* **2008**, *185*, 641-648.
[http://dx.doi.org/10.1016/j.jpowsour.2008.09.003]

[281] Burnat, D.; Heel, A.; Holzer, L.; Otal, E.; Kata, D.; Graule, T. On the chemical interaction of nanoscale lanthanum doped strontium titanates with common scandium and yttrium stabilized electrolyte materials. *Int. J. Hydrogen Energy,* **2012**, *37*, 18326-18341.
[http://dx.doi.org/10.1016/j.ijhydene.2012.09.022]

[282] Sudireddy, B.R.; Blennow, P.; Nielsen, K.A. Microstructural and electrical characterization of Nb-doped $SrTiO_3$-YSZ composites for solid oxide fuel cell electrodes. *Solid State Ionics,* **2012**, *216*, 44-49.
[http://dx.doi.org/10.1016/j.ssi.2011.11.025]

[283] Ling, Y.; Chen, L.; Lin, B.; Yu, W.; Isimjan, T.T.; Zhao, L.; Liu, X. Synthesis and characterization of a $Sr_{0.95}Y_{0.05}TiO_{3-\delta}$-based hydrogen electrode for reversible solid oxide cells. *RCS Adv.,* **2015**, *5*, 17000-17006.

[284] Chen, G.; Kishimoto, H.; Yamaji, K.; Kuramoto, K.; Horita, T. Interfacial reaction phenomenon between $La_{0.25}Sr_{0.75}TiO3$ and ScSZ. *J. Power Sources,* **2014**, *249*, 49-54.
[http://dx.doi.org/10.1016/j.jpowsour.2013.07.066]

[285] Carter, J.D.; Appel, C.C.; Mogensen, M. Reactions at the calcium doped lanthanum chromite-yttria stabilized zirconia interface. *J. Solid State Chem.,* **1996**, *122*, 407-415.
[http://dx.doi.org/10.1006/jssc.1996.0134]

[286] Yamamoto, T.; Itoh, T.; Mori, M.; Mori, N.; Watanabe, T.; Imanishi, N.; Takeda, Y.; Yamamoto, O. Chemical stability between NiO/8YSZ cermet and alkaline-earth metal substituted lanthanum chromite. *J. Power Sources,* **1996**, *61*, 219-222.
[http://dx.doi.org/10.1016/S0378-7753(96)02369-5]

[287] Sfeir, J.; van Herle, J.; McEvoy, A.J. Stability of calcium substituted lanthanum chromites used as SOFC anodes for methane oxidation. *J. Eur. Ceram. Soc.,* **1999**, *19*, 897-902.
[http://dx.doi.org/10.1016/S0955-2219(98)00340-9]

[288] Matsuda, M.; Nowotny, J.; Zhang, Z.; Sorrell, C.C. Lattice and grain boundary diffusion of Ca in polycrystalline yttria-stabilized ZrO_2 determined by employing SIMS technique. *Solid State Ionics,* **1998**, *111*, 301-306.
[http://dx.doi.org/10.1016/S0167-2738(98)00163-5]

[289] Sfeir, J. $LaCrO_3$-based anodes: stability considerations. *J. Power Sources,* **2003**, *118*, 276-285.
[http://dx.doi.org/10.1016/S0378-7753(03)00099-5]

[290] Mims, C.A.; Bayani, N.; Jacobson, A.J.; van der Heyde, P.A.W. Modes of surface exchange in $La_{0.2}Sr_{0.8}Cr_{0.2}Fe_{0.8}O_{3-d}$. *Solid State Ionics,* **2005**, *176*, 319-323.
[http://dx.doi.org/10.1016/j.ssi.2004.08.014]

[291] Vashook, V.; Vasylechko, L.; Zosel, J.; Müller, R.; Ahlborn, E.; Guth, U. Lanthanum-calcium chromites-titanates as possible anode materials for SOFC. *Solid State Ionics,* **2004**, *175*, 151-155.

[http://dx.doi.org/10.1016/j.ssi.2004.09.060]

[292] Sauvet, A.L.; Fouletier, J. Electrochemical properties of a new type of anode material $La_{1-x}Sr_xCr_{1-y}Ru_yO_{3-\delta}$ for SOFC under hydrogen and methane at intermediate temperatures. *Electrochim. Acta,* **2001**, *47*, 987-995.
[http://dx.doi.org/10.1016/S0013-4686(01)00811-8]

[293] Marrero-López, D.; Peña-Martínez, J.; Ruiz-Morales, J.C.; Pérez-Coll, D.; Martín-Sedeño, M.C.; Núñez, P. Applicability of $La_2Mo_{2-y}W_yO_9$ materials as solid electrolyte for SOFCs. *Solid State Ionics,* **2007**, *178*, 1366-1378.
[http://dx.doi.org/10.1016/j.ssi.2007.07.008]

[294] Corbel, C.; Mestiri, S.; Lacorre, P. Physicochemical compatibility of CGO fluorite, LSM and LSCF perovskite electrode materials with $La_2Mo_2O_9$ fast oxide-ion conductor. *Solid State Sci.,* **2005**, *7*, 1216-1224.
[http://dx.doi.org/10.1016/j.solidstatesciences.2005.05.007]

[295] Kostgloudis, G.Ch.; Ftikos, Ch.; Ahmad-Khanlou, A.; Naoumidis, A.; Stöver, D. Chemical compatibility of alternative perovskite oxide SOFC cathodes with doped lanthanum gallate solid electrolyte. *Solid State Ionics,* **2000**, *134*, 127-138.
[http://dx.doi.org/10.1016/S0167-2738(00)00721-9]

[296] Zhang, X.; Ohara, S.; Okawa, H.; Maric, R.; Fukui, T. Interactions of a $La_{0.9}Sr_{0.1}Ga_{0.8}Mg_{0.2}O_{3-\delta}$ electrolyte with Fe_2O_3, Co_2O_3 and NiO anode materials. *Solid State Ionics,* **2001**, *139*, 145-152.
[http://dx.doi.org/10.1016/S0167-2738(00)00833-X]

[297] Huang, J.; Wan, J.H.; Goodenough, J.B. Increasing power density of LSGM-based solid oxide fuel cells using new anode materials. *J. Electrochem. Soc.,* **2001**, *148*, A788-A794.
[http://dx.doi.org/10.1149/1.1378289]

[298] Kharton, V.V.; Figueiredo, F.M.; Navarro, L.; Naumovich, E.N.; Kovalevsky, A.V.; Yaremchenko, A.A.; Viskup, A.P.; Carneiro, A.; Marques, F.M.B.; Frade, J.R. Ceria-based materials for solid oxide fuel cells. *J. Mater. Sci.,* **2001**, *36*, 1105-1117.
[http://dx.doi.org/10.1023/A:1004817506146]

[299] Eguchi, K.; Setoguchi, T.; Inoue, T.; Arai, H. Electrical properties of ceria-based oxides and their application to solid oxide fuel cells. *Solid State Ionics,* **1992**, *52*, 165-172.
[http://dx.doi.org/10.1016/0167-2738(92)90102-U]

[300] Yokokawa, H.; Sakai, N.; Horita, T.; Yamaji, K.; Brito, M.E.; Kishimoto, H. Thermodynamic and kinetic considerations on degradations in solid oxide fuel cell cathodes. *J. Alloys Compd.,* **2008**, *452*, 41-47.
[http://dx.doi.org/10.1016/j.jallcom.2006.12.150]

[301] Filonova, E.A.; Dmitriev, A.S.; Pikalov, P.S.; Medvedev, D.A.; Pikalova, E.Yu. The structural and electrical properties of $Sr_2Ni_{0.75}Mg_{0.25}MoO_6$ and its compatibility with solid state electrolytes. *Solid State Ionics,* **2014**, *262*, 365-369.
[http://dx.doi.org/10.1016/j.ssi.2013.11.036]

[302] Li, J.; Zhong, C.; Meng, X.; Wu, H.; Nie, H.; Zhan, Z.; Wang, S. $Sr_2Fe_{1.5}Mo_{0.5}O_{6-\delta}$ - $Zr_{0.84}Y_{0.16}O_{2-\delta}$ materials as oxygen electrodes for solid oxide electrolysis cells. *Fuel Cells (Weinh.),* **2014**, *14*, 1046-1049.
[http://dx.doi.org/10.1002/fuce.201400021]

SUBJECT INDEX